Library of
Davidson College

A Perspective on Infantry

John A. English

PRAEGER SPECIAL STUDIES • PRAEGER SCIENTIFIC

Library of Congress Cataloging in Publication Data
English, J. A. (John Alan)
 A perspective on infantry.

 Bibliography: p. 297
 Includes index.
 1. Infantry. 2. Infantry drill and tactics.
I. Title.
UD145.E53   356'.1   81-5230
ISBN 0-03-059699-8   AACR2

Published in 1981 by Praeger Publishers
CBS Educational and Professional Publishing
A Division of CBS, Inc.
521 Fifth Avenue, New York, New York 10175 U.S.A.

All rights reserved

© 1981 Praeger Publishers

3456789 052 98765432
Printed in the United States of America

*This work is respectfully
dedicated
to those Canadian infantrymen
who fell in foreign fields
far from the snows of home*

Contents

FOREWORD	vii
PREFACE	ix
LIST OF MAPS AND TABLES	xi
LIST OF ABBREVIATIONS	xiii
PROLOGUE: ON INFANTRY AND WAR	xv
1. AN EPOCH-MAKING CHANGE: THE DECENTRALIZATION OF INFANTRY TACTICS	1
2. HANDYMAN OF BATTLE: ENDURING THE SHOCK OF ARMOR	37
3. THE BLINDNESS OF A MOLE: COMPARATIVE TACTICS AND INFANTRY BETWEEN WARS	61
4. FAIR-WEATHER WAR: OF INFANTRY AND BLITZKRIEG	86
5. THE BACKBONE OF LAND FORCES: INFANTRY OF THE EASTERN FRONT	110
6. CUTTING EDGE OF BATTLE: SECOND-FRONT INFANTRY	155
7. CINDERELLA OF THE ARMY: INFANTRY EAST OF SUEZ, WEST OF PEARL	201
8. BEYOND DETERRENCE: THE CONTEMPORARY REQUIREMENT FOR INFANTRY	240
9. FOUL-WEATHER WARRIORS: TOWARD A PERFECTION OF INFANTRY	282
BIBLIOGRAPHY	297
INDEX	325
ABOUT THE AUTHOR	346

> Look after the little things,
> and the big things will look after themselves.
>
> — Infantry Platoon Commander's
> Aide Memoire (1959),
> Royal Canadian School of Infantry

Foreword

Study of military history is essential to the formation of any soldier. Although new technology, more sophisticated armaments, and indeed the new geopolitical implications of major conflicts have demanded changes in the art of warfare, no one can afford to ignore what has been done in the past. Whatever the changes in methodology and tactical concepts, basic principles that have found their roots in the evolution of warfare itself remain very much the same. It is therefore from the sound knowledge of former battles, from the study of the evolution of military thought, that one can refine one's judgment, develop one's skills, and have a basis for developing the new tactical concepts necessary to the modern battlefield.

The Canadian Army has not fought on a battlefield since Korea; it is therefore doubly essential that, in peacetime, we learn our job not only in the actual drills of the battlefield but also through in-depth study of history. We must understand the raison d'être in the evolution of warfare and learn to apply well-tested principles to ever-changing tactical concepts.

I have thus read with much pleasure Major English's book <u>A Perspective on Infantry.</u> The subject itself, small-unit tactics and basic infantry training, appealed to my infantry upbringing. I was more than glad to discover that the author had the particular talent of turning a study that could have been dry and factual into a most fascinating, novel-like work. What is more important than the mere pleasure of reading this book, however, are the lessons learnt through this detailed study of "things infantry." The scope is wide, as stated by the author himself: "will include tracing the development of the infantry combat arm from 1866 to the present." Yet when I closed my book I realized that the author had made an extensive <u>tour d'horizon</u> on the evolution of warfare during the <u>last century</u>, not limiting himself to infantry tactics.

Major English is hence a significant contributor to helping us understand the essence of infantry. His work deserves to be given the widest distribution to serving soldiers or to anyone interested in a profession as old as humanity.

Lieutenant-General J. J. Paradis, CMM,CD
Commander, Mobile Command (Canadian Army)

<u>St-Hubert, Québec</u>
<u>2 March, 1981</u>

Preface

I have always been fascinated by small-unit tactics and basic infantry training. Tours of duty with five different infantry battalions--three Canadian and two British--have over the years served to intensify this interest. While employed with the Infantry Department at the Combat Arms School, however, I was struck by the dearth of consolidated works that one could actually consult to gain a broader understanding of the fundamentals of infantry. Comprehensive studies abounded on armor and artillery, but when it came to infantry, one was often sorely pressed to find even a reasonable bibliography. Many good infantrymen consequently never became acquainted with such classic infantry studies as <u>Men Against Fire</u> or <u>The Future of Infantry.</u>

Although my foremost purpose in originally attempting this work was to increase my personal understanding of things infantry, it soon became evident that the results of such endeavor could doubtless provide some bibliographical guidance to other interested individuals. I have at least been assured by a number of serving comrades in arms that they would like to retain copies for reference. I refer, in particular, to my fellow field officers John Trethewey, Dick Graham, John Selkirk, Douglas Bland, Wayne Dehnke, Mike Paré, and George Oehring, all of whom were kind enough to proofread chapters and proffer advice. For further technical insight and much appreciated criticism, I am beholden to Captains Bob Anglin, Pat Paterson, and Mitch Kryzanowski.

In meeting the specific demands of preparing for publication, however, I must acknowledge the enlightened academic guidance and solid support provided by Dr. B. D. Hunt of the department of history, Royal Military College of Canada. A former infantryman himself, he gave freely of his valuable time to read and correct my proofs, displaying, I might add, great tolerance and forbearance in the process. I am likewise similarly indebted to Lieutenant-Colonel J. H. Allan, Deputy Commandant of the Canadian Land Forces Command and Staff College, who, in commenting on the substance and quality of my original drafts, also rendered invaluable assistance and advice. For consistent encouragement and helpful counsel throughout this pleasant ordeal, of course, I remain most grateful to my former Commandant at the Combat Arms School, Colonel D. A. Nicholson, Fort Garry Horse.

Since being posted to the National Capital Region, I have fortunately received additional enthusiastic countenance from several other quarters. Lieutenant-General J. Chouinard, Colonel Commandant of Canadian Infantry, Colonel C. B. Snider, Director of Infantry, and Lieutenant-Colonel J. E. L. Gollner were particularly instrumental in bringing my work to the attention of the Canadian infantry. Lieutenant-General J. J. Paradis, Commander, Mobile Command, kindly agreed to write the Foreword and Major-General G. G. Brown, Colonel of the Regiment, Princess Patricia's Canadian Light Infantry, lent further encouragement. Colonels I. S. Fraser and R. L. Cowling, former and currently serving commanders of the Canadian Airborne Regiment, respectively, were also kind enough to lend support to my efforts.

I doubt very much that this book could ever have been completed, however, without the voluntary assistance provided by Sergeant-Major J. G. D. Dupuis and the staff of the Office of the Deputy Chief of the Defense Staff. I remain most grateful for their collective and concerted efforts in helping me to produce final manuscript copy. I am especially appreciative of the travail of two of their number, Elaine McHugh and Master Corporal M. C. Roy, both of whom worked tirelessly to see me through.

For graphic design and library support second to none, I would like to express my appreciation to Roger Frenette, Jack Bernard, and the staffs of the RMC Massey Library and the National Defence Headquarters Library. Mr. Keith Crouch and Mr. Cliff Watt of the Massey Library were particularly helpful in assisting me with my research. The latest information on the nuances of "Hutier Tactics," of course, was graciously provided by Dr. Harold C. Deutsch of the U. S. Army War College.

Finally, to my wife, Valerie, I remain forever indebted for her unwavering support and great depth of understanding during the long period of my almost total commitment to this endeavor. Together with my Irish setter, Cromwell, she established a secure and firm base for operations. Fortunately, both were steady under fire.

List of Maps and Tables

Table 1-1	Prussian "Company Column" Deployment	3
Table 1-2	Extended-Order Field Deployment	5
Table 1-3	British Infantry Battalion Organization and Deployment (Circa 1914)	8
Table 1-4	The Decisive Attack	12
Table 1-5	German Triangular Divisional Organization in the Great War	16
Table 1-6	Elastic Defense	20
Table 1-7	Improved <u>Gruppe</u> Organization	25
Table 2-1	Liddell Hart's Principles of Tactics	44
Table 2-2	The "Expanding Torrent" Along Lines of Least Resistance	46
Table 3-1	German Standard Infantry Regimental Organization (Circa 1920)	66
Table 3-2	French Standard Divisional Organization (1920-30)	68
Table 3-3	French <u>Groupe de Combat</u> (Circa 1920)	69
Table 3-4	Italian Infantry Battalion Organization (Circa 1920)	72
Table 3-5	British Battalion Organization (1937)	77

xii LIST OF MAPS AND TABLES

Table 4-1	German Infantry Division Organization, (1939)	90
Table 4-2	Panzer Division Organization (1938-39)	92
Table 4-3	<u>Schwerpunkt</u> and <u>Aufrollen</u>	94
Map 4-4	Plan Yellow	97
Map 5-1	Western Russia	119
Table 5-2	Red Army Infantry Regimental Organization (World War II)	125
Table 6-1	American Infantry Squad Organization	166
Table 6-2	American Infantry Regimental Organization (1942)	167
Table 7-1	Japanese Rifle Section	205
Table 7-2	Japanese Infantry Regimental Organization (World War II)	207
Table 7-3	U.S. Marine "Fire Team" Squad Organization (World War II)	213
Table 7-4	U.S. Marine Infantry Regimental Organization (World War II)	216
Table 7-5	U.S. Army Infantry Platoon Organization (Korean War)	225
Table 8-1	Sharon's Tactics for Seizing a Stronghold	243
Table 8-2	Infantry Antitank Weapons	256

List of Abbreviations

AA	Antiaircraft
Adm	Administration/Administrative
APC	Armored personnel carrier
AR	Automatic rifle
Armd	Armored
Arty	Artillery
Aslt	Assault
ATGM	Antitank guided missile
Atk	Antitank
BAR	Browning automatic rifle
Bde	Brigade
Bn	Battalion
Bty	Battery
Coy	Company
Def	Defense/defensive
Div	Division
Elm	Element
Engr	Engineer
Fd	Field
Inf	Infantry
HMG	Heavy machine gun
How	Howitzer
HQ	Headquarters
Hy	Heavy
IFV	Infantry fighting vehicle
Lt	Light
LMG	Light machine gun
MG	Machine gun
MMG	Medium machine gun
Mech	Mechanized
Med	Medical
MC	Motorcycle
MICR	Mechanized infantry combat vehicle

LIST OF ABBREVIATIONS

MOBA	Military operations in built-up areas
Mor	Mortar
Mot	Motorized
MOUT	Military operations in urban terrain
NCO	Noncommissioned officer
Offr	Officer
Org	Organization
Pl	Platoon
PLA	Peoples' Liberation Army
Pnr	Pioneer
Pz	Panzer
Pzkw	Panzerkampfwagen
Recce	Reconnaissance
Regt	Regiment
RL	Rocket launcher
Sect	Section
Sgt	Sergeant
Sigs	Signals
SMG	Submachine gun
SP	Self-propelled
Sp	Support
Str	Strength
Sup	Supply
Svcs	Services
Tk	Tank
TOW	Tube-launched, optical-tracked, wire-command link missile
U.S.A.A.F.	United States Army Air Forces
Wpn	Weapon
Yds	Yards

Prologue

On Infantry and War

This work is about infantry. Specifically, it is about infantry in war, its primary focus on the fundamentals of infantry operations and training. The basic organization, equipment, weapons, and tactics of several national infantries have accordingly been compared and contrasted at some length within these pages. Combat experiences and training philosophies have also been examined in an attempt to ascertain what qualities or characteristics seem to distinguish good infantry from bad. The principal argument emanating from this essentially historical analysis is that groups of foot soldiers remain to this day among the most powerful and influential forces on the battlefield. An important corollary to this thesis, however, is that the overall operational effectiveness of the infantry arm is determined most fundamentally by the performance of its smallest units and their leaders. To think of a division as but 30 companies, then, is to gain a clearer understanding of the fighting essence of infantry.

As might be expected, of course, it is somewhat difficult to examine infantry in isolation since it long ago came to share pride of place as a means of combat decision with other more technologically and materially endowed arms. The introduction of the tank and airplane and the development of quick-firing artillery made the battlefield a much more lethal place for the man on foot. The infantry arm alone was less likely to prevail; in fact, a preponderance of foot soldiers within an army has been generally regarded as a sign of military backwardness. Yet, on all modern battlefronts, the infantryman has endured. Ironically, the tremendous growth in military motorization and mechanization did not result in a corresponding decrease in infantry requirements during World War II. According to John Keegan, the battles in Burma and in the Pacific "were fought almost wholly without benefit of armour . . . [and] almost without intervention of aircraft." The long retreat up the boot of Italy was fought by the Germans without air cover and with few tanks. In Russia, after the first four months, the campaign remained up to the last year of the war essentially a clash between mighty infantry armies, a struggle of "shoe leather and horseflesh--to be eaten when times were hard."(1)

The tendency developed nonetheless, especially in Western industrialized circles following the fall of France, "to magnify the role of the machine in war while minimizing the importance of large forces of well-trained foot soldiers."(2) The reason behind such thinking was rooted in the popular misconception that battlefield mobility was related almost entirely to speed; ergo, the marching pace of the infantryman rendered him superfluous. Yet, as Western armies painfully began to discover, a profusion of machines did not necessarily guarantee an improved battlefield mobility in all circumstances and terrain. In the "turtle and hare" juxtaposition in Korea, the Red Chinese infantry achieved a surprising degree of battlefield mobility by moving slowly but steadily by night. In Vietnam, the helicopter provided the Americans with a superior technological mobility, but the North Vietnamese clearly retained a superior tactical mobility by staying light on their feet.(3) Though American infantry units and subunits literally "commuted" to war, they were often left to flounder pathetically in the immensity of the Vietnamese landscape, borne down by the weight of helmets, entrenching tools, rations, and flak jackets as well as ammunition. As one veteran who had been shot at by the silent, invisible enemy related:

> . . . with the aircraft gone we were struck by the utter strangeness of this rank and rotted wilderness. . . . I . . . never felt so exhausted, and yet I had walked only three miles, less than one-tenth the distance I had marched at Quantico. . . . Helmets bouncing against our heads, canteens against our hips, rifle slings and bandoliers jiggling, we sounded like a platoon of junkmen.(4)

In the judgment of Brigadier-General S. L. A. Marshall, true mobility could never issue primarily from machines; if such were the case, he argued, then one "could turn the whole problem over to the Ordnance and the Transportation Corps." In his view, mobility was less a matter of "traction and speed" than it was the ability to "stand against fire and to deliver it."(5) Historically, he was on rather firm ground, for even in Napoleonic times marching armies were occasionally able to cover as many as 15 miles a day for two or three weeks on end. Few Allied armies in World War II were able to sustain a faster rate of advance in face of relatively equivalent opposition. The average speed of General George Patton's Third Army after its breakout from Normandy was but 15 miles per day; at its best, it covered 30 miles in one day. The simple fact was that the speed and range of modern transport had been largely canceled out by the huge consumption of modern armies.(6) As more troops and greater quantities of supplies began to clutter up rear areas, armies became increasingly road bound. Growing specialization and sophistication meant, in turn, that modern mechanized armies were more, not less, sensitive to variations in terrain, weather, and resupply. In such circumstances, mobility was as much a matter of mind and spirit as of technical equipment. Thus, the infantryman, moving at a marching pace across all types of ground in all weathers and striking at unexpected times, retained a significant offensive potential. Sadly, however, the advent of the machine failed to decrease by a single pound the weight an individual foot soldier, particularly an Allied one, was compelled to carry on his back.(7)

While the foregoing argument is not intended to decry the obvious merits of mechanization, it does serve to illustrate that tactical mobility incorporates as many conceptual as technological dimensions. Mechanization per se was neither a panacea for improved military performance nor an excuse for ignoring the real problems associated with battlefield mobility. According to Captain B. H. Liddell Hart, the "locomobility" of infantry, its ability to move throughout every locality and fire from every position, remains the foundation stone of infantry's value and no doubt an important reason for its continued existence as a viable combat arm.(8) Offering the smallest target and endowed with the best battlefield computer yet devised--the human brain--the infantryman has played a prominent, if not dominant, role in the more than 30-odd major military engagements(9) that have occurred since World War II. The ability of infantry to adapt to changing conditions has, in fact, been nothing short of phenomenal. There is, indeed, much irony in the realization that in the postwar world, in the very shadow of weapons of mass destruction, the most effective means of altering the status quo has turned out to be a little man armed with nothing more than a rifle in his hand and an idea in his head.(10)

It is, of course, a matter of historical record that, like Horatius at the bridge, small groups of determined infantrymen on a road to Moscow or on a hill in Korea have been able to influence the fate of nations out of all proportion to their numbers. In World War II, to paraphrase Marshall, it was demonstrated time and again that a handful of men at a certain spot at a given hour exerted a more powerful influence on a battle than ten times that number 24 hours later. By prompt and imaginative action, lone riflemen and companies sometimes diverted whole enemy corps, while a machine-gun squad at a roadblock began the defeat of an armored division. In short, though mass was there somewhere in support, many great victories pivoted upon the fire action of a very few. For the infantry soldier, the major lesson of World War II minor tactics was "the overpowering effect of relatively small amounts of fire when delivered from the right ground at the right hour."(11) Since this discovery, the infantryman, rarely out of step with technology for long, has seen his arsenal of firepower increase dramatically.

Significantly, the enhanced importance of the small infantry group on the battlefield derived from the relentless decentralization of tactical control necessitated by the increased range and lethality of modern weapons. Whereas in 1800 roughly 20,000 men were required to hold a mile of battle front, this figure progressively dropped to 12,000 by 1870, to 2,500 by 1917, and to less than 1,000 today.(12) Rather than a glorious panorama of color, the battlefield became instead a foreboding, desolate place in which combatants maintained a high degree of invisibility. One indication of this was that the American company that made the most progress on D-day saw only six live Germans.(13) The trend for the future, whether in nuclear or conventional scenarios, points toward an even greater dispersion and independence of minor units "operating on a mission guided basis . . . so [they] can . . . be expected to continue . . . should crippling damage befall higher echelons to the rear."(14)

Thus, while the infantryman's principal battlefield focus in former times was fixed mainly upon the brigade or regimental line, from which it took a brave man to run away, the center of an infantry soldier's life in action now appears to be the section.(15) To be sure, ever since close-order drill was relegated to the parade square, "minor tactics" (as they were regrettably categorized by Baron de Jomini) assumed a far greater importance in what General Carl von Clausewitz termed the "friction" of war, "the force that makes the apparently easy so difficult."(16)

Unfortunately, there is reason to suspect that not enough attention is being paid to "the neglected area of small unit tactics which are the basis of every great battle."(17) To quote Gabriel and Savage, the present-day schooling of the average American officer is:

> . . . far too staff-oriented at far too high a level and only remotely connected with the details of small-unit combat. Few officers . . . genuinely comprehend the details and complexities of squad-, platoon-, or company-sized battle. With the stress on staff training, there has been a deemphasis of the true skills of the soldier.(18)

The scope of this study will include tracing the development of the infantry combat arm from 1866 to the present, this author being in perfect accord with the late Major General J. F. C. Fuller's epigram "Looking back is the surest way of looking forward." The infantry tactical revolution that occurred during the Great War will consequently be covered in some depth. The role played by the infantry arm in later German blitzkrieg operations will also receive reasonably extensive treatment. While other facets of World War II, Korea, and the Arab-Israeli wars will be considered at length from the standpoint of infantry, it is obvious that some selectivity will have to be exercised. For this reason, one mainly of scope, primary attention throughout this work will generally be directed toward one national infantry at a time; for instance, the Japanese infantry in Burma, Marine infantry in the Pacific, and Chinese Communist infantry in Korea. Only desultory reference will be made to the struggle in Indo-China. This complex conflict, though an obviously important military watershed, remains for the moment too enigmatic and enshrouded in controversy to permit truly objective and detailed examination, which, in any case, would doubtless constitute a book in itself. It is hoped nonetheless that the substance of this work will be sufficient to provide a more proper perspective on infantry today.

NOTES

1. John Keegan, The Face of Battle (New York: Viking, 1976), pp. 285-6. At Stalingrad, soldiers boiled their belts. Ibid.

2. S. L. A. Marshall, Men Against Fire (New York: Morrow, 1947), p. 15.

3. Dave Richard Palmer, Summons of the Trumpet (San Rafael, Calif.: Presido, 1978), pp. 92, 97, and 102.

4. Philip Caputo, A Rumour of War (New York: Holt, Rinehart and Winston, 1977), pp. 84, 89, 95, 111, and 115.

5. S. L. A. Marshall, The Soldier's Load and the Mobility of a Nation (Washington: The Combat Forces Press, 1950), pp. 1-2.

6. Seymour J. Deitchman, Limited War and American Defence Policy (Cambridge: The M.I.T. Press, 1964), p. 159; and Martin van Creveld, "Supplying an Army: An Historical View," R.U.S.I. Journal, 123 (1978): 61. The Germans in Greece in 1941, and the North Koreans in the first few weeks of 1950 also averaged 15 miles per day. Advances of 100 miles a day were made by the British "Blade Force" in Africa in 1942 and by Malinovsky's Russian armor in Manchuria in 1945 (against almost negligible resistance), but sustained advances of over 40 miles per day were generally the exception in the recent past. Neville Brown, Strategic Mobility (London: Chatto and Windus, 1963), pp. 205-6. Creveld states that no army has been able to sustain a pace of 75 miles per day for more than two or three days on end, and that even under ideal conditions 40 miles per day will be difficult to attain. Creveld, "Supplying an Army," p. 61.

7. Marshall, The Soldier's Load, pp. 2, 6, and 111.

8. B. H. Liddell Hart, The Future of Infantry (London: Faber and Faber, 1933), pp. 31-2.

9. Deitchman, Limited War, p. 15.

10. Klaus Knorr, On the Uses of Military Power in the Nuclear Age (Princeton: University Press, 1976), p. 144.

11. Marshall, Men Against Fire, pp. 2 and 68.

12. Brown, Strategic Mobility, p. 199.

13. Marshall, Men Against Fire, p. 68.

14. Colonel G. C. Reinhardt and Lieutenant-Colonel W. R. Kintner, Atomic Weapons in Land Combat (Harrisburg: The Military Service Publishing Company, 1953), p. 162.

15. According to John Baynes in his study of the Second Scottish Rifles in the Battle of Neuve Chapelle in 1915, "A private soldier in action finds that his section becomes the centre of his life. He finds his platoon and company important as well, and as far as reputation is concerned he thinks occasionally about the battalion and division he is in. But the small groups are the vital ones." John Baynes, Morale: A Study of Men and Courage (London: Cassell, 1967), p. 102.

16. Carl von Clausewitz, On War, ed./trans. Michael Howard and Peter Paret (Princeton: Univerisity Press, 1976), p. 121. The perceptive Clausewitz wrote: "Everything in war is simple but the simplest thing is difficult. The difficulties accumulate and end by producing a kind of friction that is inconceivable unless one has experienced war. . . . Countless minor incidents--the kind you can never really foresee-- combine to lower the general level of performance, so that one always falls short of the intended goal. . . . The military machine . . . is composed of individuals, every one of whom retains his potential for friction." Ibid., p. 119.

17. Lieutenant-General Anthony Farrar-Hockley, Infantry Tactics (London: Almark, 1976), p. 5 (introduction to the mechanics of war).

18. Richard A. Gabriel and Paul L. Savage, Crisis in Command (New York: Hill and Wang, 1978), pp. 137-8.

"CANADIAN SOLDIER"

After a cartoon by Augustus John.

1. An Epoch-Making Change

The Decentralization of Infantry Tactics

In the intervening years between the Crimean War and the end of the Great War, a major transformation took place in the realm of basic infantry tactics. Triggered by the introduction of the rifle, it remained largely evolutionary at first, the improved range and accuracy of the rifle merely increasing from a depth of 150 meters to 500 and more the "impenetrable zone" a defender was able to draw across his front. At the same time, the danger from outflanking movements increased in direct proportion to the effective range of rifle fire, thereby forcing an extension of the fighting line and an associated decentralization in tactical control. However, though ranges had increased and fewer troops were required to hold the same frontages, casualties inflicted on the enemy by aimed rifle fire from extended order were considerably less than the numbers inflicted in former times during line-versus-line close-order engagements.(1) A convoluted argument thus developed against open-order tactics. It was reinforced, moreover, by command and control considerations and the "accepted axiom of tactics, based on the experiences of generations, that troops in close order . . . [were generally capable of withstanding] heavier punishment without losing their forward momentum than when extended."(2) Such argument was, of course, eventually doomed to disintegrate under the concentrated fire of Great War machine guns, which, in themselves, speeded a more revolutionary transformation in infantry tactics.

As history records, the military rifle was first generally used in the Crimean War, its breech-loading version appearing later during the Danish War of 1864. While rifles, even repeaters, were available in limited

1

quantities to both sides during the American Civil War, the rifled musket remained by far the most commonly used infantry weapon in that conflict, the lessons of which in any case were largely ignored by Europeans. The truly significant nature of the rapid-fire rifle was not fully appreciated, therefore, until Napoleonic-style mass and élan were literally blown away by its devastating fire effect during the 1866 Austro-Prussian War. In this conflict, the dense formations of the cold-steel-oriented Austrian army, still mesmerized by the successful bayonet attacks of their French opponents in the North Italian war of 1859, were cut to pieces by the Prussian infantry's breech-action "needle-gun," which could be loaded lying down. The improved rate of fire and reduced target of the Prussian foot soldier presaged the eventual ascendancy of infantry over cavalry and defense over offense, though not all observers were quick to grasp this fact.(3)

During the early stages of the Austro-Prussian War, the Prussians employed a thick skirmish line based on the battalion column, which formation the great military critic Antoine Henri Jomini had always held to be the best order of attack. This formation was soon found to be too dense, however, and a more extended line of company columns was consequently adopted. Commanded by a captain, company column was considered by Count Helmuth von Moltke (the Elder) to be optimal formation for both attack and defense.(4) To form such column, a Prussian company was first drawn up in three close-order ranks, tallest soldiers in the first, best shots in the third. The company thus ordered, which normally paraded a frontage of 72 files, was then divided into two divisions of 36 files each. On the command "form company column," the three ranks of the divisions closed together, one rank behind the other, making a column six ranks deep. The column so formed consisted, practically, of three platoons in double rank, one behind the other, each commanded by a lieutenant. The third platoon (of the third ranks of the original divisions) was designated the "shooting sub-division"; on closing with the enemy, it was employed in skirmishing formation, the first rank or section 100 to 150 yards in front of the second, which was normally 100 yards in advance of the remainder of the company column. Difficulty in maneuvering skirmishers, who usually attacked by a series of "rushes," was overcome by encouraging them to act in sections or "swarms."(5) That the Austro-Prussian War was sometimes referred to as "the Captain's War" reflected the impact of such company columnar tactics.

In the Franco-Prussian War of 1870-/1, troops of both sides learned through experience that although

TABLE 1-1. PRUSSIAN "COMPANY COLUMN" DEPLOYMENT

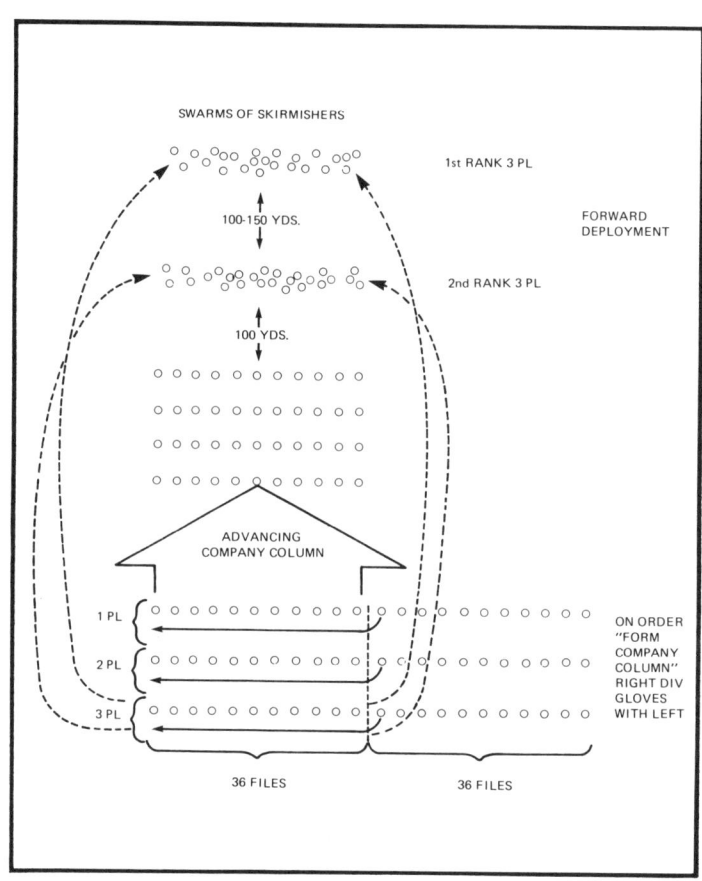

defensive rifle fire inflicted serious losses on attackers at long ranges, the efficacy of such fire did not increase as the attackers came closer; instead, the effect of defensive fire fell off noticeably at ranges below 600 meters. Though not initially well understood, the "Theory of the Rigid, Constant Cone of Misses,"(6) or as it is known today, the deadly beaten zone of falling shot, slowly came to be accepted by practicing soldiers. To be sure, the demonstrated effectiveness of such indirect fire in France was quite enough for certain Prussian field officers to recommend that henceforth the "fighting formation of . . . infantry is that of a cloud of skirmishers."(7) The Germans, in particular, had been compelled to adopt greater dispersed order since close formations like company column were found to be highly susceptible to saturation fire from the French chassepot rifle, which ranged to 2,000 yards. German infantrymen, with their "needle-guns" sighted to but 600 yards, often had to advance as much as 1,400 yards under fire without replying.(8) Massed manpower had, in fact, surrendered its position of superiority on the field of battle to concentrated firepower.

As previously intimated, however, the evolution of tactical fighting units on the battlefield was affected as much by control considerations as by developments in weapons technology. Open-order tactics tended to be disorderly and inherently prone to desertion; then, too, there were fears, best expressed by General Colmar von der Goltz, that such methods "might produce skirmishers, but not soldiers i.e. men whose devotion to duty surpasses their fear of death." There thus developed within most armies a tendency to cling to classic columnar doctrines, wherein even the battalion remained the basic tactical and fire unit. It was not until 1888, for example, that extended-order tactics were definitely accepted in the German army. With the adoption of open order, of course, and the recognition of the skirmish line as the principal combat formation of the infantry,(9) more up-to-date control measures had to be introduced. Accordingly, tacticians began to think of deploying infantry into extended-order firing lines, support lines, and reserve lines. By building up the firing line and urging volley firing, it was hoped that attack formations could increase the density and improve the control of fire. In this tactical system, the task of "supports" was to reinforce and replace casualties in the firing line as it advanced by "rushes" toward the enemy. Such methods of fighting, advocated in French army regulations of 1875,(10) would ultimately form the basis of Great War infantry tactics.

Quite obviously, the increased dispersion necessitated by the effect of the rapid-fire rifle made it impossible for one voice, or commander, to control the detailed movements of a battalion. Although generally considered at the time as being too weak to carry out a battlefield mission in action, the infantry rifle company gradually came to be regarded as the proper "fighting unit" of open warfare. The battalion remained the "tactical unit," that is, the smallest

AN EPOCH-MAKING CHANGE 5

TABLE 1-2. EXTENDED-ORDER FIELD DEPLOYMENT

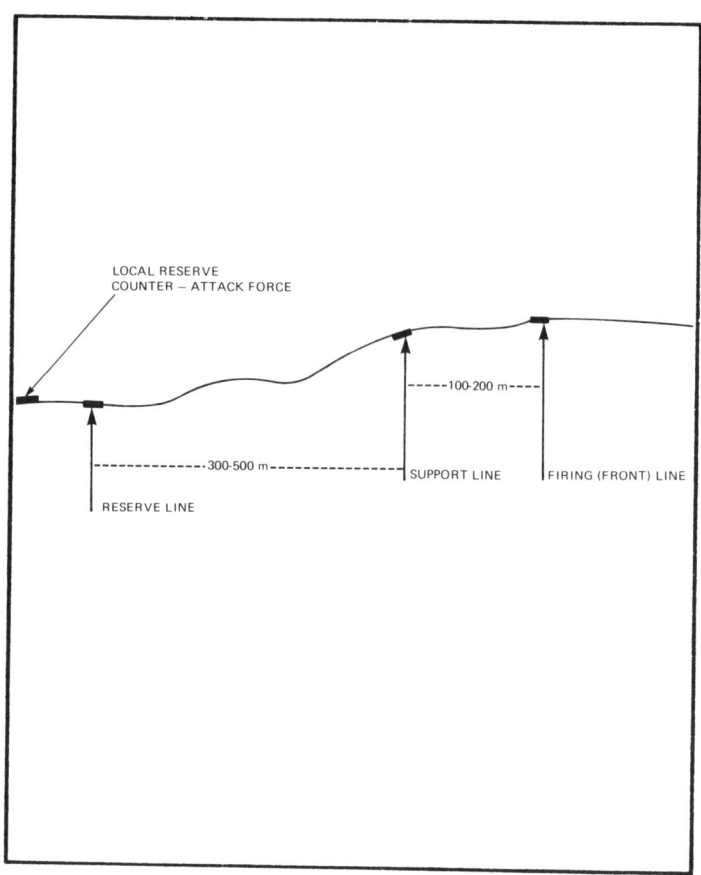

body of men that could be safely employed independently.(11) This tactical decentralization meant, of course, that the number of companies per battalion was ever fluctuating as national armies strove for optimum composition. At the time of General Emory Upton's famous trip to the armies of Asia and Europe, this number was thought to be four, that is, in every army but the British, which, in 1821, had fixed its battalion establishment at eight companies.(12)

The size of infantry companies in the larger Continental armies of the period ranged from around 200 to 250 men. However, in the opinion of Colonel William Balck, "it seemed hardly practical to exceed a strength of 150 men, as this . . . [was] about the greatest number in which a relation based upon personal influence of the leader on his subordinates . . . [could] still be obtained." It appears, therefore, that in armies in which companies consisted of more than 200 men, the extra manpower was intended as a margin of safety designed to maintain the company at strength after casualties.(13) Such a margin could also have compensated for personnel left out of battle to rest and repair.

Lack of manpower was not, of course, a major problem for the British Empire in the Boer War (1899-1902), although deficient British marksmanship and British reluctance to dig almost made it one. A "marked feature of the first part of the War . . . [was]," according to Major General Sir Ernest Swinton, the "combination of ignorance as to the value of trenches and ingrained aversion to the inglorious drudgery of digging."(14) Coupled with a failure to recognize the value of extended-order tactics, it was to cost the British dearly. At the Modder River on November 28, 1899, the devastating fire of the Boers--opened at 1,000 meters--tended to destroy the physical and moral powers of the British to the extent that not even the best regiments could be induced to advance. At Biddulphsberg, a mere 18 Boers, using ground and shooting straight, defeated two battalions of Guards in their long unarticulated lines. Too much close-order drill had, in the opinion of Liddell Hart, developed in the British soldier a sort of "tactical arthritis." The Boer tactic of crawling to within 300 meters of British positions to deliver a withering fire, which the British were unable to return effectively on the small, prone targets of the Boers, was often enough to make whole units consider surrender as the only way out (as the British did in large number at Nicholson's Nek on October 29, 1900).(15) Smokeless powder and stalking skill had all but eliminated the visible enemy from the modern battlefield.

Many of the tactical lessons and conclusions of the Boer War were confirmed in the Russo-Japanese War of 1904-5. Long firing lines and open-order tactics again characterized the battlefield, the face of which was even more pockmarked by a proliferation of entrenchments. Though the conflict cost both sides close to a quarter of a million casualties each, the Japanese appear on the whole to have employed qualitatively superior combat methods. Having apparently absorbed most of the lessons of the Boer War, they at least recognized the necessity for night attacks and making maximum use of the spade. According to Russian observers, the Japanese firing line normally advanced by "short, alternating rushes, the men then throwing themselves down and intrenching."

Japanese combat methods in general were marked by close infantry-artillery cooperation, surprise, speed in launching attacks, prompt reorganization, and no pursuit. Rushes were never made for more than 100 meters or less than 30 to 40 meters; nor were they made by units smaller than a platoon. Rushes made by companies were usually executed in wide extension since constant motion was regarded as the best protection against the fire of artillery.(16)

As fought, the Russo-Japanese clash foreshadowed the debacle of the Great War, for it clearly demonstrated the paralyzing power of machine guns, the futility of frontal attacks, and the immobilizing effect of barbed wire. These indicators, though noted by some, completely passed over the heads of the many. Having observed that an entrenched enemy under shrapnel bombardment could continue to provide effective small-arms fire through loopholes in parapets, a British observer, Lieutenant-General Ian Hamilton, felt moved to offer the caustic comment that the only good use cavalry had been put to was to "cook rice for the infantry." For this utterance, he was thought by many professional officers to be quite mad; General Douglas Haig's sanity, on the other hand, was hardly questioned on his equally topical assessment that "artillery was only effective against demoralized troops."(17) In the final analysis, the Russo-Japanese War was destined to have little more impact than the American Civil War on established European military thought. According to Major M. F. de Pardieu, battles in "entrenched camps as occurred at . Mukden [would] . . . never take place in a war with the French Army."(18)

Not surprisingly, volley fire had to be dispensed with during the Russo-Japanese War, as long firing lines made it extraordinarily difficult to transmit orders. Only whistle signals could, in fact, be heard above the din of battle. Interestingly, the German army had much earlier, in the 1870s, ceased to advocate volley firing, though only after some argument with the forces of conservatism. The French, on the other hand, had charged their 1875 instructions with "looking too exclusively to fire effect"; new instructions issued in 1884 consequently stressed the offensive and even recommended volley fire by section, notwithstanding the Napoleonic dictum that independent firing was the "only kind of fire practicable in war." The French appear to

8 A PERSPECTIVE ON INFANTRY

TABLE 1-3. BRITISH INFANTRY BATTALION ORGANIZATION AND DEPLOYMENT (CIRCA 1914)

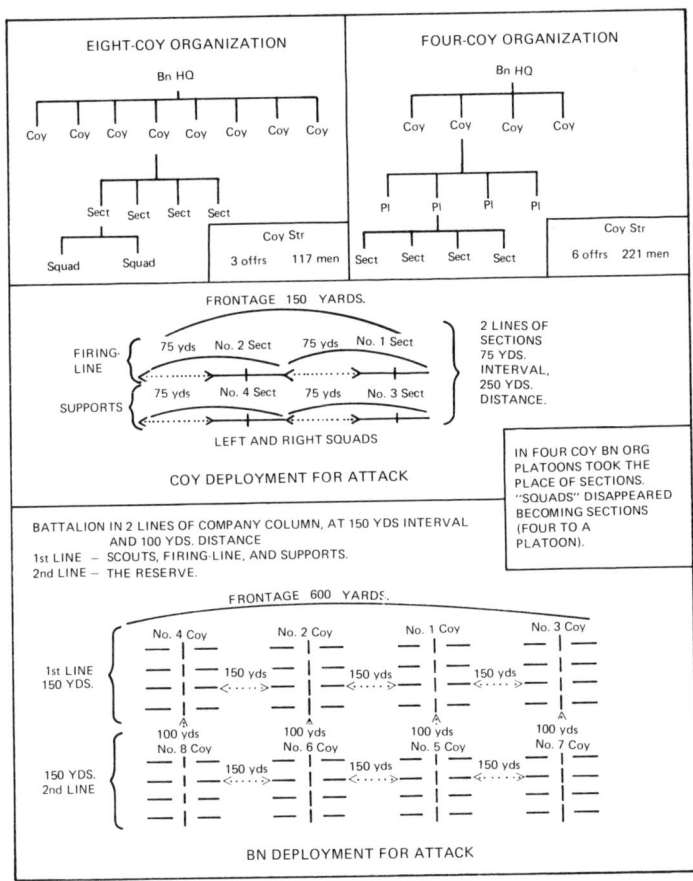

have been influenced at this time by that indefatigable Russian writer and war veteran General M. I. Dragomirov, who maintained that "national character" must be taken into account in the conduct of war. Russian soldiers, he claimed, preferred fighting shoulder to shoulder; good troops did not engage in "long-distance lead pumping," but rather husbanded their cartridges and fired volleys. Resolution in the attack was all that was required! Shock action would prevail over firepower, and what the bullet failed to do, the bayonet would. The Russian army on the eve of the Russo-Japanese War was steeped in Dragomirovian doctrine;(19) French army doctrine just before the Great War would push this idea to the extreme.

Fortunately for the British, their poor showing in South Africa forced a reexamination of tactics and training to the point where the Great War found them with the "best trained and best shooting infantry in Europe."(20) Though building up the firing line was still stressed, renewed emphasis was placed on elasticity and adapting movement to ground, utilizing such cover as was available and offering as small a target as possible to the enemy. This was not necessarily the case in Germany, where pamphlets published after the Russo-Japanese War placed additional stress on bayonet action and stated that the use of ground was limited to the preservation of direction.(21) In general, however, it was acknowledged that much greater dispersion was required in infantry ranks than that afforded by the venerable company column, which, in any case, soon passed the way of the phalanx. Though even short rushes made by long lines eventually tended to become too costly in lives to be executed, it was in the meantime widely recognized that smaller groups were able to move with startling rapidity.(22)

On the eve of the Great War, four-company battalion organization was introduced alongside eight-company organization within the British army. This move involved few changes tactically since new battalion field operations were merely tailored to fit existing deployment concepts; companies accordingly replaced companies, with new platoons of four sections supplanting the old section "fire unit"(23) of two squads. British infantry battalion organization was thus brought more into line with Continental establishments. In the number of platoons per company, however, the British remained at variance with certain Continental armies, whereas the Austrian, Russian, and French armies built "square" companies of four platoons like the British, the German, Japanese, and Belgian armies built "triangular." There had been, nonetheless, universal agreement for some time that the infantry company rather than the battalion constituted the tactical unit. It was "the smallest element of a body of troops capable of sustaining an action independently, or performing a simple combat task . . . and the elements of which . . . are personally known to the leader . . . [and] small enough [for him to exercise voice control] . . . ." The infantry company between 1905 and 1914 occupied a battle frontage of roughly 200 meters; a battalion of four companies could, therefore, expect to defend a frontage of 800 meters with a firing, support, and, on occasion, reserve line.(24)

Clearly, an infantry force organized and deployed in the foregoing fashion had little to fear from the formerly dominant cavalry even on level ground. The rapid-fire rifle had by this time established an unquestionable infantry ascendancy on the battlefield; in fact, one of the more fallacious conclusions drawn from the Boer War was that rapid fire by trained riflemen was more reliable than machine-gun fire, which depended heavily on transport support and the uninterrupted supply of water and ammunition. The "musketry" standards established by most armies of this period were, of course, tremendously impressive. Indirect firing was taught, and range estimation received great emphasis (all German soldiers, for example, being taught to judge distance to 800 meters). A good marksman was expected to score a hit on any human target within 250 meters, on a single kneeling opponent within 350 meters, on a kneeling file of men within 500 meters, and on a standing file within 600 meters.(25) A 400-meter field of fire was considered the minimum acceptable in a defensive position, and trenches were to be invisible at 3,000 meters when looked for through field glasses. These standards were probably considered quite normal, however, during a period when 600 meters and under were regarded as "close" range for a rifle and 600 to 1,400 meters "effective" range, the latter no doubt reflecting a recognition of the beaten zone principle.(26)

The infantry defense proper was based on the entrenched firing line with supports and reserves. Digging a hasty fire trench 5 feet long by 3 feet wide and 3 feet deep with a 1.5 foot parapet was the task of one man. If time permitted, the trench would be deepened by 1.5 feet with head protection, 9 to 12 inches of earth being the recommended standard. In 500 meters of prepared position, one would normally find about 500 men in the trenches of the firing line (one man per yard), with perhaps 100 to 200 in support assigned to join the firing line in case of attack. Another 500 to 700 men would be in the reserve, the purpose of which was to counterattack, not reinforce

AN EPOCH-MAKING CHANGE 11

the firing line. Normally, a battalion would provide only firing and support lines, with reserves--divided into local (battalion) and general (brigade)--coming from other battalions. Companies in defense were usually broken into firing line and supports, with the local reserve coming under direct control of the battalion commander. A brigade could be expected to detail one-quarter of its forces for firing line and supports, another quarter for local reserves, and one-half for the general reserve or counterattack force. A battalion on its own would usually divide into 25 percent firing line, 25 percent supports, and 50 percent local reserves.(27)

In the attack scenario, an enemy would first be encountered by advance scouts, who would immediately be reinforced by the firing line. The firing line would then be built up to the necessary strength to gain superiority of fire. The object was to close to within 500 to 800 meters and "hammer all along the line," the essence of infantry tactics being to "break down the enemy's resistance by the weight and direction of fire, and then completing his overthrow by assault." This was termed the "developing attack,"(28) and it generally aimed at building up to a concentration of one to three men per meter of front in depth. In a rifle company, this could mean three sections (later platoons) in the firing line and one in support. (Section frontage was normally 75 meters, divided into left and right squads.) In the British army before 1914, the firing line was always commanded by a "half-company" commander, whose task it was to supervise the section commanders' application of fire. The company commander normally remained with the support line.

During the firefight process, the firing line advanced by a series of rushes, entrenching as necessary to consolidate ground gained. Once superiority of fire was achieved by the firing line, the general reserve would move as secretly as possible opposite a point chosen by the responsible commander and, supported by artillery, assault in a "decisive attack." In German doctrine, the "decisive attack" was invariably directed toward a flank as the primary attack, with the frontal or "developing attack" relegated to secondary holding action. This was clearly not the case in French doctrine, which held that the frontal attack could be decisive since it could "not be divined beforehand."(29) In executing such assaults, most armies fixed bayonets 200 meters from an enemy position and attempted the final rush at 100 meters. It was generally agreed that a decisive attack launched by a force smaller than three to five men per yard of frontage in depth would rarely prove successful.(30)

Although firepower was becoming more and more

12   A PERSPECTIVE ON INFANTRY

TABLE 1-4.   THE DECISIVE ATTACK

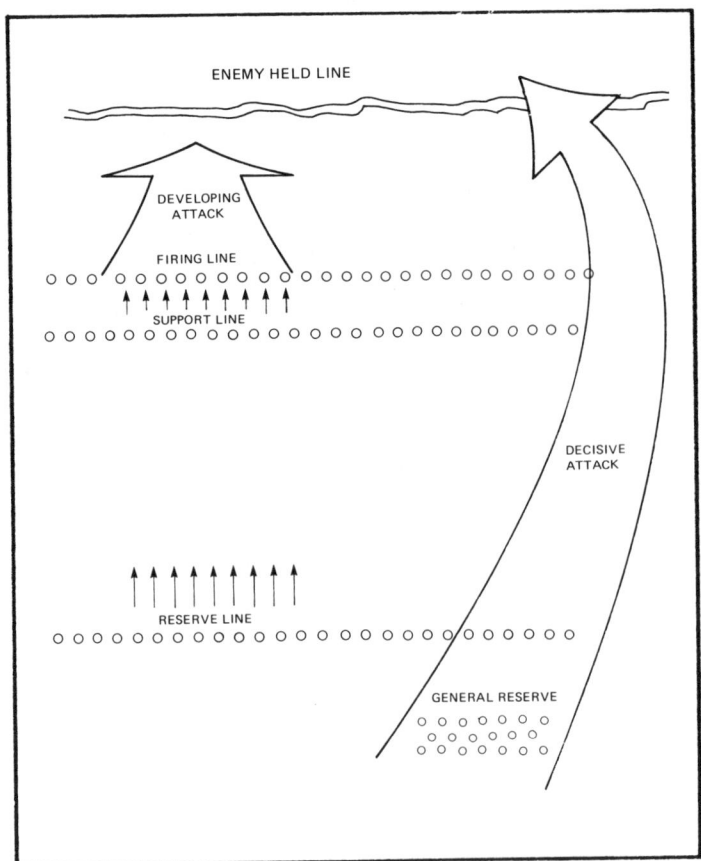

important, the theory of mass(31) continued to dominate much European military thought. When wedded to the doctrine of the offensive, despite warning signs to the contrary, it gained a particularly dangerous and mystical momentum. A French army règlement of 1887 reduced the attack frontage of a division from 2,800 to 1,600 meters and stated: "Brave and energetically commanded infantry can march under the most violent fire even against well-defended trenches, and take them."(32) A principal exponent of mass, Ferdinand Foch, argued that if 2,000 men armed with rifles capable of firing once a minute attacked 1,000 men, the "balance in favour of the attack" would be 1,000 bullets; any improvements in firearms, such as increasing the rate of rifle fire to ten rounds per minute, would merely multiply tenfold or more the strength of the offensive.(33) Belief in the absolute superiority of the massed attack eventually dominated French military thinking to the extreme, as the ideas of Foch and his star pupil, Louis de Grandmaison, prevailed. As late as 1911, the latter would declare that "there exists no other means but attack, immediate and total. . . . It is always necessary in battle to do something which would be impossible for men in cold blood: For example, march under fire."(34) Opposition to such thinking was largely interpreted as evidence of moral weakness and unfitness for command.(35)

The offensive aphorism that a "battle won is a battle in which one will not confess oneself beaten" was to exact a terrible toll in that awful morass that came to be termed the Western Front. And the toll was to be exacted primarily in terms of infantry flesh. The "idea with a sword," the offensive à outrance of Colonel Grandmaison, was to founder under the syncopated hammer of machine guns, which in eight seconds were to mathematically disprove French Field Regulations calculations that an infantryman could in a dash of 20 seconds cover 50 meters before enemy infantry could shoulder their weapons, aim, and fire. The notion that le cran, sheer guts, or will power alone could prevail was also quickly dispelled, just as attacker's bullets speeding ineffectively over the heads of entrenched enemy infantrymen disproved Fochian mathematics. The only bright light on an otherwise dismal military horizon was the much-improved marksmanship of the "contemptible little army," which, in delivering such volume of wickedly accurate, rapid fire upon the Germans during the Battle of Mons (August 23, 1914), convinced the latter that all British soldiers were armed with machine guns.(36)

The fact of the matter was, however, that the employment of mass armies paralyzed the ability of units and individuals to maneuver on the battlefield. To be sure, railways gave the infantry a new and greater strategic mobility, but sheer numbers and the weight of personal equipment worn by the rifleman crippled that arms' tactical mobility. In 1914-15, for example, an infantry battalion with all four companies in line (as was often the case) covered a frontage of 1,200 to 1,500 yards--almost one man every 10 yards. To complicate matters further, the absolute minimum weight carried by the infantryman was 60 pounds, which included his totally inadequate "wretched little entrenching tool."(37) As the number of machine guns increased each year, from six to eight per regiment or brigade in 1914 to roughly 144 in 1917, the heavily laden infantry of the firing line grew more helpless. The machine guns were the real "queens of battle," and all armies were held in the firm grip of Hiram Maxim.(38)

By the end of 1914, prewar military fixations that the contest would be quickly decided by large-scale mobile operations supplemented by sweeping cavalry actions, beginning and end, had abated before the inexorable crush of industrial technology. As the soldier dug deeper to escape the death-dealing fire of quick-action artillery and machine guns, the face of the battlefield changed. Flanks, long sought after as areas for forcing decision, virtually disappeared as scattered trenches were linked to form continuous front lines from Switzerland to the sea. The artillery arm, though also compelled to seek protection in the bowels of the earth or behind the cover of hills, became more and more dominant as the range of guns increased and a more scientific indirect method of firing was adopted. As the fire of machine guns tended to bring most advances to a halt, so a greater weight of artillery fire was gradually brought to bear to smash defending lines. It was in these circumstances, almost that of a grand artillery duel, that Marshal Henri Philippe Pétain coined his famous dictum "L'artillerie conquiert, l'infanterie occupe."(39) Infantry was relegated to "mopping-up."

Though the Germans had no more expected protracted trench warfare than the Allies, they reacted to it more sensibly by going over to the defensive and digging in after their retreat from the Marne. Unlike the Allies, they apparently sensed the futility of attempting to revive a war of movement.(40) The Germans were, of course, somewhat better equipped for seige warfare, as they had begun the war with careful plans for reducing Belgian fortresses; they had accordingly included in their weapon inventory heavy howitzers, trench mortars, and grenades. The roles of the belligerents on the Western Front were thus reversed in 1915, with the French and British committed to the offensive in accordance with the wishes of the French General Staff. Although Allied Generals were completely surprised by the German decision to dig, the wisdom of this action was subsequently borne out statistically, as the Allies never succeeded in breaching the German front, even with local tactical superiorities of five to one. (They had 140 divisions initially against 90 German in 1915,

160 against 120 in 1916, and 180 against 140 in 1917.)(41) German tactics would eventually prove, as well, that they were more sparing of soldiers' lives than those of the Allies. Even in trench design the Germans maintained a superiority.(42)

Under the spell of offensive doctrine, Allied tactics unfortunately appeared to be little else than unimaginative attempts, in the words of General Swinton, "to fight the rifle with the target."(43) Initially, artillery preparation was often overlooked, with infantry expected to advance after only the most desultory shell fire. As late as September 26, 1915, two new British divisions were committed to attack under such circumstances in the area of Loos. Twenty minutes of bombardment, which appears to have caused the Germans no casualties, was followed by a pause of about half an hour. Then 10,000 men in 12 battalions advanced up a gentle slope toward enemy trenches still protected by unbroken barbed wire. At 1,500 yards' range, the British advance met with a storm of machine-gun fire, which, in roughly three and one-half hours, killed 385 officers and 7,861 men. As remnants of the British infantry staggered away from "Leichenfeld von Loos," the Germans stopped firing in compassion. Their casualties in the same time had been nil.(44)

During the battles of the Somme in 1916 the British army introduced and perfected the "creeping barrage." Preceding the infantry at the rate of 100 yards in four, six, eight, or more minutes, according to the condition of the ground and degree of enemy resistance, it moved forward by lifts of 50 or 100 yards at a time. The basic infantry tactic, fathered by the firing-line mentality, was to advance behind the barrage in "waves" of men, two to three yards apart, each wave followed by another 50 to 100 yards in rear. A four-company battalion normally advanced in two or four waves, Lewis machine gunners (two to a gun) preponderantly placed to the front. Foremost infantry were urged, at all costs, to keep within 50 yards of the creeping barrage, which on attainment of the objective moved forward 200 to 300 yards beyond to become a "standing" protective barrage. A major advantage of the creeping barrage, in the British view, was that the infantry could move forward at a walking pace instead of being compelled to advance by short rushes.(45)

Although French small-unit tactics by the time of the Somme battles emphasized advancing by small-group rushes, the British tended to regard such "fire and movement" as difficult to teach, at least "too difficult to be taught to the Kitchener divisions." Thus, the British continued to attack in waves up to the Battle of Passchendaele (1917), giving certain

16   A PERSPECTIVE ON INFANTRY

TABLE 1-5.   GERMAN TRIANGULAR DIVISIONAL ORGANIZATION
IN THE GREAT WAR

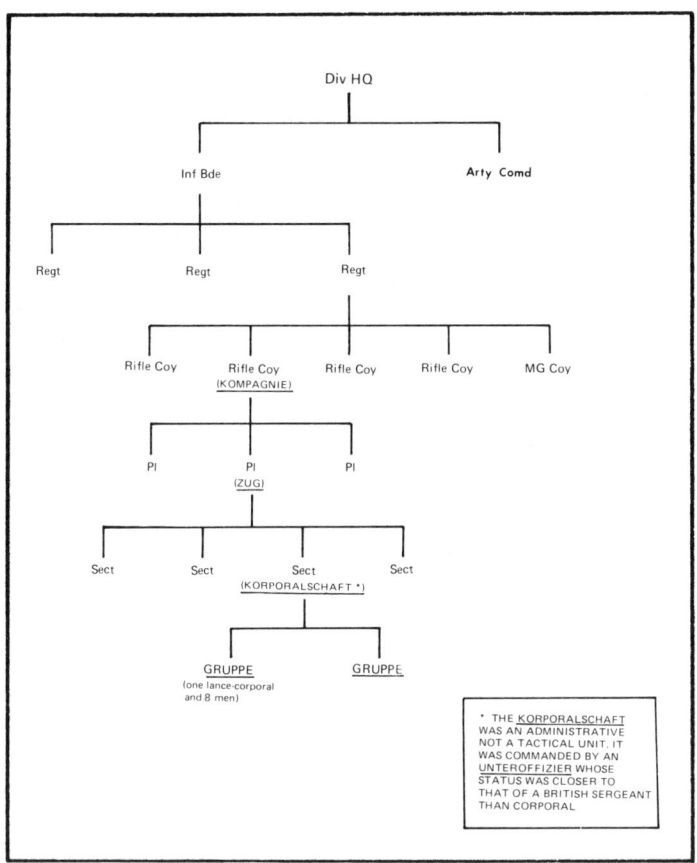

German regiments the distinct impression that the
tactical handling of British infantry was even worse
than that of Russian infantry.(46) The trouble with
such mass tactics, of course, was that there was little
room initially for employment of small tactical units
like the platoon. This weakness was unfortunately

compounded by certain major shortcomings of the creeping barrage. Quite often, there were left enemy pockets of resistance that artillery fire alone could not destroy or even neutralize. These were machine gunners and infantrymen who hid in dugouts or pillboxes or shell holes and somehow managed to survive until after the barrage had passed. They then reappeared to engage the following waves of attacking infantry with deadly effect. The latter troops, by now floundering in the mud and churned-up fields created by the barrage, could do little else than go to ground and watch their rigidly timed protective fire literally roll away. To better illustrate the double-edged nature of massive artillery bombardment, the 4-million-odd shells expended in front of the British army at Passchendaele merely served to create a self-made obstacle of "liquid mud" that doomed the offensive before it was launched.(47)

From the battles of the Somme to March 1918, the German army on the Western Front again adopted the defensive. Reorganizing their "square" infantry divisions of four regiments (in two brigades) into more maneuverable and tactically flexible organizations of three infantry regiments, the Germans commenced to reexamine their defensive doctrine. In the course of the Somme battles, the Germans had learned the necessity for absorbing the shock of assault rather than attempting to break it on one or two trench lines. On many occasions, they had succeeded in accomplishing the latter, but normally at great expense in manpower since it meant concentrating masses of troops forward, in close formation, and exposing them to almost certain destruction by enemy trench-breaking artillery bombardment. It was recognized, on the other hand, that if the infantry was held far enough back to avoid hostile bombardment and the artillery echeloned in greater depth, it would be impossible to prevent enemy-infantry penetration. The suggestion was made, however, that this would not really matter, as the opportunities for successful counterattack were greater; not only would the enemy infantry have to fight on ground more familiar to the defender, it would likely have to do so at the end of its artillery support.

A new system of defense based on counterattack was accordingly introduced within the German army. Organized in a number of defensive zones rather than in a series of easily identifiable rigid lines, as was hitherto the case on both sides, it incorporated an "elastic" or mobile capacity for defending in depth. Whereas before commanders had to hold a line, they could now resist and retire in an offensive-defensive mode as the situation allowed; knowing their own ground better than their opponents, they could even prepare and practice on it. The deeper an enemy attack penetrated, therefore, the more likely it was to encounter the unforeseen and unexpected. As a whole, the defensive battle was envisioned as a series of engagements characterized by counterattacks on a continually larger scale, the objective of every endeavor being to retain possession of all ground held originally.(48) To educate the German army in this system, a booklet entitled "The Defensive Battle" was issued to all German divisions in December 1916. Written by two operations staff officers, it had been officially sanctioned by none other than General Erich Ludendorff. Intensely interested in basic tactical techniques, Ludendorff had even personally intervened to advocate the new methods.(49)

Not surprisingly, the new doctrine espoused soon sparked a heated tactical controversy, which Ludendorff himself appears to have encouraged. Though the old Prussian maxim <u>Halten was zu halten ist</u> ("hold what you are ordered to hold") was gradually forsaken for the concept of defense-in-depth, it still influenced tactical thinking on how the battle for the foremost line should be fought. While the group of bright young staff officers(50) who first perceived the advantages of the elastic defense thought it might preferably be fought behind the foremost line, more conservative and battle-experienced officers like Colonel Fritz von Lossberg maintained that no "invitation-to-walk-in" should ever be offered to an enemy. Front trench garrisons, von Lossberg argued, should only be permitted to sidestep or go forward; to "yield elastically" was to run the risk of almost total tactical disorganization. Ludendorff, displaying great flexibility and vision, not only allowed von Lossberg to circulate his contradictory doctrine to all divisions but also had his ideas incorporated into a new "Manual of Infantry Training for War." In the course of the debate that followed, however, von Lossberg was eventually converted with but minor reservations to the school of his juniors. Ironically, he was the first to put the elastic defense into practice, by which act he ultimately gave his name to it.(51)

Von Lossberg's change of heart represented more than mere acceptance of the principle of "parry and thrust" or even the replacement of a rigid passive defense with an active mobile one. It was, in fact, a significant acknowledgment of the necessity for an increased decentralization in tactical control. Though von Lossberg had been instrumental in broadening the tactical scope and power of battalion commanders, he was reluctant to decentralize control below that level.

AN EPOCH-MAKING CHANGE   19

The greater dispersion and isolation forced on fighting units, however, plus the liberal permission granted them to retire, all conduced to make control of the battle more and more difficult for battalion and even company commanders. Fortunately for them, the Germans had an abundant supply of first-class peacetime-trained noncommissioned officers (NCOs),(52) traditionally the backbone of all conscript armies. For these and other associated reasons, the <u>gruppe</u> of one NCO and eleven men now became the "official tactical battle unit." Ludendorff, as fearful of disorder as von Lossberg, referred to this increased decentralization as "a risky business,"(53) but being a revolutionary thinker in his own right nonetheless supported it:

> . . . a new system was devised, which, by distribution in depth and the adoption of a loose formation, enabled a more active defence to be maintained. It was of course intended that the position should remain in our hands at the end of the battle, but the infantryman need no longer say to himself: "here I must stand or fall," but had, on the contrary, the right, within certain limits, to retire in any direction before strong enemy fire. Any part of the line that was lost was to be recovered by counter-attack. The group, on the importance of which many intelligent officers had insisted before the war, now became officially the tactical unit of infantry. The position of the N.C.O. as group leader thus became more important. Tactics became more and more individualized.(54)

As established, the von Lossberg style of defense consisted of a forward outpost zone, a main battle zone, and a rear battle zone. The outpost zone varied in depth from a mile to a mile and a half, depending on the lay of the land, and it contained but 10 to 12 percent of the total forward troops deployed. These were normally disposed in checkerboard pattern of mutually supporting, defended localities and machine-gun casemates, their purpose to give warning while at the same time delaying the enemy as <u>Wiederstandsnester</u> ("nests of resistance"). The outpost zone could be held elastically, with troops permitted to withdraw if heavily pressed. The main line of resistance, sited on a <u>Hinterhang</u> (reverse slope), formed the back of the <u>outpost</u> zone and the front of the battle zone that lay behind it. It had the double function of defending artillery observation posts and representing the objective of the counterattack if the enemy penetrated the position. The main battle zone, in which most enemy attacks were to be stopped, was about a mile and a half in depth and bounded on the rear by

20 A PERSPECTIVE ON INFANTRY

TABLE 1-6. ELASTIC DEFENSE

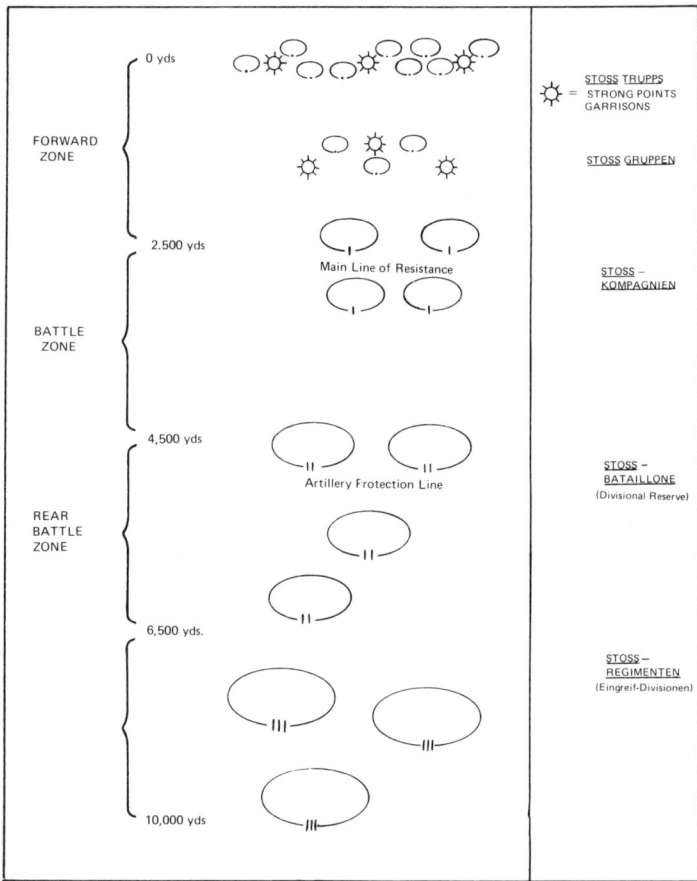

an artillery protection line, back of which the bulk of the field guns were distributed. Behind the main battle zone lay the rear battle zone in which were husbanded the bulk of the field guns and critically important Eingreif-divisionen counterattack forces. Counterattacks in the outpost zone were carried out by

Sturmtruppen and in the main battle zone, by Sturm-bataillone.(55)

The advantages of the elastic defense soon proved obvious. In providing passive protection from hostile artillery fire, it was much superior to the "plane defense," that series of lines, each comprising front, support, and reserve entrenchments, that had emerged as an alternative to a one main line of resistance. Compared by stormtrooper Ernst Junger to a "steel sinew," it was originally intended to bend and give way to an attack, then snap back, sweeping spent attackers before it. However, the easily recognizable and indeed conspicuous lines of the "plane defense" became veritable death traps for the soldiers manning them, as they were subjected to an ever-increasing weight of artillery fire. In escaping to the safety of shell holes, which provided no recognizable artillery target, on the other hand, the infantry regained to a certain degree its freedom of movement simply because it was no longer tied to one spot. Under the refinement of elastic defense, of course, positions became even more widely dispersed and consequently less vulnerable to enemy shellfire; they were thus better able to separate attacking waves of infantry from their creeping barrages. Machine guns formerly sited to break up assaults were now sited to support counterattacks.(56)

Termed by the British the "blob" system of defense, from the innumerable pillboxes and strongpoints that dotted its landscape, the elastic form was eventually to be adopted by both sides. The French version was made up of a certain number of connected "centers of resistance," formed by a combination of "strongpoints," the most basic unit for the organization of the ground being constituted by the group corresponding to the automatic weapon, or "combat group." The French considered the formation of the "combat group," the very skeleton of their defensive system, to be the most significant innovation. The British, however, took a slightly opposite view, alluding in their Official History to the British soldier's aversion to fighting in isolated "bird cages." There was, of course, some doubt as to whether the British fully appreciated the vital necessity of having counterattack forces when conducting a defense in depth. Their method of attacking such a system also differed from the German approach inasmuch as it involved ponderously "leap-frogging" troops from one secured objective to another; in this manner, the creeping barrage was prevented from running away.(57)

The German approach to breaking the impasse of the trenches was much more imaginative than that of either the British or the French. As previously intimated, to quote Ludendorff, a "vigorous intellectual life was observable in all branches of the army" around this time. Recognizing that fire had compromised surprise and shattered the chain of command in the attack as well as in the defense, the Germans began turning the decentralized tactics of the elastic defense into new tactics for the advance. Methods used in the counterattack were accordingly applied to offensive operations generally. Instead of selecting limited objectives to be "conquered by artillery and occupied by infantry," the Germans expanded the infantry's role from the mere occupation and holding of ground (but one facet of infantry operations in the resistance-riposte system of von Lossberg) to having it conduct battle through its own means to break enemy resistance. In Ludendorff's view, the German infantry in defense had grown "flabby," no longer "able to hold the enemy off and to fight from a distance." The infantryman had forgotten how to deliver accurate fire with his rifle; his main weapon had become the grenade.(58) The German Frontschwein were not alone in this regard, however, as poilus and Tommies had also entered the world of the troglodyte in the trenches and mining operations of 1915 and 1916. They, too, had embraced the club as of more use than the rifle-bayonet.(59)

In their search for a tactical rather than technical solution to the riddle of the trenches, the Germans were aided by the experiences of a French officer, Captain André Laffargue, who had participated in an attack on Vimy Ridge in May 1915. Noting that two machine guns had held up battalions of advancing troops and that artillery could not get forward to neutralize them, Captain Laffargue recommended in a pamphlet entitled L'Etude sur l'attaque that light mortars accompany the infantry up front. He further advocated sending forward from each platoon of line infantry specially trained groupes de tirailleurs armed with light machine guns or automatic rifles; their tasks were to "infiltrate" enemy lines and destroy machine-gun emplacements from the rear. Laffargue's suggested system differed from the normal practice of the time in attacking trenches, which was to spread out laterally, moving from fire bay to fire bay, bombing one's way along from traverse to traverse. A costly and cumbersome procedure, this method also took some momentum out of the attack.(60)

Unluckily for the Allies, Laffargue's ideas were largely ignored in the French army, and his pamphlet was never translated into English. A copy was captured by the Germans, however, and considered by them to correspond so closely to their own tactical thinking(61) that it was immediately translated and issued as an official training manual. There is even evidence to indicate that it was used as a basis for

Captain Geyer's handbook on "The Attack in Trench Warfare." To be sure, the combination of movement and fire proposed by Laffargue and directed toward countering the decisive defensive weapon, the well-posted machine gun, was certainly the mark of 1917-18 German infiltration tactics. Unlike British and French infantry to this point, infiltrating German infantry did not merely "mop up" the effects of fire; they were, in fact, supported by fire and thus represented a return to a doctrine that all previous armies had abided by, namely, fire effect combined with movement.(62)

The essence of German infiltration tactics as developed was their high degree of decentralization of command. Instead of attacking "limited" hard objectives in waves, the advancing infantry flowed in small groups along lines of least resistance, seeking out "soft spots" through which to penetrate enemy defenses. In order to achieve maximum surprise, the attack was opened by an intense preliminary bombardment that lasted for hours instead of days. Designed and perfected by Colonel Georg Bruchmuller on the Eastern Front, the Feuerwalze had as its object not the smashing of field fortifications but rather the paralysis of the enemy's communications and artillery. While also providing close indirect fire support to attacking troops, it had the further advantage of leaving intact the ground over which the infantryman had to pass.(63)

The infantry attack proper was spearheaded by special assault Stosstruppen or Sturmtruppen (storm troops), whose task it was to bypass strongpoints and advance boldly by infiltrating small groups until they reached the enemy's artillery. Like an increasing tide, they were to flow always forward and round by paths of least resistance. As the assault units pushed through, follow-up infantry "battle units" and reserves were committed at the point where the progress was greatest to strengthen the breakthrough. The task of the battle units, composed of infantry, machine gunners, trench-mortar teams, engineers, field artillery, and ammunition carriers, was to reduce enemy strongpoints and defeat counterattacks. No obstacle was to hold them up for long, however, as they were to keep as close as possible to advancing storm troops. Other reserves were to destroy any resistance they bypassed. Surprise was the key to these new German tactics, and troop concentrations were therefore kept secret, with most movement being executed at night. The enemy was further confused by feints and deceptions.(64)

Though improvised ad hoc organizations, storm-troop groups were made up of the youngest, fittest, and most experienced soldiers from ordinary units.(65)

Well versed in Ludendorff's new doctrine, enunciated by Geyer and others, they set the standard for the rest of the infantry. Trained essentially to overcome the elastic defense through adherence to the theory of the "unlimited" objective, they required an even higher standard of junior leadership than that demanded in the implementation of that same defense. Like the remainder of the German infantry, their basic tactical unit was the group of 10 to 12 men with its own base of fire in a light machine gun (or automatic rifle) and light mortar.(66) The advent of the light machine gun in particular had given infantrymen a most valuable weapon both to attack with and to defend themselves with if counterattacked. The realization of this double purpose of the light machine gun, now "the chief firing weapon of the infantry," appears to have prompted the Germans to radically reorganize their smallest infantry battle elements. In Ludendorff's view, companies had to be "provided with new light machine-guns, the serving of which [was to] be done by the smallest possible number of men"(67):

> Previously it had been the squad (Trupp) of seven men under a leader; it now was to be the group (Gruppe) containing a light machine gun Trupp of four men (two with the gun and two ammunition carriers) and an assault (Stoss) Trupp of seven riflemen, each Trupp under a leader. The Trupps of a Gruppe might be separated or be grouped with those of other groups, as circumstances demanded, but the important point was that this smallest molecule of the German infantry arm contained within it the essential requirements for both defence, the parry by the light machine Trupp, and for offence, the thrust by the Stosstrupp.(68)

The Germans first actually experimented with infiltration techniques in the Battle of Verdun where surprise had been achieved through a shorter than normal artillery bombardment. In this action, instead of hundreds of thousands moving toward the enemy in long, unbroken lines, small groups of infantrymen had gone forward as infiltration teams, probing for weak spots. This tactic was only executed on a narrow front, however, and never exploited.(69) Interestingly, General Sir Arthur Currie was sent to examine the Verdun battlefield in January 1917 to study and determine what lessons could be learned. On return, he recommended that the policy of attacking in waves be discontinued and that the smaller and less vulnerable platoon be established as the basic maneuver unit and used to spearhead attacks. He further advocated that platoons and sections should be assigned easily

TABLE 1-7. IMPROVED GRUPPE ORGANIZATION

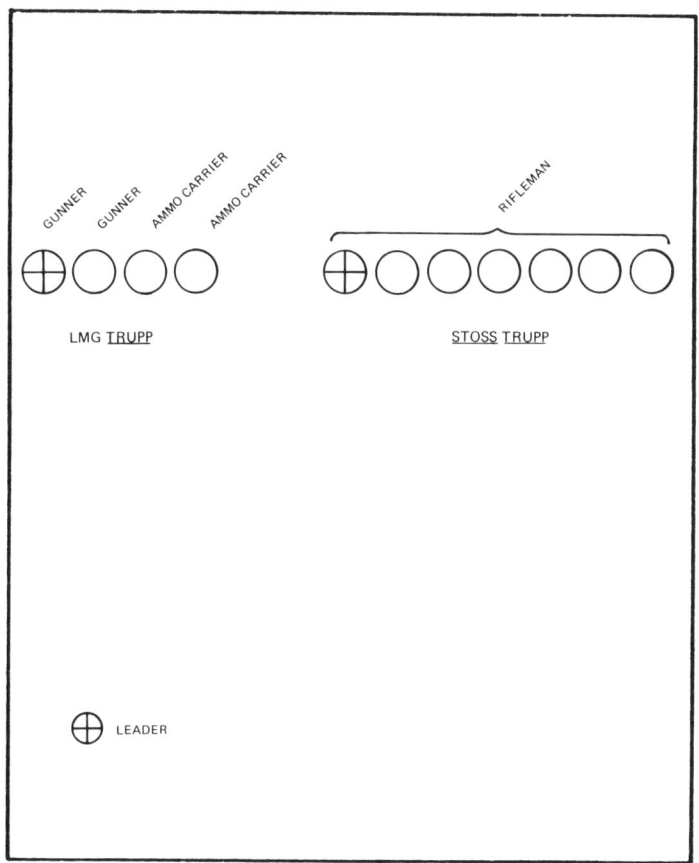

recognizable objectives in the attack and that fire and movement drills should be used to attain them. Unfortunately, however, most of Currie's progressive ideas were not immediately adopted. In pioneering the development of "soft spot" infiltration tactics, the Germans definitely held the edge.(70)

On September 1, 1917, the German Eighth Army under

General Oskar von Hutier(71) suddenly attacked Riga using storm troop "soft spot" tactics; the city fell to him in two days. The German General Staff was now confident that it had found the solution to the impasse of the trenches. Similar tactics were used with resounding success at Caporetto(72) in October, and again on November 30, 1917, to blunt the British tank attack at Cambrai. Finally, in March 1918, the Germans made their last attempt to win the war. The new tactics succeeded far beyond expectation, particularly in the sector of Hutier, "the apostle of the surprise attack," where his army advanced without check. He took 50,000 prisoners, crushed the Fifth British Army, and drove a wedge between British and French forces. Fortunately for the Allies, Ludendorff at this point demonstrated a singular lack of strategic grasp. Of the three armies he had employed in the offensive, only Hutier's continued to make progress; yet Ludendorff failed to exploit this success by redirecting the other two along Hutier's axis. Had Ludendorff done so, he could possible have split the French and British armies and perhaps gained a Sedan 22 years earlier.(73)

According to Liddell Hart, there were two possible means of reviving movement on the battlefield in the Great War: one was to make men bulletproof by putting them in armored vehicles; the other was to revive stalking methods. While the British pioneered the first, the Germans unquestionably led in development of the second. The giving of the power of maneuver to the smallest infantry units even when forming but a segment of the fabric of a large-scale offensive was in Liddell Hart's opinion, an "epoch-making change." Though convergent attack from two directions simultaneously had always been a traditional guiding aim of major tactics, its application to minor tactics really dates from the Great War. For centuries, battalions and companies forming part of a large attacking force had but line or frontal roles. Under new circumstances of dispersion, however, penetration between centers of resistance was made possible. A gap found or opened by a small section could now lead to bigger things. Reserves could accordingly be used for their intended purpose: the exploitation of success rather than the reinforcement of failure. Moreover, the issue of battles depended as never before on the tactical knowledge and ability of junior officers and NCOs.(74) And herein lay the greatest change of all.

NOTES

1. In the eighteenth century, armies in direct contact suffered extremely heavy casualties, often as

high as 50 percent when they decided to fire for effect. J. F. C. Fuller, The Conduct of War (London: Eyre and Spottiswoode, 1961), pp. 20 and 22-3.

2. Colonel F. N. Maude, Notes on the Evolution of Infantry Tactics (London: William Clowes, 1905), pp. 77, 107, 147, and 154-6.

3. T. Miller Maguire, Notes on the Austro-Prussian War of 1866 (London: Hugh Rees, 1904), pp. 62-3; Colonel E. M. Lloyd, A Review of the History of Infantry (London: Longmans, Green, 1908), pp. 252-3; Captain A. F. Becke, An Introduction to the History of Tactics, 1740-1905 (London: Hugh Rees, 1909), pp. 40, 44, and 45; and Theodore Ropp, War in the Modern World (New York: Collier, 1962), pp. 163-9 and 190-6. The dominance of the infantry firearm was not admitted by all since the army possessing the inferior infantry firearm prevailed in 1859 (the French generally winning the field by bayonet assault), and again during the Franco-Prussian and Russo-Turkish wars. Becke, History of Tactics, pp. 37 and 53. Nonetheless, the Prussians had already recognized the dominance of infantry over cavalry inasmuch as, by 1866, the former arm was advised not to bother forming square to resist the latter. Lloyd, History of Infantry, p. 265.

4. Lloyd, History of Infantry, pp. 232 and 239; and Becke, History of Tactics, pp. 45 and 48.

5. Major General Emory Upton, The Armies of Asia and Europe (New York: D. Appleton, 1878), pp. 27-75.

6. Colonel William Balck, Tactics, trans. Walter Krueger (Fort Leavenworth: U. S. Calvalry Association, 1911), Vol. 1, pp. 173-6 and 182. The lesson was driven home in the Russo-Turkish War where "long range infantry fire [even unaimed]" was found to do "much damage if the slope of the ground . . . [was] favourable." Becke, History of Tactics, p. 51. Maude, on the other hand, continued to believe that the "ratio of unintentional hits to rounds fired must be far less in the future than in the past." Maude, Evolution of Infantry Tactics, p. 147.

7. Jay Luvaas, "European Military Thought and Doctrine," The Theory and Practice of War, ed. Michael Howard (London: Cassell, 1965), p. 83.

8. Maude, Evolution of Infantry Tactics, pp. 51, 61, and 82.

9. Balck, Tactics, Vol. 1, pp. 102-5; and Lloyd,

History of Infantry, pp. 267 and 273-4. William I of Prussia and his advisers were "unwilling to accept . . . [the] loss of control" resulting from skirmish tactics. Others were opposed to skirmishing because it "spoil[ed] the men's clothes" and because "the teaching of grown-up men to play hide-and-seek seem[ed] in itself to be antagonistic to the average soldier's conception of his duty." Maude, Evolution of Infantry Tactics, p. 5.

    10. Lloyd, History of Infantry, p. 267; Luvaas, "European Military Thought and Doctrine," p. 84; and Balck, Tactics, Vol. 1, p. 75.

    11. Balck, Tactics, Vol. 1, p. 33.

    12. The number previously varied from 10 in Cromwell's time to 13 under Marlborough to 11 under Wellington. Each company from 1821 consisted of four sections of two squads each. Company strength was about three officers and 117 men under conditions of active service. A battalion of infantry thus consisted of roughly 1,000, all ranks. Lieutenant Colonel H. J. Kinsman, Tactical Notes. (Dublin: E. Ponsonby, 1914), pp. 37, 47, and 60; and Lieutenant Colonel R. J. S. Langford, Corporal to Field Officer: A Ready Reference for All Ranks in Peace and War (Toronto: Copp Clark, 1941), p. 5. By Upton's time, some Russian line battalions consisted of five companies. In both the British and Indian armies there were eight companies per battalion. The French, following the Franco-Prussian War, reduced the number of companies per battalion from six to four. This brought them in line with German organization, as well as with the armies of Austria and Japan, all of which had four companies to a battalion, each commanded by a captain. Upton, The Armies of Asia and Europe, pp. 9, 65, 149, 162-3, 192-3, 227, 251, 270, and 287.

    13. Balck, Tactics, Vol. 1, pp. 34-5.

    14. Major General Sir Ernest Swinton, Eyewitness (London: Hodder and Stoughton, 1932), p. 114. Swinton wrote a pamphlet on defensive digging entitled, The Defence of Duffer's Drift by "Backsight Forethought"; by any measure, it is a classic tactical document, as relevant to today's battlefield as it was to Swinton's.

    15. Deneys Reitz, Commando: A Boer Journal of the Boer War (London: Faber and Faber, 1929), pp. 39 and 42; B. H. Liddell Hart, Thoughts on War (London: Faber and Faber, 1944), pp. 90, 93, and 282; and Balck, Tactics, Vol. 1, pp. 88 and 132. Reitz records: "I saw

soldiers [at Nicholson's Nek] looking over their defences to fire, and time after time I heard the thud of a bullet finding its mark." Reitz, Commando, p. 42.

16. Lloyd, History of Infantry, p. 287; and Balck, Tactics, Vol. 1, pp. 90-1, 321, 341-2, and 392. The Japanese found in their attacks that at ranges from 150 to 75 paces hostile rifle fire had no effect. Balck, Tactics, Vol. 1, p. 179. Wounds in the Russo-Japanese War were, on the basis of a 100 hits, distributed as follows: lower limbs, 39.5; upper limbs, 25.4; abdominal region, 16.5; chest, 15.5; spinal column, 15; and head, 11. Wounds caused by the small 6.5-millimeter caliber Arisaka '97 rifle healed more rapidly than those caused by the Russian 7.62-millimeter '91 model; Japanese flesh wounds healed in ten days, Russian in four. Balck, Tactics, Vol. 1, pp. 127-8

17. Barbara W. Tuchman, The Guns of August (New York: Macmillan, 1962), pp. 189 and 232; and Sir John W. Hackett, The Profession of Arms (London: Times Publishing, 1963), pp. 50-1. As late as 1915 Haig would also write the machine gun off as a much "overrated weapon" and recommend two per battalion as sufficient.

18. Major J. F. de Pardieu, A Critical Study of German Tactics and of New German Regulations, trans. Captain Charles F. Martin (Fort Leavenworth: U.S. Cavalry Association, 1912), p. 117.

19. Balck, Tactics, Vol. 1, pp. 34, 341-2, and 392; Lloyd, History of Infantry, pp. 272-3 and 287; and Becke; History of Tactics, p. 87.

20. Liddell Hart, The Future of Infantry, p. 24.

21. Major M. F. de Pardieu, German Tactics, pp. 22-3 and 102; and Luvaas, "European Military Thought and Doctrine," p. 86. Emphasis on bayonet action was surely retrograde, as the great Moltke had in 1861 stated: "In principle when one makes a bayonet charge, it is because one supposes that the enemy will not await it." Liddell Hart, The Future of Infantry, p. 23. The myth of the decisiveness of the bayonet attack was to die hard.

22. Balck, Tactics, Vol. 1, p. 87; and Lloyd, History of Infantry, p. 274. The French army following the Boer War gave a freer hand to captains of companies and entrusted section commanders with fire control. A system of advance by groups was also substituted for a chain of skirmishers. Lloyd, History of Infantry, p. 283.

23. Langford, Corporal to Field Officer, p. 5.

24. Balck, Tactics, Vol. 1, pp. 32 and 47.

25. Balck, Tactics, Vol. 1, pp. 168, 236, and 302. Machine guns were viewed as "not adapted to carry on protracted fire actions." Ibid., p. 236. This opinion was only based on mobile warfare experience, however; static warfare conditions quickly proved it to be fallacious. For convincing proof of the tremendous emphasis on individual shooting, see War Office, WO 2227, Musketry Regulations, Part 1, 1909 (London: HM Stationery Office, 1909, amended 1914).

26. Charles Edmund Carrington, Soldier From the Wars Returning (London: Hutchinson, 1965), p. 25; Kinsman, Tactical Notes, pp. 87, 171, and 179; and Balck, Tactics, Vol. 1, p. 143. Long range was defined as between 1,400 and 2,000 meters and distant as 2,000 to 2,800 meters. War Office, WO 2232, Field Service Regulations, Part 1: Operations 1909 (London: HM Stationery Office, 1909, amended 1914), p. 17. Again, much of this fire must have been effective only because of the beaten zone effect. Magazine rifles issued to the German army in 1890-1891 could send their bullets 3,000 meters, which was not much short of the effective range of shrapnel; the rifle thus threatened to overtake the field gun as a long range weapon of engagement. Lloyd, History of Infantry, p. 275.

27. Most detail here is taken from British references, but Continental armies in principle generally employed similar tactics. War Office, WO 2052, Infantry Training: Company Organization (London: HM Stationery Office 1914), pp. 148-57; Field Service Regulations, pp. 145-54; and Kinsman, Tactical Notes, pp. 86, 101, 143, 147-8, 154, 157, and 163.

28. Infantry Training, pp. 133-47; and Kinsman, Tactical Notes, pp. 101, 109, and 147-8.

29. de Pardieu, German Tactics, pp. 130-1 and 140-1. The Germans consistently advocated the envelopment or wing attack; always they attempted to hold the enemy on his front, turn him, and crush him. Major de Pardieu considered it a mistake to "claim that the wing attack is always preferable to a frontal attack." Ibid., pp. 140-1.

30. Infantry Training, pp. 133-47; Field Service Regulations, p. 19; Becke, History of Tactics, p. 57; and Kinsman, Tactical Notes, pp. 84-6, 100, 104, 107, 113, and 163.

31. According to Liddell Hart, this monster was the "child of the French Revolution" and the "midwife who brought it into the world . . . the Prussian philosopher of war, Clausewitz, cloudily profound." Liddell Hart, Thoughts on War, pp. 33.

32. Correlli Barnett, The Swordbearers (London: Eyre and Spottiswoode, 1963), pp. 245-6.

33. Mass was the central term of Foch's three characteristics of modern war: preparation, mass, and impulsion. Ropp, War in the Modern World, pp. 196-7 and 218. According to Liddell Hart, Foch was launched upon his "successful" military career when he organized a great review at the end of annual maneuvers. Foch supposedly told the story: "At eight o'clock there wasn't even a cat on the parade ground. At ten minutes past there were 100,000 men." To Liddell Hart it indicated the "paradox in military affairs that the only way to obtain license for intellectual ideas is to prove oneself expert in conventional practices. Having proved himself a super-drillmaster, Foch was considered fit to guide the higher study of war." Liddell Hart, Thoughts on War, p. 102. Thus was introduced to the military mind the "bullet-counting" method of working out tactical doctrines, as fallacious a process as counting chicken entrails as a form of safe prediction.

34. Barnett, The Swordbearers, pp. 247-8.

35. Ropp, War in the Modern World, p. 222. The brilliant French military writer and critic Captain Mayer attacked Foch's ideas and prophesied a seige warfare of years. He had little influence and less promotion. Barnett, The Swordbearers, p. 247.

36. Tuchman, The Guns of August, pp. 32-3, 232, and 256; Ropp, War in the Modern World, p. 218; and Brig. General J. L. Jack, General Jack's Diary, ed. John Terraine (London: Eyre & Spottiswoode, 1964), p. 29.

37. Terraine, General Jack's Diary, pp. 34 and 82; and Robert Graves, Good-bye to All That (London: Jonathan Cape, 1929), pp. 146-7. The entrenching tool was quite adequate when digging was in its infancy, which was precisely the point at which it was introduced; suitable for tactics of the firing line, it was useless when it came to serious digging, as General Jack so clearly indicted.

38. Liddell Hart, The Future of Infantry, pp. 25-6; and Miksche, Atomic Weapons and Armies (London:

32 A PERSPECTIVE ON INFANTRY

Faber and Faber, 1955), p. 51. The Germans were initially armed with the same gun as the British, the Vickers-Maxim (made under license in Germany), but they had a greater appreciation of its value. Swinton, Eyewitness, p. 181.

39. Carrington, Soldier From the Wars Returning, pp. 25, 83, and 89; Miksche, Atomic Weapons and Armies, pp. 49-50; and Shelford Bidwell, Modern Warfare: A Study of Men, Weapons and Theories (London: Allen Lane, 1973), p. 62.

40. Leon Wolff, In Flanders Fields (New York: Viking, 1958), p. 6. German strategy, of course, called for holding on the Western Front, as they were heavily engaged in the East.

41. Terraine, General Jack's Diary, p. 73; Carrington, Soldier From the Wars Returning, p. 83; Eric J. Leed, No Man's Land; Combat and Identity in World War I (Cambridge: University Press, 1979), p. 98. In the autumn of 1918, the Allies, in spite of a threefold superiority, succeeded only in pushing the Germans out of successive defense lines but not in breaking them. Otto Heilbrunn, Conventional Warfare in the Nuclear Age (New York: Praeger, 1965), p. 71.

42. S. L. A. Marshall, Blitzkrieg (New York: 1940), p. 82. The German trenches were far better constructed, as they had dug in to stay. According to Charles Carrington, the Germans were "far cleverer than we at trench design and repair" and capturing a German trench meant moving from discomfort to comfort. Charles Edmund Carrington (pseud. Charles Edmonds), A Subaltern's War (London: Peter Davies, 1929), p. 215; and Alan Lloyd, The War in the Trenches (New York: David McKay, 1976), p. 49.

43. Swinton, Eyewitness, p. 115.

44. Hackett, The Profession of Arms, p. 51. One machine gun fired 12,500 rounds that afternoon. It was "the birthday of the idea of siting heavy machine guns at the back of . . . [a defensive] position." Captain G. C. Wynne, If Germany Attacks: The Battle in Depth in the West (London: Faber and Faber, 1939), pp. 76-7.

45. "Infantry Tactics, 1914-1918," R.U.S.I. Journal, 64 (1919): 463-5; Terraine, General Jack's Diary, pp. 144-5, 225-6, and 272; Carrington, A Subaltern's War, pp. 63 and 137; and Guy Chapman, ed., Vain Glory (London: Cassell, 1937), p. 44. The "creeping" barrage was first used by General Walter

AN EPOCH-MAKING CHANGE 33

Congreve's Thirteenth Corps on July 1, the first day of the Somme. Moving forward by 100-yard lifts intimately covering the advance of infantry, it was a refinement of the 1915 "lifting" barrage that shifted from objective to objective.

46. John Keegan, The Face of Battle, p. 226; and Wynne, If Germany Attacks, pp. 277-8.

47. Tom Wintringham, Deadlock War (London: Faber and Faber, 1940), pp. 147-8; Carrington, A Subaltern's War, p. 137; and John Swettenham, To Seize the Victory; the Canadian Corps in World War I (Toronto: Ryerson, 1965), pp. 148, 181, and 197.

48. Wynne, If Germany Attacks, pp. 152-3, 155-8, and 222.

49. General Erich Ludendorff, My War Memories (London: Hutchinson, 1919), Vol. 1, p. 387 and Vol. 2, p. 573.

50. Major (Later Colonel) Bauer, Major Bussche, and Captains Geyer and Harbou. Of the four, only the last had front-line experience.

51. Wynne, If Germany Attacks, pp. 88-9, 132, 149, 157-61, 188, 202, 206-7, and 292.

52. By way of comparison, the French army had only one-third of the number of NCOs on strength of the German army. Barnett, The Swordbearers, pp. 41, 210, 244, and 249.

53. Wynne, If Germany Attacks, pp. 86, 125, 157, 160, 209, 251, 291, and 295.

54. Ludendorff, My War Memories, Vol. 1, p. 387

55. Falls, Ordeal by Battle, pp. 105-7; Ludendorff, My War Memories, Vol. 1, p. 387; Lieutenant Colonel Lucas, The Evolution of Tactical Ideas in France and Germany During the War of 1914-1918, trans. Major P. C. Kieffer (Paris: Berger-Leorault, 1923), pp. 122-4 and 156; and Wynne, If Germany Attacks, pp. 148, 150-1, and 185. During the battle for Verdun in May 1915, the Germans had encountered a form of defense in depth built around mutually supporting "centres of resistance." As described by one combatant, Verdun was "a battle of very small units, each directing its own fate." Lucas, Evolution of Tactical Ideas, pp. 71 and 75-9; and Leed, No Man's Land, p. 104.

34   A PERSPECTIVE ON INFANTRY

56. Ludendorff, My War Memories, Vol. 1, pp. 271 and 273; and Leed, No Man's Land, pp. 102-4; and Wynne, If Germany Attacks, pp. 121-3 and 209.

57. Brig. General Sir James Edmonds and Major A. F. Becke, British Official History of the Great War: Military Operations, France and Belgium, 1918 (London: Macmillan, 1937), pp. 477-8; Swinton, Eyewitness, p. 153; Lucas, Evolution of Tactical Ideas, p. 124; and "Infantry Tactics," R.U.S.I. Journal, pp. 466-8.

58. Ludendorff, My War Memories, Vol. 1, pp. 267, 272, and 388. Brig. General Jack, in order to maintain shooting skills in his battalion while commanding, ordered every man in a forward trench to fire two shots a day at tin cans hung on their protective wire. Terraine, General Jack's Diary, p. 170. By 1918, many soldiers preferred not to shoot for fear of drawing fire. Chapman, A Passionate Prodigality, p. 224.

59. Leed, No Man's Land, pp. 108 and 138-9. This is not surprising, as opposing trenches were on average but 100-400 yards apart, with advance posts closed to within 10-to 12 yards of each other. Carrington, A Subaltern's War, pp. 215 and 217; and Michael Moynihan, ed., A Place Called Armageddon: Letters From the Great War (London: David and Charles, 1975), p. 126.

60. Wynne, If Germany Attacks, pp. 54-9 and 147-8; Tom Wintringham; Weapons and Tactics (London: Faber and Faber, 1943), pp. 166-8; and Lucas, Evolution of Tactical Ideas, p. 38. German machine guns were usually protected by steel plates and concrete or sandbag cover from frontal fire. Wynne, If Germany Attacks, p. 56. In the trench system, straight lines facing the enemy constituted "bays" and the diagonals (or perpendicular kinks) "traverses." A standard bombing party might consist of two bayonet men, two bomb throwers, an officer with periscope, and a tail of understudies and bomb carriers. By 1917, bomb fighting had given way to shell-hole warfare, as the intensity of artillery fire had made trench lines untenable. Carrington, Soldier From the Wars Returning, p. 175; and A Subaltern's War, pp. 217-18. On trench design, see Barrie Pitt, 1918; The Last Act (London: Cassell, 1962), pp. 7-14.

61. In the summer of 1915, Captains Rohr and Reddeman under the supervision of Colonel Bauer had worked out the technical details of a special assault Sturmbataillon, which was intended to be a model on which the entire German army was eventually to be equipped and trained. Wynne, If Germany Attacks, p.

147. Rohr's troops adopted the Death's Head as their badge. Roger A. Beaumont, Military Elites (Indianapolis: Bobbs-Merrill, 1974), p. 21.

62. Wynne, If Germany Attacks, pp. 57-8; Wintringham, Weapons and Tactics, pp. 159 and 168; Major General D. K. Palit, War in the Deterrent Age (London: Macdonald, 1966), p. 93; and Brian Bond, Liddell Hart: A Study of His Military Thought, (London: Cassell, 1977), p. 25.

63. Dr. Laszlo M. Alfoldi, "The Hutier Legend," Parameters, 5 (1976): 72-3; and Swettenham, To Seize the Victory, p. 196. Colonel Bruchmuller was nicknamed "Durchbruch" ("breakthrough") Muller." Wynne, If Germany Attacks, p. 294. The Germans also designed a signals system whereby forward junior Commanders could control the "creep" of their supporting barrage. Pitt, 1918: The Last Act, p. 43.

64. Alfoldi, "The Hutier Legend," p. 72; Barnett, The Swordbearers, pp. 291-2; Colonel T. N. Dupuy, A Genius for War: The German Army and the General Staff, 1807-1945 (London: Macdonald and Jane's, 1977), pp. 169-70 and 172; Major General G. M. Lindsay, The War on the Civil and Military Fronts (Cambridge: University Press, 1942), p. 5; and Pitt, 1918: The Last Act, pp. 42-7.

65. Barnett, The Swordbearers, p. 254. In 1918, the Germans divided their front-line troops into "shock" and "trench" divisions. Ibid., p. 351. Ludendorff regretted this decision. Ludendorff, My War Memories, Vol. 2, p. 583.

66. Wintringham, Deadlock War, pp. 164, and 169-73; and Bond, Liddell Hart, p. 25.

67. Ludendorff, My War Memories, Vol. 2, p. 272.

68. Wynne, If Germany Attacks, p. 295. A German platoon originally consisted of two sections (Korporalschaft) of two Gruppes, each of eight men and one lance corporal. The section was an administrative unit. German Army Handbook, April 1918 (London: Arms and Armour Press, 1977), pp. 43-4 and 61.

69. Alastair Horne, The Price of Glory: Verdun 1916 (London: Macmillan, 1962), p. 336; and Dupuy, A Genius for War, p. 170.

70. Herbert Fairlie Wood, Vimy! (London: Corgi, 1972), pp. 76, 85-7, 110-11, and 141; and Swettenham,

To Seize the Victory, pp. 147 and 149. Currie also recommended putting maps in the hands of front-line troops. This was done for the battle of Vimy Ridge. Wood, Vimy!, pp. 86 and 105.

 71. Colonel Bruchmuller was Hutier's artillery commander at Riga. The tactics applied by Bruchmuller and Hutier have been inappropriately referred to by many authors, Addington and Dupuy among them, as "Hutier Tactics." However, as Dr. Alfoldi points out, there is no evidence to support that Hutier was primarily responsible for their development. In fact, if anyone deserves credit, it is probably Ludendorff. Alfoldi, "The Hutier Legend," pp. 70-1 and 73.

 72. At Caporetto, the Italians lost more than 300,000 men; the defeat almost took Italy out of the war. It was at Caporetto that Rommel made a name for himself, his experiences there providing the basis for his book The Infantry Attacks.

 73. Swettenham, To Seize the Victory, pp. 196-9; Barnett, The Swordbearers, pp. 258-9 and 293-9; Dupuy, A Genius For War, pp. 171-2; and Alfoldi, "The Hutier Legend," p. 69.

 74. Liddell Hart, The Future of Infantry, p. 27; and Thoughts on War, p. 275.

2. Handyman of Battle

Enduring the Shock of Armor

The Great War tactical reorganization of German infantry, though revolutionary, was overshadowed by the appearance on the battlefield of the mechanical "machine gun destroyer." The infantry's preeminence had thus no sooner been reestablished than it was once again challenged. In the long run, however, the revitalized employment of infantry turned out to be of greater military significance than the mere technical application of the internal-combustion engine. The new system of infantry tactics devised by Liddell Hart, for example, bore an uncanny resemblance to the later German development of blitzkrieg,(1) which tactical process many continue to regard as simply the unleashing of masses of tanks.

Liddell Hart's infantry ideas, like those of Captain Laffargue, not only added to the general compendium of German military thought but also served to confirm the widely held conviction that infiltration tactics by all arms still possessed merit. This assertion is not meant, of course, to discount the revolutionary impact of either the tank or airplane, both of which had undeniably and permanently joined the order of battle by this time. In the military watershed that was the interlude between the wars, however, the roles of these new arms had to be properly determined and that of the more ancient infantry adjusted accordingly. In short, the foot soldier had to learn to live with mechanization. During the controversy that subsequently raged on this and associated topics, the most productive and lasting military thought emanated from but a few military theorists, the most important of whom would exercise a profound influence in the development of modern tank and infantry arms. To gain a sharper perspective on infantry, then, it would be most worthwhile to examine the weight of infantry theory in the aftermath of the Great War, particularly as it was affected by the coming of the tank.

The arch-apostle of mechanization and, indeed, real "father of blitzkrieg" was the soldier-theorist J. F. C. Fuller. Believing that "weapons, if only the right ones can be discovered, form 99 per cent of victory,"(2) he seized upon the tank as the weapon of the future. In May 1918, in a military "novelette" entitled "Plan 1919"(3) he laid the foundation stone for the champions of mechanized warfare. This revolutionary plan, to be executed by an armored force, had as its objective the moral deterioration or psychological dislocation of the German high command. Through a surprise stroke by some 5,000 medium and heavy tanks supported by artillery, motorized infantry, and air, the Allies hoped to drive deep into the enemy rear to capture, destroy, or paralyze enemy head-quarters and communication centers.(4) It was the prototype of blitzkrieg, and had it been executed before the war ended, the Allies may have known better how to deal with the events of 1939.(5)

As Fuller envisioned it, the 1919 offensive would take place in three distinct phases: breakthrough, disorganization, and pursuit. The immediate aim was to attack the brain rather than the body of the enemy field army. In hunter's parlance, the beast was to receive a head shot; but more than that, it was to be an assault on its "cybernetic system"; the tanks and planes were but means to this end. To better effect the desired disorganization, Fuller even suggested not destroying communications as it was "important that the confusion resulting . . . should be circulated by the enemy. Bad news confuses, confusion stimulates panic."(6)

Obviously an innovative military thinker, Fuller was perhaps the "profoundest intellect which [had . . been applied to military thought in this century."(7) He was certainly the first British military theorist "who ever made the heads of Continental armies look to England for professional guidance."(8) Thinking in terms of tactical elements or function--protection, offensive action, and movement ("guarding, hitting, and moving")--rather than the conventional arms (cavalry, artillery, and infantry), Fuller saw that the tank combined within itself these three tactical elements to a higher degree than any other arm. Clearly, the tank was now the dominant force on the battlefield; "armour . . . [had] completely defeated the rifle bullet."(9) In pondering the question of what infantry could do on this same battlefield, Fuller deduced "nothing outside playing the game of interested spectators." Rather than "queen of battle," infantry would become "queen of fortresses," responsible for holding tactical points as it had traditionally done in the era when the armored knight held sway over Europe.(10)

That Fuller had an excellent grasp of infantry tactics there can be no doubt. Stating that the chief concern of a soldier in war is not to kill but to live, he defined the two main duties of infantry in battle as "holding by fire and occupying by movement"--in short, to provide the continuous secure base for armored action. He distinguished between the frontal assault, the frontal threat, and the frontal holding attack; the object of the second being to compel the enemy to assume the defensive and of the third to force him to maintain it and to pin him to a locality. In a masterly bit of reasoning, he determined that to train infantry for the assault was outdated, "the old idea of the assault . . . [having] long vanished from practical tactics."(11) Recent history proved, he insisted, that such assaults were always costly, but with the bullet "eliminated" as "one of the principal weapons," they were now indefensible. To him, the true attack took the form of a flank or rear maneuver. Even in purely infantry attacks in mountains or forested areas, he recommended that:

> . . . riflemen reconnoitre for the machine gunners. They move forward, search the ground, select positions and then, as the machine gunners come up, move to the flanks, the machine gunners fixing the enemy with their fire. Under cover of this fire the riflemen should once again move forward and be followed by the machine gunners, until the enemy is hedged round by a circle of bullets and is definitely pinned down. When this fixing is accomplished, the final act is not to assault him, but to interpose a force between him and his line of retreat and so compel him to surrender.(12)

While appreciating that light infantry operating in such manner would always be required under certain conditions, Fuller still inclined toward denigrating the infantry's role. For the old adage that "artillery conquers, infantry occupies," Fuller substituted "tanks conquer, infantry holds." Envisioning two classes of troops, mechanized and motorized, he saw the first as being the "protective sword" of the second and the second as being the "protective shield" of the first. In his thinking, for offense or defense, the ideal mechanized force should consist of both a tank and an antitank wing (the natural usurpers of the old infantry-artillery combination), with offensive power measured in shells and armor-piercing projectiles rather than bullets. The antitank wing he saw as being primarily composed of artillery and engineers, not infantry. Interestingly, Fuller did not consider light infantry equipped with antitank weapons to be true infantry, referring to them later as "antitank foot." In any event, apart from light infantry, obviously necessary for fighting in areas unsuited for tanks, Fuller foresaw but two other specialties as required for "second line" infantry tasks. The first, field pioneers, would be armed with antitank weapons and transported in cross-country vehicles to establish the defenses of the army of occupation, "the base of the mechanized and motorized troops." The second, field police, would be armed with machine guns, rifles, and possibly nonlethal gas to occupy and organize conquered areas and territories. Fuller also toyed with the idea of "motorized guerillas" or scouts, originally basing them on cavalry but ultimately on infantry as well.(13)

On the whole, however, Fuller was skeptical of the infantry's role on the mechanized battlefield. "To combine tanks and infantry," he argued, "is tantamount to yoking a tractor to a draught horse. To ask them to operate together under fire is equally absurd."(14) Employing the tank as simply a "machine gun destroyer" to facilitate the advance of slower-moving infantry was to fail to exploit the full measure of its battlefield potential. What was required, therefore, was not the "infantry tank" so commonly advocated by many military authorities at that time but rather a vehicle to protect "tank infantry," soldiers specially trained to support tanks.(15) Because of his extreme emphasis on the machine, Fuller came to be lumped into that peculiar British grouping known as the "all-tank" school of military thought. This trend had been foreshadowed in 1916 by Captain (later Lieutenant General) Gifford le Q. Martel in a paper entitled "The Tank Army," which first suggested "the employment of an army consisting entirely of fighting vehicles."(16) Originally inspired by the "landship concept," which visualized the tank as the direct equivalent of the warship and land warfare in terms of fleet actions, the necessity of combining with other arms was not recognized. The "all-tank" trend was to continue long after the landship concept went out of fashion,(17) however, and adherents such as Major General H. Rowan-Robinson were able to write as late as 1934 that "in its existing forms [infantry] has no great scope in continental warfare of the more advanced type." He even charged that the "fiction that infantry is still the Queen of battle is of continental concoction."(18) The last statement was probably directed against the French, who, in reading the lessons of the Great War, decided to worship firepower over mobility and defense over offense, eventually tying their tanks to infantry.(19)

At the same time that the "all-tank" school was developing, Liddell Hart was approaching the "riddle of the trenches" from a slightly different angle. His theories, ultimately to exert profound influence on Commonwealth (and American) infantry doctrine, were rooted in his personal experiences of war. A subaltern in the King's Own Yorkshire Light Infantry, he had participated in the 1916 Somme offensive, during which struggle his battalion was destroyed and he was badly gassed. His experience in this and previous actions prompted him to seek, like the Germans, a tactical rather than technical solution to the trench impasse. Eventually invalided to the half-pay list and finally out of the army, Liddell Hart was in the interim to make himself an expert on infantry small-unit tactics and training. In fact, his entire professional reputation as a military critic was founded on his radical reformulation of infantry tactics.(20)

While serving as adjutant to volunteer units in Stroud and Cambridge during the last two years of the war, Liddell Hart published numerous pamphlets on discipline, training, and tactics. Of particular importance was "The 'Ten Commandments' of the Combat Unit--Suggestions on its Theory and Training," in which he stressed that the platoon had now become a "combat unit" in the sense that it contained "several sub-divisions, each capable of separate manoeuvre."(21) He also developed a "battle drill," a new and simplified method designed to facilitate movement and to exercise formations that could be applied in battle and practiced on the parade square. However, it was in attempting to devise a new tactical training system for the defense, attack, and counterattack that he came to doubt the validity of tactical precepts and practices that had been taught before and during the Great War. In 1920, encouraged by a broad-minded patron, General Sir Ivor Maxse, he published his "Man in the Dark" theory of war and "A New Theory of Infantry Tactics." These articles were so well received that he asked to assist in the preparation of the postwar <u>Infantry Training</u> manual. In the course of doing so, he devised his "expanding torrent" system of attack for breaking through a defense in depth. He presented his expanded theories in lectures to the Royal United Service Institution (RUSI) and Royal Engineers, and they were subsequently published in several journals and books. This was indeed fortunate because by the time the army manual was circulated through the War Office and Staff College for comment, "the basic ideas had been watered down."(22)

Like Fuller, Liddell Hart thought in terms of tactical functions, and he used the example of a man fighting another in the dark as his model for framing fundamental principles. The analogy of personal combat was to be the "soldier's pillar of fire by night," a framework of elementary tactical principles on which the junior officer and NCO could base his battlefield actions.(23) Hoping to work "upwards from the elementary, instead of downwards from the complexities of large operations," he described the fighter's actions as follows:

> In the first place . . . the man stretches out one arm to grope for his enemy, keeping it supple and ready to guard himself from surprise (principle of protective formation). When his outstretched arm touches his enemy, he would rapidly feel his way to a highly vulnerable spot such as the . . . throat (reconnaissance). The man will then seize his adversary firmly by the throat, holding him at arm's length so that the latter can neither strike back effectively, nor wriggle away to avoid or parry the decisive blow (fixing). Then while his enemy's whole attention is absorbed by the menacing hand at his throat, with his other fist the man strikes his opponent from an unexpected direction in an unguarded spot, delivering out of the dark a decisive knock-out blow (decisive manoeuvre). Before his enemy can recover the man instantly follows up his advantage by taking steps to render him finally powerless (exploitation).(24)

Noting that the actions of the "man in the dark" could be simplified to two, guarding and hitting, Liddell Hart extracted from the five foregoing "battle principles" two "supreme governing" principles: security and economy of force. In applying these to deployment of infantry platoons, companies, battalions, and brigades (all of which were "square" at the time) he deduced that only three dispositions were possible: one fixing, three maneuver; two fixing, two maneuver; or three fixing, one maneuver. Stressing that maneuver "is useless unless the enemy is first fixed in another direction," he at the same time intimated that the maneuver body, thought not necessarily the largest, is often "the last straw which breaks the camel's back." Old infantry terminology was then called into question with devastating logic:

> . . . the recognized terms "firing line" and "supports" . . . are no longer applicable to modern infantry tactics. In order to instil a correct doctrine into the minds of the average

officer and NCO it is advisable to eschew misleading terms, and not try to reconcile modern ideas with out-of-date phraseology.

The term "firing line" does not convey the idea of the outstretched arm or of distribution in depth. It suggests a broad frontal attack with no attempt to make use of covered ways of approach or to find the soft spots.

In the case of "supports" both the word and the idea are dangerous. It does not inculcate the essential idea of manoeuvre, but rather the obsolete and unsound idea of reinforcing frontally troops who are held up, which means piling the dead in front of the enemy's strongest points.(25)

In developing his argument for fixing and maneuvering "right down to the scale of the smallest combat unit," Liddell Hart emphasized that weight of force in modern war was related to weight of fire and not merely numbers of men. He further stressed that though in large actions it might appear that infantry units have been confined to a purely frontal role, the wider dispersion forced on combatants by the increased effectiveness of modern weapons rendered possible penetration by fire units (sections) between enemy defense posts. The role of the section, platoon, and company commander in such circumstances is to exploit "their penetration to change their sector of the battle from a mere bludgeon fight into a manoeuvre combat."(26)

The infantry in attack Liddell Hart likened to a "human tank, comprising both offensive power and protective armour," the former in its weapons and legs (maneuver power) and the latter in its field formation, "which prevents more than one of the sub-units being surprised by the enemy." To ensure the efficient operation of subunits, he recommended adoption of sensible open formations such as diamond, square, and section arrowhead--and the use of field signals. He further recommended the use of section scouts moving ahead by bounds. Maneuver groups and fixing (forward) groups at all levels were to move in the same manner. He finally stipulated that though attacking battalions should be divided into maneuver, forward, and reserve bodies, companies and platoons need only be distributed in the first two.(27)

For breaking through a series of layered enemy positions in depth, such as the elastic defense used by the Germans in the Great War, Liddell Hart devised the "expanding torrent" method of attack. Again, he based his tactics on the "intelligent manoeuvre of firepower," likening them to the flow of a strong water current exploiting a breach in an earthen dam. The

44 A PERSPECTIVE ON INFANTRY

TABLE 2-1. LIDDELL HART'S PRINCIPLES OF TACTICS

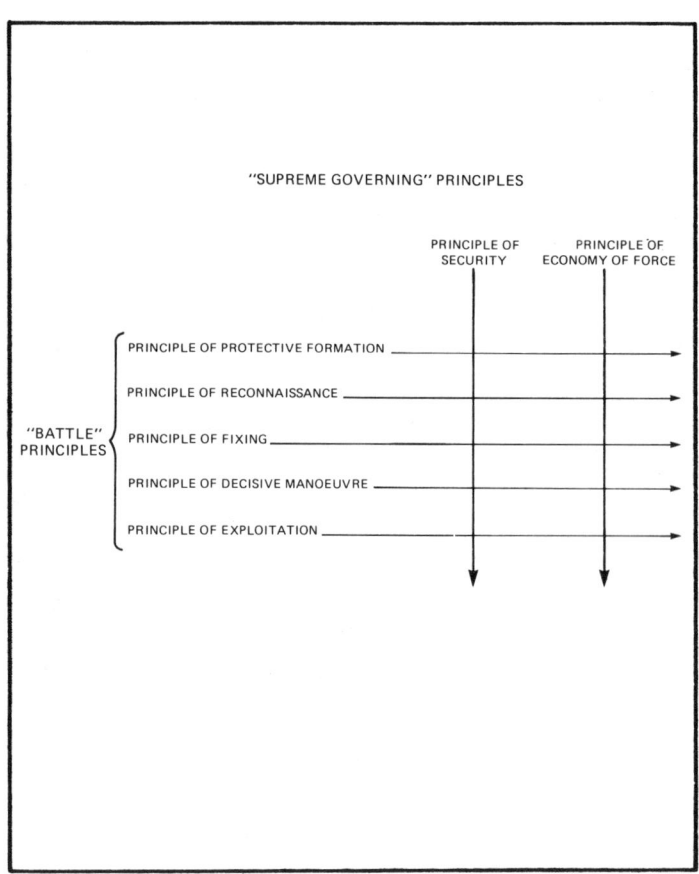

essence of his "scientific system of attack" was the combination of "speed with security." Rejecting as a "sheer waste of force" the notion that an attack should be pressed equally at all points, he advocated feeling and testing an enemy position everywhere to find or make a weak spot, and then exploiting it to the fullest with the use of reserves.(28)

As for method, the first forward (fixing) subunit to effect a breach was to continue to advance straight ahead so long as it was still backed up by the maneuver element of its parent unit. Forward units held up on the flanks of the breach would then dispatch their respective maneuver elements through the breach to attack the enemy in the flanks and so widen the gap. Meanwhile, units in the rear would pass through the gap and take over the frontage and lead in place of the units engaged in fighting. With the reduction of the enemy, the latter units would then follow on as maneuver units to support the new forward units. This flowing system of attack was, in Liddell Hart's opinion, universally applicable to all units and formations from the platoon upwards. He cautioned platoon commanders, however, that sections were not tactical units "composed of interdependent fighting parts and so capable of fixing and manoeuvring simultaneously." They possessed no maneuver body of their own. Therefore, if a forward section of a platoon was held up, the platoon commander would have to use his maneuver sections to extricate it before continuing the platoon advance.(29)

The keynote of Liddell Hart's system was simplicity. Its main advantages were that the pressure of the advance was maintained and its control rested with the "immediately superior commander," thus ensuring the preservation of tactical unity "instead of an unorganized dog fight to get forward, with each unit playing for its own hand." He also foresaw the battle tactics of the infantry becoming more automatic and less dependent on receiving fresh orders from superiors in the rear. In fact, he wanted to drill his framework into junior leaders until they acted on it instinctively.(30)

As regards defense, Liddell Hart merely turned the "expanding torrent" into an infiltration-defeating "contracting funnel," explaining that the defense was simply the attack halted. Platoons, companies, and battalions were still to be distributed into fixing (forward), maneuver, and (in the case of the last) reserve elements. For a "square" organization, two subunits forward and two for maneuver would be the norm. He advocated allotting each unit and subunit an area to hold as opposed to a point; thus, "if the enemy takes a defense post at a disadvantage, by crushing shell fire, smoke or manoeuvre, the commander should use his initiative to quit the post and take up a fresh position on the flanks of the post so he can out-manoeuvre the attacking infantry." Such optional action was recommended as much for forward section commanders as for platoon commanders.(31)

While noting that defense was incapable of beating

TABLE 2-2.   THE "EXPANDING TORRENT" ALONG LINES
             OF LEAST RESISTANCE

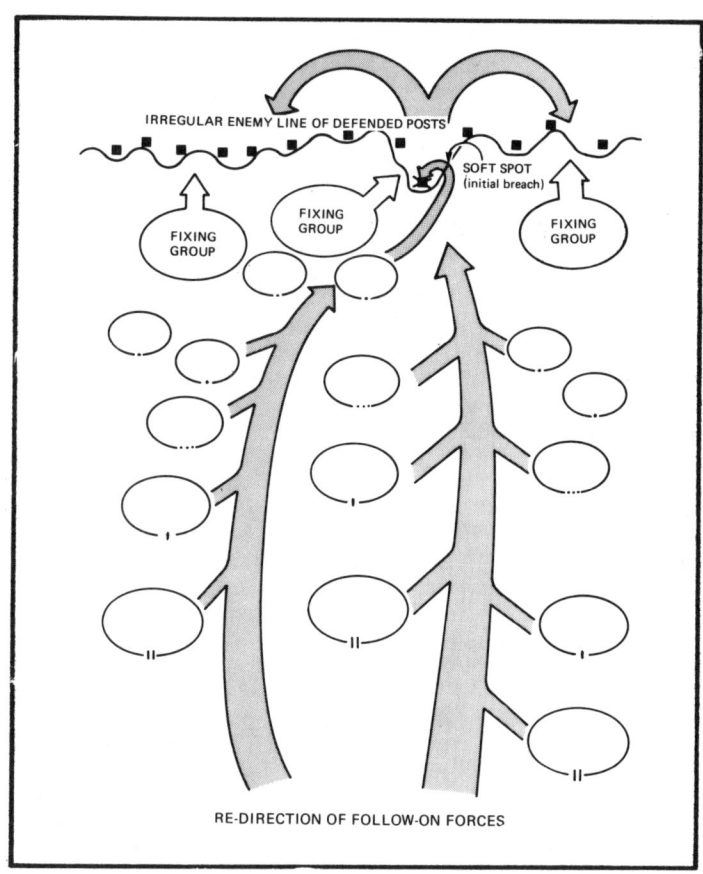

an enemy, Liddell Hart nonetheless recognized that the best way to blunt an attack and reseize the initiative was to take it in the flank. As he saw it, the function of maneuver element commanders would be to effect this, either through enfilading counterfire or even assault. One great advantage of the defense was that maneuver elements could be positioned beforehand in critical areas or rehearsed in movement to those areas. The "sheer waste of force" of a counterassault that fails was, of course, emphasized, and counterattack by "any unit smaller than platoons" was not recommended. In his opinion, the best course of action for a maneuver section was to apply fire and not counterassault. Great stress was also placed on mutually supporting fire and active patrolling, the ultimate aim of all actions being to draw the enemy into a funnel raked by flanking fire--in short, a mobile defense executed from the lowest tactical level up.(32)

How much Liddell Hart was influenced by others in developing his theory is difficult to establish. He admitted in his Memoirs that he had examined the infiltration techniques evolved by the Germans and imitated by the Allies in the Great War. However, in 1925, he argued that his "expanding torrent" idea differed from the infiltration method of "soft spot" attack in the important respect that it spelled out a controlled system for the coincident progressive expansion of the penetration and the maintenance of the original breadth and pace of the advance. In other words, it explained more specifically how to use reserves to breach the defense zone in depth.(33)

How much the infantry theories of Liddell Hart influenced others is equally difficult to determine. We do know, however, that his book The Future of Infantry (1933), which contrasted the German tactical (infantry) solution to the "riddle of the trenches" with the British technical (tank) solution, found an extremely receptive audience in Germany. Fuller, recently returned from observing German army maneuvers in 1935, told Liddell Hart that one German military spokesman had ordered 5,000 copies of The Future of Infantry for his particular group. As this book also explained in detail the theory of the "expanding torrent," it may well have exerted a significant influence on German military thought generally. We know for certain that Heinz Guderian read it.(34) There is even evidence to believe that it was better received than Lieutenant Colonel Erwin Rommel's comparatively successful work, The Infantry Attacks (Infanterie Greift an).(35)

The genesis of The Future of Infantry was rooted in Liddell Hart's membership on a War Office committee examining the lessons of the Great War. While so engaged, he came to believe that no attack in modern war could succeed against an enemy in position unless his resisting power was paralyzed either through surprise or preponderating fire. He therefore began to look upon the night attack as a means of attaining tactical surprise. Though subsequent investigation revealed that the night attack was a rarity in the

Great War, he was struck by the frequency with which there had been foggy conditions at the start of a number of successful offensives:

> Fog had cloaked the French attack close to the Somme on the morning of July 1, 1916. Fog had cloaked the remarkably successful French offensives at Verdun that autumn, which regained in two quick bites most of the ground lost during the prolonged German offensive from February to July. Fog had cloaked the British surprise stroke at Cambrai in November, 1917. Fog had cloaked all three breakthroughs by the Germans in 1918, but was absent in their three abortive offensives of March 28, June 9, and July 15. It had shrouded the battlefield for the benefit of the British on the morning of August 8, 1918, and had come to the aid of the 46th Division when it broke the Hindenburg Line of September 26.(36)

In <u>The Future of Infantry,</u> Liddell Hart advocated night actions and the development of new fighting aids such as artificial moonlight and artificial fog. He also propounded numerous other suggestions for increasing the tactical effectiveness of the infantry arm. Maintaining that the "mounted arm had always and naturally been, in the true time sense, the decisive arm [because it could move quicker]," he stated that the "true function of infantry" was to disorganize resistance and prepare the way for a decision. Arguing that one could not "expect mobility on the battlefield unless the man who fights on foot is given the chance to be mobile," he recommended splitting infantry into heavy and light grades; the former tailored only for defense or garrison duties, the latter for a modern mobile role as part of a mechanized-motorized army. In short, he urged the revival of light infantry.(37)

As explained by Liddell Hart, the revived infantryman would be "<u>tria juncta in uno</u>--stalker, athlete, and marksman." Equipped with a lighter rifle, to alleviate ammunition supply, and trained to a high standard in fieldcraft, he would be capable of destroying enemy machine-gun and antitank positions through stealth and deadly accurate small-arms fire. Unlike his Great War counterpart, he would not be a beast of burden carrying 70 pounds of personal kit; rather, he would carry but one-third of his own weight. Dressed as an athlete and "light of foot," he would also be "quick of thought" and capable of acting on his own or as part of an independent team. The elastic chain of little groups would replace the traditonal infantry line.(38)

Through <u>The Future of Infantry,</u> Liddell Hart made

specific, practical suggestions for infantrymen. He advocated adoption of the "framed rucksack" as the individual load carrier, and he suggested that in selecting webbing "there should be nothing that dangles, especially below the hips." As regards drill, he charged that the parade square fostered an instinct for "alignment" instead of developing "the vital instinct of direction." He was therefore convinced that tactical training, using sandtable instruction, should commence before "square bashing" began so that a soldier's innate tactical instincts could be built upon. Significantly, he advocated basing the tactics of the new mobile infantry on infiltration "in its more developed form." Stressing that on the battlefield "success is usually uneven and progress unequal," he explained how this circumstance could be turned to advantage "by using the unevenness of the enemy's front to apply flanking fire as a lever to loosen the projecting 'corner posts'(39) . . . developed by the 'expanding torrent' method of exploitation." He went to great lengths in the book to explain how this method "would work on the battlefield, as only a shadowy indication was contained in the new Infantry Training manual that [General] Montgomery had produced."(40)

In addition to his theory of the "expanding torrent," Liddell Hart spelled out three specific methods or forms of attack. The first, the "stalking attack," involved infiltration utilizing patience and craft; the second, termed the "masked attack," took advantage of darkness and fog, natural or induced. The third form, and one used regularly by the Germans with great effect, was the "baited attack" whereby an opponent was lured to a repulse and his recoil exploited by a riposte. Such systematizing of tactics was no doubt linked to Liddell Hart's strong belief that it was a fallacy to consider the infantry the "most easily trained arm." To him, the infantry was "essentially the tactical arm": to train it was an art, whereas the training of technical arms was a science.(41)

The Future of Infantry was generally well received in both the British and American armies, though obviously many of Liddell Hart's more progressive ideas were not adopted. Significantly, the translated edition was greeted far more enthusiastically in Germany. This may have been partly due, of course, to the praise the author accorded the German infantry for its outstanding performance in the Great War. On the other hand, the book must have possessed substantial merit since it received a wide circulation within the army. This was perhaps best indicated by the fact that it was used in the training of motorized infantry regiments of panzer divisions formed in 1935 and adopted as a textbook for the Waffen S.S.(42)

50 A PERSPECTIVE ON INFANTRY

Whether or not the Germans were profoundly influenced by the infantry ideas of Liddell Hart is difficult to prove absolutely. In a keen, well-trained army officered by but 4,000 professionals, however, a book of such circulation must have had a substantial impact.(43) By outlining the "expanding torrent" theory, it might even have helped convince the Germans of the value of infantry in the all-arms team. In any event, they ultimately lent more emphasis than the British to the integration of infantry into mechanized land operations. This may, of course, also have been due to a number of other considerations, in particular, an earlier-established German belief that the infantry arm was an effective instrument of offense in its own right.

For his part, Liddell Hart sometime previously had begun to appreciate the necessity for mechanizing the modern army. As he had been from the beginning essentially dedicated to the idea that infantry was still queen of the battlefield, however, he naturally encountered the conflicting views of Fuller. In an exchange of views on tactics, the latter suggested that "if the enemy produces tanks your infantry tactics will prove useless, for it is impossible to pit infantry against machines." Although Liddell Hart had never really contended that infantrymen alone were a match for machines, it was not long before he was converted to the mechanical-warfare school. Asked to defend the infantry as "queen of battle" in modern war for an Encyclopaedia Britannica article in 1921, he found that he could not logically "uphold the infantry against the inroads of mechanical warfare." In fact, he began to suspect that the infantry "is more likely to endure because of conservatism, financial and official, than its own inherent merits." By 1924, he was arguing that the "tank is likely to swallow the infantryman, the field artilleryman, the engineer, and the signaller, while mechanical cavalry will supersede the horseman." He shortly requested a transfer to the Tank Corps.(44)

In the 1925 publication Paris, or The Future of War, statements such as infantry "may find use . . . as caretakers . . . the only feasible role for infantry in mobile warfare of the future" clearly showed Liddell Hart leaning toward the Fuller camp. He did nonetheless make brief mention of "land marines" to be used "to clear fortifications and hill defenses under cover of the fire from the tank fleet."(45) In his book The Remaking of Modern Armies, however, he suggested mounting individual infantrymen in Martel-Morris and Carden-Loyd tankettes, machines originally championed by Martel. Several years later, he went further, advocating the creation of a corps d'élite of light "mounted" infantry organized in motorized battalions for mobile operations. Always mindful of the basics, he placed great stress on individual light-infantry skills such as "stalk and skirmish" methods, infiltration techniques, improved marksmanship, and reducing individual combat loads. He also made the particularly logical point that since infantry could not compete in volume of fire with mechanized troops, riflemen should concentrate on accuracy of fire and fieldcraft to enable them to pick off machine-gun crews and tank commanders. Not surprisingly, he railed against the infantryman's entering the Great War as a rifleman and returning as a "human Christmas tree" armed with hand grenades, rifle grenades, smoke grenades, heavy machine guns, light machine guns, trench mortars, trench daggers, and clubs.(46)

> The result was not to increase the offensive power of infantry--which depends not only on fire-power, but on protection and mobility--but to limit such offensive value as he still retained. For the value of infantry in the present Army, as of tank-marines in the future, rests solely in their loco-mobility--their ability to move over every sort of ground and to clear every yard of any locality.(47)

Despite his apparent conversion, then, it would appear that Liddell Hart did foresee a greater role for the infantry on the modern battlefield than did Fuller, although it would be an error to say that the latter did not see any use for infantry at all. Both preferred to think in terms of battlefield functions as opposed to arms, Fuller rightly or wrongly coming to be identified with the "all-tank" school and Liddell Hart dimly seeing infantry operating as a force in its own right threatening the enemy from another direction. The latter, slightly more balanced view would not prevail in the British army unfortunately, and for various reasons--the antagonism of conservative infantry and cavalry officers to mechanization per se and the perhaps understandable extremism of the tank advocates--British armored forces during the twenties and early thirties consequently tended to be "all-tank" in their outlook.(48) It should be stressed here, of course, that the "young Turks" of mechanization were not at this time a majority in the British army; only after Dunkirk would they gain sufficient seniority to put their "all-tank" views into practice.

Across the Channel, a similar schism was occurring within the postwar German army, as certain of its officers also began to grasp the revolutionary significance of the Schutzengrabenvernichtungpanzer-kraftwagen.(49) It would not be allowed to develop so disastrously as the British split, however, since most conservative officers of the German army generally chose to temper, rather than oppose outright, the more extreme ideas of the proponents of mechanization. Generals Werner Von Fritsch and Ludwig Beck,(50) for example, chose to reserve judgment on the decisive role of the tank until a host of questions dealing with weapon efficiency, fuel supply, and methods of organization and command were satisfactorily answered.(51) Beck additionally believed that the army could not do without traditional infantry for the foreseeable future because that arm still provided the necessary "mass of decision." A devout disciple of Count Alfred von Schlieffen and the elder Moltke, he wished for more than an army capable of springing encirclements; he wanted an army with sufficient "mass and defensive firepower" so that it could win the decisive "battles of annihilation" that would inevitably accompany or follow encircling maneuvers. Beck was anything but a reactionary,(52) yet he tended to be lumped with the German "antitank" school.(53) By championing balance in the German development of blitzkrieg, however, he fortuitously may have made the most critical contribution of all.

If Beck personified the "antitank" element within the Germany army of the time, Guderian was surely the most outspoken and active member of the mechanized school. How much he was influenced by foreign theorists is, of course, again difficult to establish. Although referring in his postwar publication Panzer Leader to the "books and articles of the Englishmen, Fuller, Liddell Hart, and Martel," he referred only to the writings of Fuller, Martel, and Charles de Gaulle in his 1937 recipe for blitzkrieg, Achtung! Panzer!(54) He later indicated to Liddell Hart that de Gaulle's Vers l"Armée de Metier (1933) was considered rather "fantastical" and "up in the clouds"; it was accordingly largely ignored.(55) In further conversation, Guderian mentioned that he had first read Liddell Hart's articles during the period 1923-24. According to Brian Bond, he supposedly referred so often to Liddell Hart in the presence of Beck that the latter was heard to complain that he wished "he could have six months without hearing Liddell Hart's name.(56) There still remains a suspicion, however, that except for his brilliant analysis of infantry and associated theories, Liddell Hart had far less overall impact on the Germans than did Fuller.(57) Interestingly enough, it has also been claimed--though the claim has remained largely unexplored--that Austrian General Ludwig von Eimannsberger, through the second edition of his book Der Kampfwagenkrieg, exerted an exceedingly strong influence on German military thought between wars,

even to the extent that he coined the term panzer.(58)

In April 1922, Captain Guderian was assigned as a
General Staff officer to the Inspectorate of Transport
Troops, Motorized Transport Department, in Berlin. A
signal officer by classification, he was already aware
of the potential of radio for controlling tactical
operations. By 1929, he was also "convinced that tanks
working on their own or in conjunction with infantry
could never achieve decisive importance." In his
opinion, the effectiveness of tanks would increase in
proportion to the ability of infantry and other arms to
follow them across the country.(59) In coming to this
highly significant (and early) conclusion on the
importance of "all arms," Guderian was obviously
influenced from many quarters, not the least of them,
as we shall see, the very institution of the German
army itself. According to General F. W. von Mellinthin,
German army tactical theory in that year "had
progressed beyond that of Great Britain." The German
panzer division was consequently, from its inception, a
self-contained balanced force of all arms, a state the
British would not reach until 1942. The fifty-fifty
split of tanks and infantry with panzer organizations,
though doubtless reflecting a German military
conservatism and hesitancy to place all eggs in one
basket, also mirrored a healthy respect for the
infantryman as at least the "handyman" of battle.(60)

NOTES

1. All of the ingredients of blitzkrieg were
evident in the Great War. The American Holt Caterpillar, its potentialities first recognized by Swinton,
had been transformed into the tank; the German
employment of low-strafing aircraft to support the
counterattack at Cambrai had also been hailed as "a
distinct advance on anything that had hitherto been
attempted by either side." Swinton, *Eyewitness,* pp. 32,
81, and 235. It was not until these means were coupled
with German infiltration and surprise tactics, however,
that "lightning war" was born.

2. J. F. C. Fuller, *Machine Warfare* (London:
Hutchinson, 1941), p. 74.

3. Originally entitled "The Tactics of the
Attack as Affected by the Speed and Circuit of the
Medium D Tank" and later as "Strategical Paralysis as
the Object of the Decisive Attack."

4. J. F. C. Fuller, *Memoirs of an Unconventional*

Soldier (London: Ivor Nicholson and Watson, 1936), pp. 322-36; and Charles Messenger, The Art of Blitzkrieg, (London: Ian Allan, 1976), pp. 27-28.

    5. Marshall, Blitzkrieg, pp. 94 and 103.

    6. Fuller, Memoirs, pp. 325-7 and 329-30; also detailed in "Strategical Paralysis as the Object of the Decisive Attack" in On Future War, (London: Sifton Praed, 1928), pp. 83, 86-87, 93, and 100-1; and Bidwell, Modern Warfare, p. 199; and Kenneth Macksey, Tank Warfare: A History of Tanks in Battle (London: Rupert Hart-Davis, 1971), p. 71.

    7. Brian Bond, Liddle Hart, p. 30.

    8. Jay Luvaas, The Education of an Army: British Military Thought, 1815-1940 (Chicago: University Press, 1964), pp. 374-5.

    9. J. F. C. Fuller, Armored Warfare: An Annotated Edition of Lectures on F.S.R. III (Operations Between Mechanized Forces) (Harrisburg: The Military Service Publishing Company, 1943), p. 12; and Lectures on F.S.R. II (London: Sifton Praed, 1931).

    10. J. F. C. Fuller, The Reformation of War (London: Hutchinson, 1923), pp. 158 and 161-4.

    11. Fuller, Lectures on F.S.R. II, pp. 14-16; Armored Warfare, p. 107; and The Reformation of War, p. 164. Fuller states: "We are told that 'the main object of infantry . . . is to close with the enemy and destroy him.' Then we are told that this demands superiority of fire; that numbers are unlikely to produce this superiority, and that the infantry in the defence are stronger than infantry in the attack. In brief, that infantry cannot close unless strongly assisted by artillery or tanks. When such a happy climax to the attack as closing with the enemy does take place, as the last war proved very clearly, either the enemy has been slaughtered, or he has withdrawn, or he surrenders without a struggle; consequently there will seldom be any need to destroy him, for to all intents and purposes he will already be destroyed." Lectures on F.S.R. II, p. 14. In this regard, Fuller's ideas closely approximated those of Moltke and other German military thinkers who tended to view the assault as merely "presenting the cheque for payment." Liddell Hart, The Future of Infantry, p. 18.

    12. Fuller, Armored Warfare, pp. 16-17 and 106-7.

HANDYMAN OF BATTLE 55

13. Fuller, Lectures on F.S.R. II, p. 88; Armored Warfare, pp. 16-17, 70, and 106-7; and Machine Warfare (London: Hutchinson, 1941), p. 93. Fuller also noted that "guerilla wars may run concurrently with great ones, wars of the first magnitude, as happened in the Franco-Prussian war." Lectures on F.S.R. II, p. 3.

14. Fuller, Armored Warfare, p. 16.

15. Fuller, Machine Warfare, p. 55.

16. Lieutenant General Sir Gifford Martel, An Outspoken Soldier (London: Sifton Praed, 1949), p. 14; and Macksey, Tank Warfare, p. 70.

17. R. M. Ogorkiewicz, Armoured Forces (London: Arms and Armour Press, 1970), p. 57. In fairness to Martel, he "was always convinced that the use of all arms would be needed for practically every kind of operation." In fact, he was greatly critical of Brigadier P. C. S. Hobart's "all armoured views" in commanding the Tank Brigade under Major General G. Lindsay, an "all arms" armored supporter. Martel, An Outspoken Soldier, pp. 67 and 124-5.

18. Major General H. Rowan-Robinson, The Infantry Experiment (London: William Clowes, 1934), pp. 36-37. A gunner, he maintained that the infantry in its "present shape" was too vulnerable and "too ineffective to justify its existence." He prophesied that "infantry is seldom likely to make attacks except in small wars" and recommended it be made an "auxiliary arm." Ibid., pp. 67 and 77.

19. Ogorkiewicz, Armoured Forces, p. 72; and Luvaas, Education of an Army, pp. 377 and 404.

20. Bond, Liddell Hart, pp. 16-17, 23 and 32-33. On the first day of the Somme the British Army suffered 60,000 casualties, the heaviest day's toll in British military history. Liddell Hart's battalion was practically wiped out; he was one of but two surviving officers. Ibid, p. 17. He had seen enough action by this time, however, to make him reflect that something was wrong with the system of tactics.

21. B. H. Liddell Hart, "The 'Ten Commandments' of the Combat Unit--Suggestions on its Theory and Training," RUSI Journal 64 (1919): 288.

22. B. H. Liddell Hart, "'Man-in-the-Dark' Theory" pp. 1-22; B. H. Liddell Hart, Memoirs (London: Cassell, 1965), vol. 1 pp. 28-29, 31-32, 34-35, and

37-48; Bond, Liddell Hart, pp. 19 and 23-24; and Luvaas, Education of an Army, pp. 378-9. The 1931 revision of Infantry Training was done by General Montgomery, and some restorations were made. Liddell Hart, Memoirs, vol. 1, p. 48.

23. B. H. Liddell Hart, "The Soldier's Pillar of Fire by Night; The Need for a Framework of Tactics," RUSI Journal, 66 (1921): 619-20 and 623-5.

24. Liddell Hart, "'Man-in-the-Dark Theory," pp. 2-3; also in identical publication "A Science of Infantry Tactics," The Royal Engineers Journal 33 (1921), pp. 169-70; and later in A Science of Infantry Tactics Simplified (London: William Clowes, 1923).

25. Liddell Hart, "Science of Infantry Tactics," pp. 170-3; and Thoughts on War, p. 128.

26. Liddell Hart, "Science of Infantry Tactics," pp. 173-7; and Memoirs, I, pp. 44-45.

27. Liddell Hart, "Science of Infantry Tactics," pp. 175-8 and 181-2. "Bounds" are tactically defensible features from which one fire element can cover the movement of another. Liddell Hart advocated a pair of scouts from each section, moving individually by bounds, the last signaling the section to come forward or not. He later changed his mind on the use of section scouts, claiming them to be "more nuisance than protection--a source of continual delay, masking fire, and breaking up the fighting unit and fire-readiness of the section." Liddell Hart, Thoughts on War, p. 185.

28. Liddell Hart, "Science of Infantry Tactics," pp. 215-16; and Bond, Liddell Hart, p. 27. It should be noted here that Liddell Hart employed the terms "maneuver" and "movement" rather interchangeably but essentially in the sense of movement. In current NATO terminology, "Maneuver" embraces both fire and movement. To avoid confusion, I will use the terms as Liddell Hart employed them. Clearly, fire and movement are separate functions; the first "the physical and psychological effects of weapons," the second "the change of position of the weapons producing fire." We can, therefore, have fire in movement (i.e., tank action), fire after movement (bounding artillery), or fire and movement (two groups). Each form will be explained, as required, in this work. Definitions are taken from Miksche, Atomic Weapons and Armies, p. 46.

29. Liddell Hart, "Science of Infantry Tactics," p. 216; and Luvaas, Education of an Army, p. 380.

30. Liddell Hart, "Science of Infantry Tactics," pp. 217-18; and "The Soldier's Pillar of Fire By Night," p. 620; and Luvaas, Education of an Army, p. 381.

31. Liddell Hart, "Science of Infantry Tactics," pp. 218-19 and 222; and Memoirs, vol. 1, p. 45.

32. Liddell Hart, "Science of Infantry Tactics," pp. 219-23; and Memoirs, vol. 1, p. 45.

33. Bond, Liddell Hart, pp. 25-27; and Liddell Hart, Memoirs, vol. 1, pp. 43-44.

34. Bond, Liddell Hart, pp. 216, 219-21; and 229. He also read The Remaking of Modern Armies and, it is suspected, "The Development of the 'New Model' Army." Ibid., p. 222. It is questionable whether the latter article really had the effect on Guderian that Liddell Hart claims it did.

35. General F. M. Erwin Rommel, The Infantry Attacks, trans. Lieutenant Colonel G. E. Kidde (Washington: The Infantry Journal, 1944), forward. The book was given "only perfunctory reviews in German military periodicals." Ibid. David Irving states that the book was a best seller in Germany and that it was "probably one of the best infantry manuals ever written." David Irving, The Trail of the Fox, (New York: Avon, 1978), pp. 35-36. Desmond Young concurs, reporting that it became a textbook in the Swiss army but, significantly, not in the German. Desmond Young, Rommel (London: Collins, 1950), p. 50.

36. Liddell Hart, Memoirs, vol. 1, pp. 212, 215-16; The Future of Infantry, pp. 53 and 55; and Thoughts on War, p. 207.

37. Liddell Hart, Memoirs, vol. 1, p. 218; and The Future of Infantry, pp. 29, 41, and 55.

38. Liddell Hart, The Future of Infantry, pp. 38-39, 41, and 62-63.

39. In World War II, the Germans attempted to decapitate Russian breakthroughs by holding on to "cornerstones," as they termed them. Heilbrunn, Conventional Warfare in the Nuclear Age, pp. 67-68.

40. Liddell Hart, The Future of Infantry, pp. 42, 48-50, and 58. He also suggested a waterproof cover for the rifle, there being no "adequate reason why a military rifle should not be treated with the same care

as a sporting rifle." Ibid., p. 37; and Thoughts on War, p. 259.

41. Liddell Hart, The Future of Infantry, pp. 53-57 and 64. He was very up-to-date in his contention that infantry training time could be reduced if it were possible to cut "those parts of the technical and tactical instruction which go beyond what . . . infantry actually need to receive." Ibid., p. 36.

42. Liddell Hart, Memoirs, vol. 1, p. 222. The Infantry School at Fort Benning as late as 1959 requested copies, as the conclusions were judged still valid. Ibid.

43. The decisive step in increasing the Reichswehr from 100,000 (officered by 4,000) to more than 300,000 effectives was taken on October 1, 1934. Herbert Rosinski, The German Army (Washington: The Infantry Journal, 1944), pp. 134-5.

44. Luvaas, Education of an Army, pp. 381-2; Anthony John, Trythall, "Boney" Fuller: The Intellectual General, 1878-1966 (London: Cassell, 1977), pp. 92-93; Bond, Liddell Hart, pp. 28-34; Messenger, The Art of Blitzkrieg, p. 43; Liddell Hart, Memoirs, vol. 1, p. 90; and "The Development of the 'New Model' Army" The Army Quarterly 9 (October 1924): 44.

45. B. H. Liddell Hart, Paris, or The Future of War (London: Kegan Paul, Trench, Trubner, 1925), pp. 75 and 88. In 1922, he envisaged an army "composed principally of tanks and aircraft, with a small force of seige artillery, for the reduction and defence of the fortified tank and aircraft bases and of mechanical-borne infantry for use as land-marines." Luvaas, The Education of an Army, p. 383.

46. B. H. Liddell Hart, The Remaking of Modern Armies (London: John Murray, 1927), pp. 61-79 and 115-16; The British Way in Warfare (London: Faber and Faber, 1932), pp. 218-24 and 244-5; Memoirs, vol. 1, pp. 217-22; and Luvaas, The Education of an Army, p. 404.

47. Liddell Hart, The Remaking of Modern Armies, pp. 61-79.

48. Bond, Liddell Hart, p. 29; Messenger, The Art of Blitzkrieg, pp. 43-46; and Ogorkiewicz, Armoured Forces, p. 386.

49. Early German descriptive word for the tank as reported by Swinton. The British, having thrown out "landship" and "land cruiser" in the interests of secrecy, agonized over "container," "receptacle,," "reservoir," and "cistern" before deciding on the simplistic "tank." Swinton, Eyewitness, pp. 186-7.

50. Beck was Chief of the Army General Staff 1935-1938. Fritsch was head of the army from 1933. In 1935, he became Commander-in-Chief of the Army.

51. Walter Goerlitz, History of the German General Staff, 1657-1945, trans. Brian Battershaw (New York: Praeger, 1953), p. 300.

52. Addington, The Blitzkrieg Era, and Goerlitz, German General Staff, p. 301. General Guderian painted him in this light, most falsely, I believe. Bond, Liddell Hart, p. 226.

53. Liddell Hart, The Other Side of the Hill (London: Pan, 1978), p. 47.

54. Macksey, Tank Warfare, pp. 86 and 92. Achtung! Panzer! itself was not much of a book, contributing little to tank theory. Len Deighton, Blitzkrieg (London: Jonathan Cape, 1979), p. 146.

55. Liddell Hart, The Other Side of the Hill, pp. 121-2. On reading this de Gaulle book, I am inclined to agree most heartily with Guderian. A lot of history and politics are included in its pages but few specifics on military theory. General Charles de Gaulle, The Army of the Future (Philadelphia: Lippincott, 1941). See also Messenger, The Art of Blitzkrieg, p. 115.

56. Bond, Liddell Hart, pp. 223-4 and 229.

57. Fuller wrote the Reformation of War in 1922. Guderian read this book plus Lectures on F.S.R. III (1932). Only 500 copies of F.S.R. III were distributed in Britain: 30,000 were sold in Germany according to Professors Preston, Wise, and Werner. (Fuller in his Memoirs mentions 30,000 were translated into Russian.) Liddell Hart's the Remaking of Modern Armies appeared in 1927 after Paris (1925). Guderian supposedly made up his mind in 1929 on the worth of all arms. Bond records "Fuller's name was particularly associated with tanks, whereas Liddell Hart was highly regarded as a military historian. . . ." Richard A. Preston, Sydney F. Wise, and Herman O. Werner, Men in Arms (New York: Praeger, 1962), p. 281; Bond, Liddell Hart, p. 235; Trythall, Fuller, pp. 93 and 163-4 and 175; and Fuller, Memoirs, p. 455.

58. F. O. Miksche, *Blitzkrieg* (London: Faber and Faber, 1941), pp. 105-6; and Kenneth Macksey, *Guderian: Panzer General* (London: Macdonald and Janes, 1975), p. 69.

59. General Heinz Guderian, *Panzer Leader*, trans. Constantine Fitzgibbon (London: Michael Joseph, 1952), pp. 18-19, 24, and 27.

60. Major General F. W. von Mellenthin, *Panzer Battles, 1939-1945*, trans. H. Betzler (London: Cassell, 1955), pp. xv, 5, and 24; Messenger, *The Art of Blitzkrieg*, pp. 79-81, 99-101, 103-4, and 114; Liddell Hart, *Thoughts on War*, p. 254; and Macksey, *Tank Warfare*, pp. 87, 93-94, and 96-98.

3. The Blindness of a Mole

Comparative Tactics and Infantry Between Wars

Progressive German military thinkers and practitioners were probably influenced as much by the extensive war experience and military history of their own forces as by foreign military theorists. The Germans, more than any other power, appear to have discovered the secret of institutionalizing military excellence within their army;(1) at least postwar military developments tend to show this to be the case. The Treaty of Versailles, which disallowed tanks, heavy artillery, and planes, also restricted Germany's total military establishment to 96,000 men and 4,000 officers, organized in seven infantry and three cavalry divisions. It was the nadir of the German army, yet its very weakness became a strength; in so small a force, no effort was spared to encourage intellectual activity and the pursuit of professional military knowledge. Quality rapidly replaced quantity in both officers and men.(2) The former, in particular, were very carefully selected, virtually all with education superior to the average German.(3)

    Equally important, under the guiding hand of General Hans von Seeckt, Chief of the Heeresleitung (army command) from 1921 to 1926, the army returned to the tradition of mobile warfare as the basic strategic idea. Seeckt had fought in the East, where, as early as 1916, a German infantry force mounted in trucks had executed a lightning strike as part of General Erick von Falkenhayn's offensive against Rumania.(4) A believer in the most intricate coordination of all arms, Seeckt saw the "whole future of warfare" as lying in the "employment of mobile armies, relatively small but of high quality and rendered distinctly more effective by the addition of aircraft."(5) While tending to think of mobile ground striking forces in cavalry rather than armored terms, Seeckt demonstrated remarkable perception in placing strong emphasis on the development of complementary offensive aviation.(6)

Under Seeckt's tutelage, the general staff of the German army, disguised as the Truppenamt (troop office),(7) began to look into the causes of the Central Powers' collapse in the Great War. Starting with an inquiry into the war plans of 1914 and their execution, the investigation ultimately covered many operational facets considered to be of significance. A major conclusion of this postwar assessment was that the Germans accepted as sound the basic precepts of the Schlieffen Plan, with its emphasis on envelopment. Flexible tactics of fire and movement were also seen as the logical application of traditional principles to improvements in rapid-fire weaponry. It was noted, however, that the initial effectiveness of "soft spot" tactics was largely negated by the sluggishness of contemporary exploitation measures. Front-line infantry tended to outrun its artillery, which, like logistic supply columns, had difficulty in traversing the pulverized morass of no man's land and beyond. The answer as Seeckt and others saw it, lay in improved battlefield mobility; they did not, however, immediately hit upon the tank as a solution. In an army planned essentially as an infantry force, and partly because of a traditional German predilection for that arm, "soft spot" tactics using airplanes to project firepower forward were favored instead. Accordingly, emphasis continued to be placed on the basic elements of these tactics: surprise and disruptive infiltration by small groups.(8)

Overriding all, it must be stressed, was the omnipotent influence of Cannae, the Prussian doctrine of Kesselschlacht pounded into German military heads by Schlieffen. Decisive victory was not to be found on the "hostile front," but "when the rear or at least one flank of the enemy . . . [was] made the objective of the attack." The "essential thing" was to "crush the flanks," which "ought not to be sought merely in front but along the entire depth and extension of the hostile formation." This doctrine, so expressed, applied to all levels of German tactical thinking since modern firepower was considered strong enough to permit units and formations to "skeletonize" their centers.(9) Unlike the British, who appeared "lacking to some degree in that 'sense of manoeuvre' which is such a vital asset in war," the Germans forever thought in terms of turning the flanks of their enemies.(10)

There is reason to believe, then, that the Germans in the development of blitzkrieg relied heavily on their own fundamental concepts. The theory of infiltration or irruption was certainly home grown, spreading from minor tactics to grand tactics and strategy, just as Kesselschlacht doctrine spread downward. There is also evidence to suggest that the Germans had long appreciated the value of psychologically disorganizing an enemy.(11) Fuller may have exerted some influence in this regard, of course, by spelling out in "Plan 1919" that the destruction of "command" paved the way for the destruction of "personnel."(12) According to Kenneth Macksey, once the Germans had embraced the mechanized theory of warfare, they consistently regarded the role of the tank as an attack on morale, whereas the British and French looked upon it as a means to attack material.(13)

The important point of all of this is that the German army between the wars was a vibrant, searching subsociety--a definitely fertile soil for the implantation of new ideas or the unearthing of matured ones. The small, highly trained Reichsheer thus reached a peak of efficiency previously unknown, with much of its attention being devoted to the study of even the smallest detail. Whereas army authorities originally focused on welding major units together for possible immediate service, they later concentrated on the elaborate instruction of the smallest units, companies, squadrons, and batteries. With the adoption of conscription in March 1935, there were many like General Beck(14) who urged a gradual expansion to preserve the combat effectiveness that had become the hallmark of the smaller force. Even under the new conditions, however, the basics were not forgotten, and the requirement for individualization and intelligent study at all levels was still stressed.(15)

One of the most difficult infantry problems that the Truppenamt addressed during the interwar period was the question of what was termed "the last 300 yards." The problem of getting the infantryman alive over this distance had grown increasingly difficult during the Great War, and mechanization appeared to have had made it worse. The German investigators, therefore, turned to an examination of the musketry training of the German and French armies in the years before 1914, during which period the former had specialized in "target shooting" and the latter in "volley fire." The inquiry disclosed that there was something to be said for both systems, much, of course, depending on the accurate sighting of the rifles used. Experiments conducted in Austria indicated that soldiers who had received little musketry practice, when given faultily sighted rifles to shoot, scored more hits than practiced marksmen firing at the same targets with the same rifles. Still further investigation revealed, however, that under the noise of supporting artillery fire, there was a natural human tendency to shoot too high, in which circumstance target firers scored more hits than volley firers. It was additionally noted that this same artillery fire tended to make attacking enemy infantrymen dash forward in the hope of evading hostile shells. Such inadvertent combination theoretically increased German chances in attacks since volley-trained French infantry would invariably be firing too high in the event. Naturally, this theory did not work entirely to German advantage in 1914, as it had no application in respect to the mechanically regulated fire of machine guns, which could be fired along predetermined or "fixed" lines. Besides, the French infantry eventually discontinued the practice of volley fire. (16)

The problem thus remained essentially unresolved. To make matters worse, by 1918, German artillery methods had improved to the point where French guns were normally silenced by preliminary fire before attack. German infantry were consequently no longer drawn forward by enemy artillery action, nor could they "run under" the fire of the French infantry as they had previously done. A conclusion made from this discovery was that it was ridiculous to stress soldierly élan, courage, and endurance in the face of such fire. Other solutions were obviously required. One suggestion put forth was to arm one-third of the attacking infantry with smaller-caliber automatic (machine) pistols; lighter ammunition meant that more could be carried forward, thereby ensuring that ample weight of suppressive fire support would be available at critical junctures during the last 300 yards. By 1939, German section commanders were armed with submachine guns, uniquely intimidating weapons at the time.(17) Even more important, however, a specially designed light machine gun, the MG 34, was introduced as the main section weapon.

Eventually, in the course of this and associated studies, the Germans also began to suspect that they had discovered an alternative solution to the problem of the "last 300 yards" in the distracting fire potential of the man-portable mortar. A 50-millimeter mortar was consequently issued down to platoon level by 1939. The Germans subsequently became expert in the handling of mortars, developing an acknowledged and deadly combat effectiveness. Despite this advancement, however, individual rifle-shooting skills, so ardently pushed by Ludendorff, continued to be stressed to a marked degree in the German army. In certain units, the names of marksmen were permanently inscribed at the entrance to barracks, while still others were presented with much-coveted written records of their shooting achievements.(18)

Progressive individual training in the German infantry was matched by an overall good sense of organization. Standard infantry regiments consisted of three battalions, supporting artillery, and a large signals element. Battalions comprised three rifle companies and a machine-gun company. Each rifle company was based on three platoons, the organization by "threes" apparently having the advantage of inculcating in a commander's mind the ideas of maneuver, concentration, and variation of effort. Organization by "fours" was thought to produce a stultifying uniformity and lack of tactical flexibility since dispositions tended always to be made two by two, thereby annuling the principle of concentration and the idea of maneuver.(19)

For all the emphasis given to infantry, however, it remains a matter of record that the Germans followed the disciples of armor further than any other army was prepared to go. Nor were they deterred when the war in Spain, which broke out in July 1936, led many observers to conclude that the antitank gun had outstripped the tank. Although the conflict confirmed the French in their view that tanks should only be used in support of infantry for maximum effect, the Germans were careful to regard the war as a technical as opposed to tactical laboratory. They thus continued to place their faith in the tank as a decisive instrument. The defeat of Italian tank and motorized forces at the Battle of Guadalajara was written off by the Germans to Italian incompetence rather than to any weakness in the new methods of warfare. Yet some German officers like General Beck remained skeptical about the worth of the tank as an arm of decision or at least reserved judgment about it until a host of questions dealing with fuel supply, weapons efficiency, and methods of organization and command could be satisfactorily answered. Fortunately for the German army, Beck insisted on changing certain of Guderian's heavily tank-biased procurement priorities; had he not done so, of course, German marching infantry divisions would have gone to war without critically important antitank weapons. Also, partly due to Beck's influence, four motorized infantry divisions were created to work closely with the new panzer divisions. German armored forces were thus not composed of just tanks alone but instead established as "all-arms" teams backed up by infantry divisions.(20)

While the German army was developing its unique "all-arms" operational doctrine, the armies of other powers were evolving along their own particular lines. The French army of this period, though reputed to be the strongest and best trained in Europe, represented a nation with a seriously declining birth rate. Virtually

TABLE 3-1. GERMAN STANDARD INFANTRY REGIMENTAL
ORGANIZATION (CIRCA 1920)

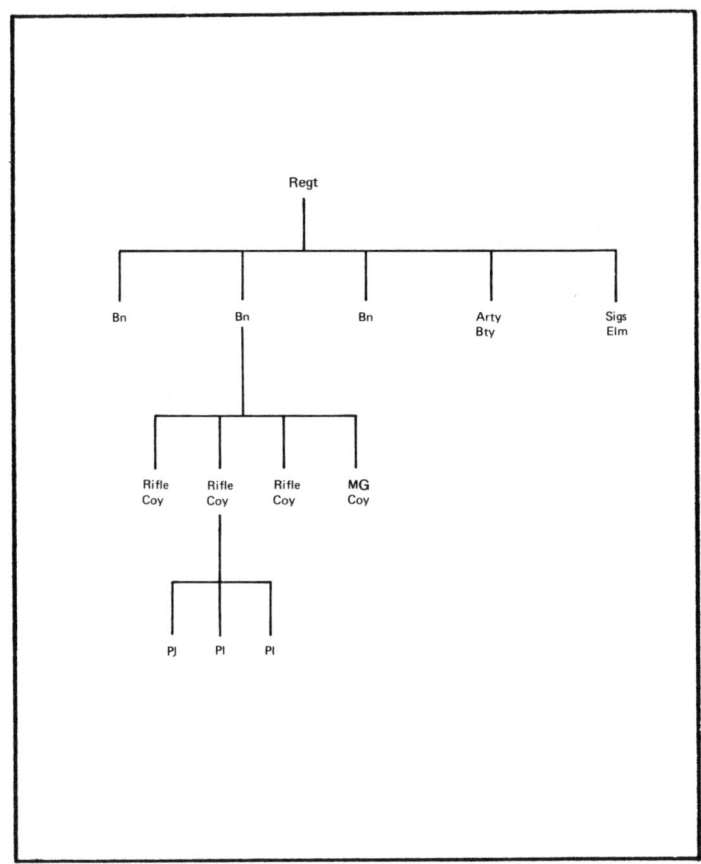

bled white by their super human efforts in the Great War, the French may also have left the mind and fighting spirit of their military leadership buried at the Mort Homme near Verdun.(21) According to General Sir John Kennedy, the French army in 1938 looked "as majestic as ever," it never entering his or other observers' heads "that their morale was rotten to the core."(22)

Despite de Gaulle's advocacy of a professional mechanized <u>armée de métier</u> of, oddly, 100,000 men, "the effective strength of the Reichswehr,"(23) the French army generally tended to think in terms of artillery dominating the battlefield. Described by some as a "clumsy army, still beladen with much equipment from . . . 1918," it was nonetheless of "immense potential, in men, tanks, and guns."(24) The dictum that the tactics of each war begin where those of the last left off made it easier, of course, for dimming memories to believe that positional linear warfare by mass armies was the trend of the future. Total renunciation of Foch's doctrine of the offensive also nurtured a widespread tendency to glean the wrong lessons from the Great War. Tactically, the French remained convinced that the artillery-infantry array was all but invincible.(25) They accordingly pinned their hopes and operational plans on the "supreme value of the defensive" epitomized by the Maginot Line, completed in 1934 at a cost of 30 million pounds sterling.(26)

The defensive orientation of the French army naturally placed great stress upon the dominance and development of firepower. "Weight of Metal" was considered paramount for both attack and defense. Of the two elements of fire and movement, the French military considered the former absolutely predominant; <u>une manoeuvre de feu</u> of automatic weapon fire was unquestionably preferred to any maneuver of the infantryman.(27) The high priest of the omnipotence of antitank defense, General Narcisse Chauvineau, constantly stressed the value of the continuous line of static positions, arguing that "like little mice which . . . [run] into their holes when threatened," infantry could evade attacking tanks and take them in the rear.(28)

French field army organization in the late twenties and thirties reflected this overpowering defensive doctrine. A standard division comprised three or four infantry regiments, each of three battalions, supported by two regiments of organic artillery.(29) An infantry battalion consisted of three rifle companies and a close support company, which included a section of antitank guns and close support mortars. Within a rifle company there were four "sections" (platoons), each of three groups. The tactical distribution of the division was effected in three echelons: the battle echelon, entrusted to the divisional infantry commander; the artillery main body, under the divisional artillery commander; and the divisional reserve. It was more a rigid fighting machine, with artillery openly regarded as the decisive arm, than a flexible instrument.(30)

TABLE 3-2. FRENCH STANDARD DIVISIONAL ORGANIZATION (1920-30)

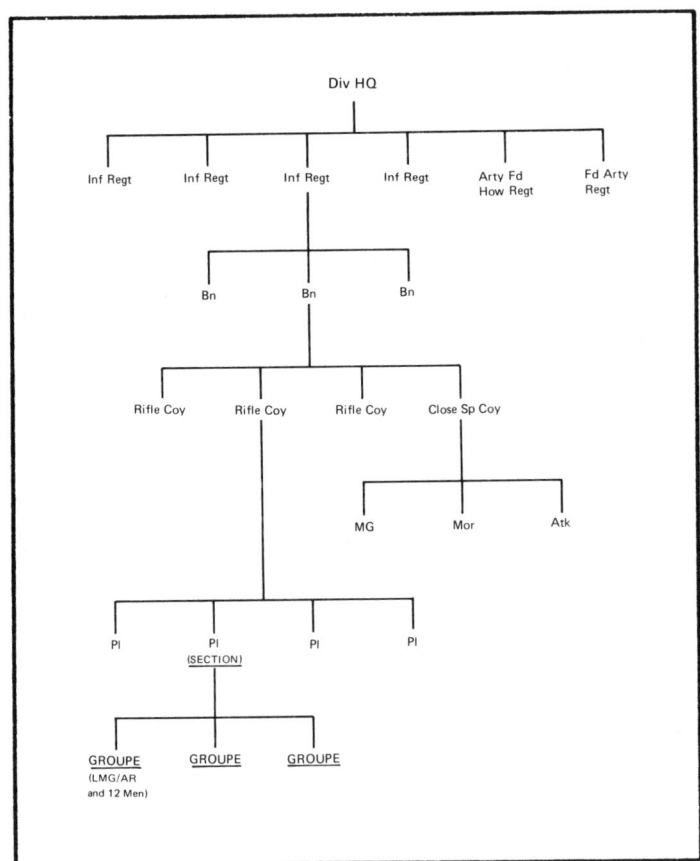

   The combat and fire unit of the French infantry was its Great War innovation, the groupe de combat, 12 strong. Its members were virtually tied to the light machine gun, the fusil mitrailleur, according to whether their individual role was "to move it, service it, feed it, or protect it." Not surprisingly, the

THE BLINDNESS OF A MOLE 69

TABLE 3-3.   FRENCH GROUPE DE COMBAT (CIRCA 1920)

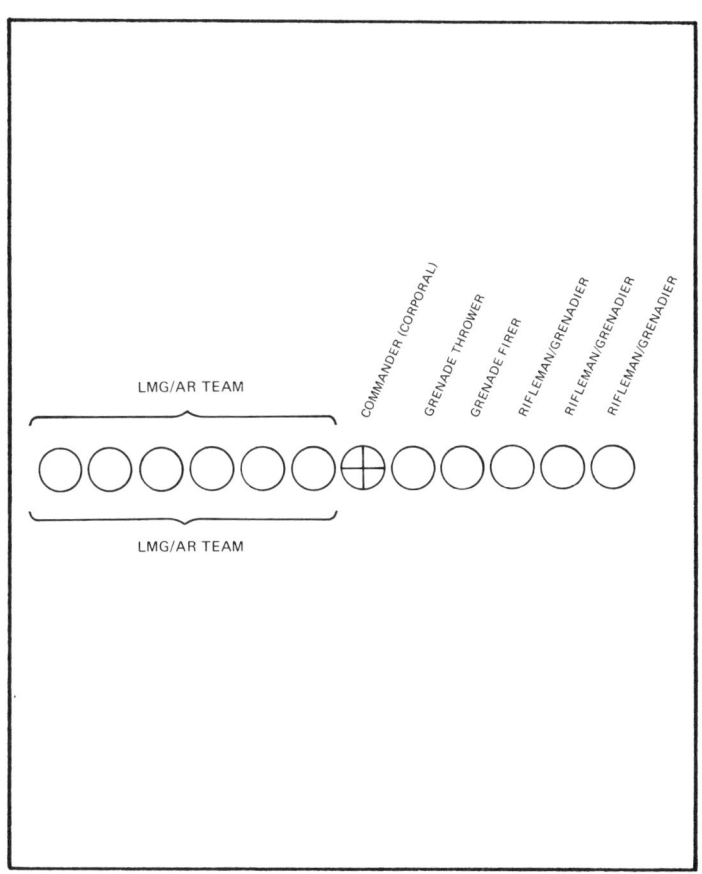

tactical training of automatic-rifle and machine-gun teams was maintained at a reasonable standard, the technique of creating a tidal wave of small-arms fire ahead of advancing infantry being developed to a particularly high pitch.(31) The rifle, on the other hand, was regarded as very much a subsidiary weapon, and the standard of shooting was universally low. The further subdivision of the *groupe* into an automatic rifle squad and a rifle team partially reflected this priority since the latter, commanded by a corporal, was not strictly a rifle team but rather a specialized bombing cell of grenade thrower, grenade firer, and three riflemen-cum-grenadiers.(32) Ideally suited for clearing the fire bays and traverses of a trench system, the French *groupe* remained essentially indivisible and untrained to maneuver within itself; whereas the German *Gruppe* operated by fire and movement, the former represented fire preceding movement. This difference in the organization of the smallest component of the French and German infantry reflected the overall doctrinal difference between the two armies.(33)

French defensive thinking, based on the preeminence of firepower, remained essentially linear in conception. Superiority in fire was thought of as superiority along a whole front rather than as in German doctrine, of superiority in a localized sense. Defensive tactics thus stressed stringent centralized control(34) and the protection of flanks. Antitank weapons, of which a division in 1940 had 52 on paper but in reality only 12, were to be placed well forward. Reinforcement of forward positions by reserves was considered normal practice, though little reliance was placed on the counterattack. Practically no emphasis was given to preparation of strong points in depth. Scarcely any French villages or cities were fortified even though these had been the backbone of German antitank defense in 1917-18. Had Paris and other major French centers been so prepared and held in 1940, they would doubtless have proven of inestimable importance.(35) To quote Fuller, "There is no anti-tank obstacle so formidable as a great town."(36)

In offensive operations, probably because the attack was considered the weaker form of warfare, the French stressed that "the offensive power of the unit . . . [was to be] maintained by the gradual fusion of the reserve into the echelon of fire." This doctrine of direct reinforcement rather than outflanking maneuver, indicated a distinct French preference for deliberate methods of attack, if such had to be made. Rejecting the entire idea of "storm troop" tactics, French military opinion advocated attacking with preponderant fire on a narrow frontage of from 300 to 800 meters, the principle being that such concentration would ensure a dense volume of fire. It was "building up the firing line" all over again. The initiative of small units was restricted to a minimum.(37)

Within this general conception of operations, the role of tanks was thought to begin at the moment the infantry were held up or reached assault distance.

Although normally controlled at a high level (that is, corps), tanks were treated essentially as a subdivision of infantry. Because of the perceived antitank threat, the 1937 manual stated that tanks should never move out of the range of artillery support. It was common practice, therefore, to attach a tank company (of three sections of five tanks each) to an infantry regiment for extra fire power. The task assigned tanks would usually be to reduce centers of resistance encountered by foot soldiers. In the assault, they would advance immediately in front of the infantry, which on "point of honour" were never to let a tank, considered as "precious," fall into enemy hands. The French tanks were thus employed, in fact, as but armored "pillboxes" on tracks," and so endowed with the power of movement at a foot-soldier's pace."(38) Such thinking continued to permeate the French army until almost the eve of World War II, when in a flurry of activity, three heavy armored divisions were formed. In the meantime, however, three already existing light cavalry <u>Divisions Légeres Mécaniques</u> continued to be employed in traditional reconnaissance and security roles, even though each remained the organizational equivalent of a panzer division.(39) Unlike the Germans, who had learned from Spain that war had changed, the French persisted in pretending that it had remained the same.(40)

If Verdun shook the French nation to its very soul, Caporetto must have come close to doing the same thing to the Italians. The army of the new Italy of Mussolini in any case, even with its spur of fascism, also reflected to a substantial degree the considerable pervasive influence of trench warfare. The mountainous frontiers of the Alps and a critical shortage of oil, of course, made the Italian situation unique, and tank development was consequently neglected initially. Emphasis was placed instead on training a sharp-shooting, agile, light infantry, a force that could also be utilized in the grand colonial designs of <u>Il Duce.</u> The tactical methods employed by Italian infantry during this period are nonetheless of interest.

On visiting Italy in the autumn of 1927, Liddell Hart gained the distinct impression that the Italian military was training "an army of human panthers," the physical training of the soldiers being "far superior to anything . . . [he had] ever seen." He described the marching endurance of the Italian soldier as "astonishing," particularly in light of their slender ration scales at the time.(41) For additional mobility, <u>Bersaglieri</u> soldiers were issued with folding bicycles that could be strapped on their backs.(42)

The Italian infantry battalion during this period consisted of three rifle companies and a machine-gun

TABLE 3-4. ITALIAN INFANTRY BATTALION ORGANIZATION
(CIRCA 1920)

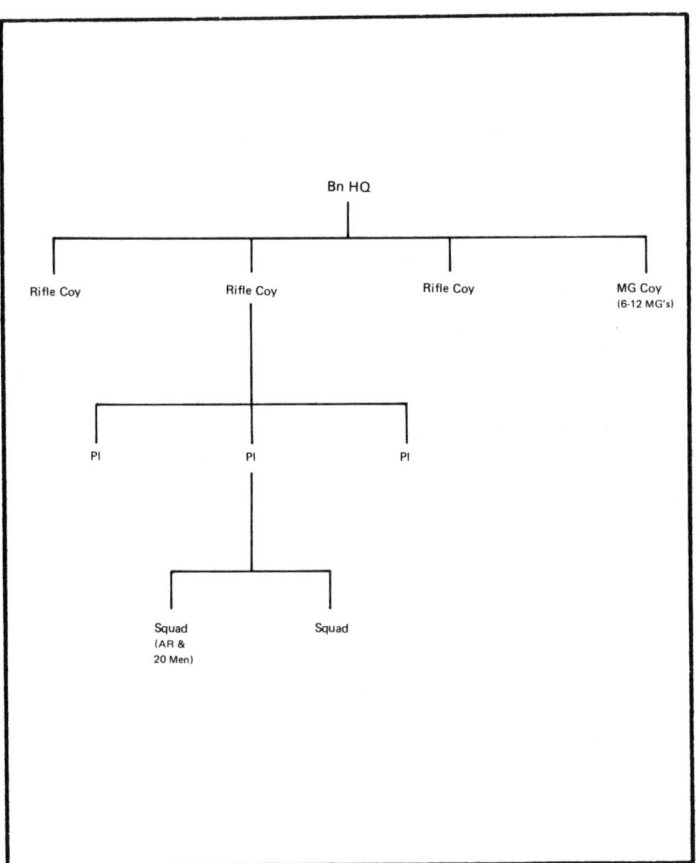

company of six guns, peace establishment, and 12 war
establishment. Each rifle company was divided into
three platoons of two squads of 20 men each. One light
automatic weapon was allocated per squad, but unlike
French army practice, the combat of the squad was not
tied so narrowly to that particular weapon. In the

advance, the Italian platoon moved forward in two long squad "worms," with the light automatic at the head of each. Upon encountering effective enemy fire, the squad riflemen would fan out to the right and left, respectively, seeking to maneuver around each flank, assaulting from both sides if necessary. The squads of 20 were further broken down into fighting groups of 3 to facilitate better control and more flexible movement. Throughout the encounter action, the squad light automatics, supported by heavy machine guns from the rear, were to keep the enemy pinned down. It was a precept of Italian operations that heavy machine-gun suppressive fire was necessary for the infantry to advance at all. Surprisingly, Italian doctrine recommended narrow attack frontages of 50 yards for a platoon and 400 yards for a battalion. Such allowances were, in Liddell Hart's opinion, bound to "have a corpse-producing . . . [effect] under modern conditions."(43)

In general, Italian tactics of the period seemed to savor strongly of Great War methods. Though slightly more elastic than those of the French, there was much less firepower to fulfill Italian tactical deliberations. Whereas a French division of nine infantry battalions also comprised one field-artillery and one field-howitzer regiment for support, an Italian division of 11 battalions (war establishment) was supported by only one regiment of mixed field artillery.(44) Tactical ability and firepower, or the lack thereof--as shall be seen--would ultimately count for more in the Italian army's future performance than any of its emphasis on better body building.

The British army, like the French and Italian, also manifested a form of trench-induced myopia during the period between wars. Despite the rather revolutionary approach to tactics urged by Fuller, Liddell Hart, and others, the shadow of the Great War fell heavily across such rays of enlightenment. The tank was passed over in favor of infantry in the British army for several reasons: tradition, financial stringency, and military policy among them. Though regarded as essentially a "weapon of offence," the tank was not necessarily rejected for this reason alone; in the eyes of the Exchequer, infantry clearly cost less than tanks and were judged to be better suited to the defensive role.(45) Moreover, the British army, even more than the Italian, was heavily involved in international policing and colonial pursuits, tasks considered by most soldiers as ideal for infantry.

In short, the British army between the wars experienced pulls from many different directions. With practically all of its reformers in the extreme "all-tank" camp, however, it would appear that the bulk of this cavalry-infantry army retrenched into conservatism. The less numerous infantry reformers, of whom Liddell Hart was the most prominent, were accordingly rebuffed. General Maxse's remarks of 1921 in this regard are most illuminating:

> I do not think there is any harm in saying that in the condition in which we are now, in regard to the infantry, we are hardly doing any tactics at all. We are employed to a great extent, as policemen, housemaids, orderlies, gardeners, and grooms; I could give you a list of other things which we are learning to do with energy. Even those who wish to practice tactics have few opportunities of doing so at present.(46)

A major infantry problem associated with devising a new system of tactics revolved around determining a proper role for the Lewis light machine gun within the infantry platoon. Unlike the Germans, who designed their 1918 offensive tactics around the light automatic weapon, the British, in typical empirical fashion, merely introduced the Lewis gun in greater and greater numbers as the need arose. Initially, specialization demanded by trench warfare resulted in the reorganization of the British platoon into a Lewis gun section, a bombing section, a rifle-grenade section, and a rifle section. With the introduction of a second Lewis gun, two light machine-gun sections were created. After the war, platoon organization was established at two Lewis gun sections and two rifle sections, each comprising a commander and six men.(47) This, of course, led to difficulties in the tactical maneuvering of the platoon. Liddell Hart's impressions of 1925 British army maneuvers are pertinent in this regard:

> Instead of the platoons acting as a combination of independently manoeuvring sections, they became so many thick clusters of men, human stop-butts had the defenders been firing real bullets. And the root cause of this return to the cannon-fodder methods of a past age lay in the too difficult tactical role of the modern platoon--due to its hybrid composition. It is illogical to expect the platoon commander, a junior officer or NCO to combine the action of four radically dissimilar sub-units, where the experienced company commander has only to coordinate four homogeneous sub-units. This anachronism means, in plain language, that the more experience an officer has the simpler is the tactical task allotted to him, under our present organization. To remove this fatal flaw it was argued, as an axiom of organization, that the

sections of each platoon should be similar in composition and armament.(48)

This was not to be the last word, however, as the organizational experts first sought to reorganize the company into three platoons of riflemen and one platoon of Lewis gunners. Such solution, in effect, merely turned the hands of the clock back, which to certain British infantrymen was undoubtedly a most desirable state of affairs. The experiment did, nonetheless, raise the question of whether the Lewis gun was to be regarded as an automatic rifle or as a light machine gun to be used farther forward than the heavier Vickers. Progressives were generally of the opinion that the proper role of the Lewis gun was that of an automatic rifle: that it was adopted as essentially a weapon of mobility in contrast to the true machine gun, a weapon of stability. In short, its intended purpose was to strengthen the mobile firepower of the platoon, to give the forward infantry greater firepower than that with which the rifle endowed them.(49) In their comprehension of this tactical idea, both the Germans and the French were ahead of the British. As late as 1937, when platoons were reorganized into three sections, the <u>Infantry Training</u> manual still mentioned that all "sections will be trained to act as rifle and/or light machine gun sections."(50)

An additional problem area for British infantry was to be found in the framework of field maneuvers, which in many cases were highly imitative of the Great War. Divisions and units were generally arrayed in one line opposing another, with little scope left for practicing minor tactical maneuvers. The aim of many schemes and exercises appeared to be to develop a smooth-working tactical process at the higher levels rather than to ameliorate the tactical art. Charging that the British commander of the day was more adept in the technique of war than in the art of war, Liddell Hart urged the cessation of such practices so that tactics could once more be elevated in position of importance above standing operating procedures. In his opinion at the time, it was the exception to find any battalion wherein correct platoon tactics were fitted into the larger framework of a battalion attack.

> The typical battalion, or larger, attack, soon becomes as linear a procession as in 1914. The sections usually try to use any cover which lies directly in their path, but so they did in 1914. They are rarely given time or space to use it as does a stalker or poacher, and, as the advance progresses to close quarters, they thicken up into a "firing line" hardly distinguishable from that

of 1914. And still more than in 1914 are they but swathes of human corn ripe for the machine gun reaper. We have foresworn the old shibboleth of weight of numbers, and yet in practice it is constantly repeated.(51)

As corollary to the foregoing shortcoming, Liddell Hart noted a tendency on the part of infantry sections to revert to the "rushing tactics" of the Great War. When infantry attacked in suicidal waves, such resort to basic fire and movement was forced, but these methods at the same time tended to catch the enemy eye and thereby ran the risk of drawing fire. A better system, as the Boers had demonstrated, was a "stealthy imperceptible advance by stalking methods." For some reason, however, the "soft-spot" infiltration tactics of the Great War appear to have been ignored during this period;(52) accordingly, the "expanding torrent" system of Liddell Hart did not really make much headway within the British army of the time. In fact, Liddell Hart's theories of tactics were anathema to many professional soldiers.(53)

Tom Wintringham, former commander of the British battalion in the Spanish Civil War, was incensed by this attitude. In his opinion, refusal "to adopt the tactics of infiltration and of elastic defence . . . [was] as out of date as to build walled castles." He claimed he detected a noticeable dislike of the concept of infiltration on the part of some British officers, likening their attitudes to those of French army authorities who, in 1918, banned the very use of the word among their war correspondents. While admitting that bits and pieces of the infiltration doctrine were being picked up by the British army, Wintringham nonetheless worried that a failure to understand concepts in their entirety often resulted in the adoption of their more dangerous and potentially fatal parts. This had actually happened during the Great War where the British army, in adopting elastic defensive measures advocated in a captured German pamphlet,(54) neglected to provide their most important element, namely, counterattack forces.(55)

A major effort was made just before World War II, however, to rejuvenate the British infantry. It began in 1936 with the replacement of the Lewis gun by the lighter, simpler Czech-designed Bren gun. Platoons were allocated three Brens in lieu of two Lewis guns and reduced from four sections to three, each with a Bren and tactical strength of corporal and seven men. Each platoon additionally received a 2-inch light mortar (with smoke ammunition only) and a 0.55-inch Boyes antitank rifle. Standard rifle companies at the same time were reduced from four to three platoons.(56)

TABLE 3-5. BRITISH BATTALION ORGANIZATION (1937)

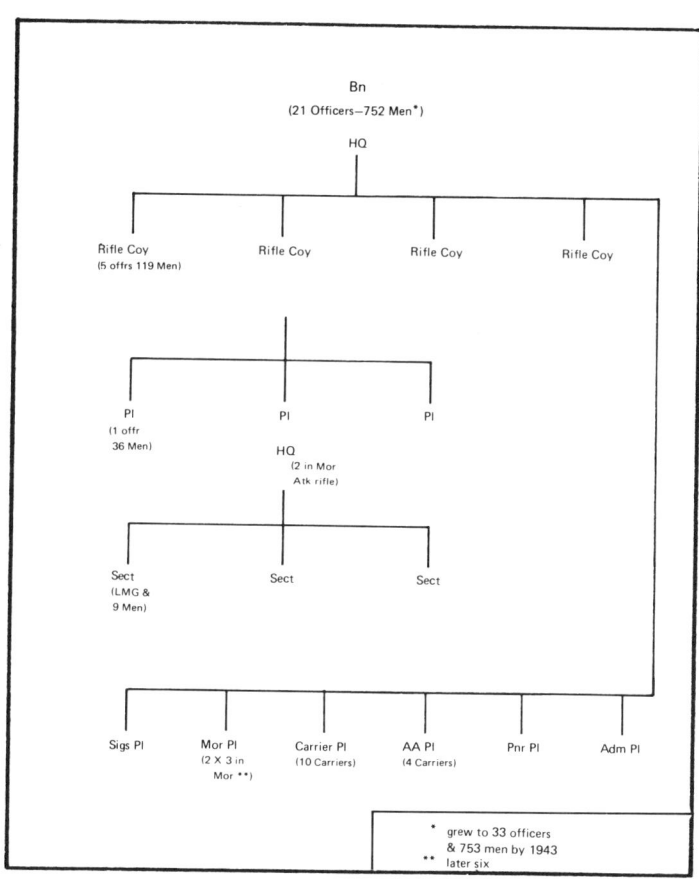

Because of a shortage of qualified junior officers, a new rank of Warrant Officer, Class III was introduced, placing Platoon Sergeant-Majors in command of two of these platoons.(57)

With the introduction of the lighter Bren, fire and movement was emphasized as never before within the platoon; "if the riflemen are moving, the Bren should be the foot on the ground," and vice versa, ran the training parody. A "carrier" platoon of ten Bren gun-tracked carriers was also added to battalion establishments. This idea, which was a specialty of British infantry, was to give the battalion commanders a reserve of firepower capable of crossing bullet-swept ground. A battalion was also allotted two 3-inch mortars. Heavy antitank guns and heavy machine guns were allocated from brigade and divisional resources.(58)

Although the Infantry Training manual of 1937 attempted to accommodate these new developments, it remained very much a compromise publication. While it stressed fire and maneuver rather than "reinforcement of troops held up in the hopes of carrying a position by weight of numbers," it also recommended "rapid and extensive digging" in defense on a Great War scale. Though alluding to depth and counterattack action in so many words, it tended to recommend in its several defensive diagrams a basically linear and frontally oriented posture.(59) Furthermore, flank security was strongly stressed in both attack and defense, and objectives were strictly limited. The stereotyped frontal attack supported by a timed barrage received as much emphasis as the flanking attack. The manual nonetheless represented a slight step forward in the modernization of British military thought.(60)

Unfortunately, however, it was a case of too little too late. As it came at virtually the eleventh hour, the British army went off to fight armed largely with weapons and equipment developed before or during the Great War. Many of its regular soldiers had never before seen most of the important weapons with which they would later fight. Worse yet, the army that finally went to war in 1939 stressed old deliberate methods of attack and defense, not even as up to date as those used by the Canadian and Australian Corps in the Great War.(61) There had, indeed, been scant time spent on collective training. Although capable of marching 20 to 30 miles a day and sticking bayonets into sacks filled with straw, the British infantry in truth was not prepared for modern war. According to General Fuller, the mentality of the times was such that it was "like living in a lunatic asylum."(62) The Great War trenches had exacted their toll; in preparing for "mole warfare,"(63) the British army, too, came to suffer from the blindness of a mole.

NOTES

1. Dupuy, A Genius for War, p. 5.

2. Rosinski, The German Army, pp. 123-8 and 141-2.

3. Gabriel and Savage, Crisis in Command, pp. 34-35. The rigorous standard for selecting officers was maintained throughout the war. In the 1939 field army, officers constituted 2.96 percent of total fighting strength. This low ratio was maintained throughout the war. Ibid.

4. Addington, The Blitzkrieg Era, pp. 28-30; and Bond, Liddell Hart, pp. 216-17.

5. General Hans von Seeckt, Thoughts of a Soldier, trans. Gilbert Waterhouse (London: Ernest Benn, 1930), p. 62. It is significant that Seeckt thought of airplanes as close support weapons and not merely instruments for reconnaissance, as Fuller originally regarded them. Lufaas, The Education of an Army, p. 362.

6. Addington, The Blitzkrieg Era, p. 30; and Rosinski, The German Army, pp. 132, 125-6, 128 and 141-2. On the mobile issue, opinions were sharply divided in Germany, many holding that trench warfare was the trend of the future. Ibid., p. 128.

7. Responsible for purely operational and training matters and advising the Chief of the Heeresleitung as appropriate.

8. Dupuy, A Genius for War, pp. 212-14, 240 and 256-7; Addington, The Blitzkrieg Era, pp. 28-30; and Rosinski, The German Army, p. 141.

9. General F. M. Count Alfred von Schlieffen, Cannae (Fort Leavenworth: The Command and General Staff School Press, 1936), pp. 4, 15 and 200-1. Kessel means "cauldron," kettle or container; Schlacht means "battle." The Field Service Regulations that the Germans went to war on in 1914 stressed the doctrine of envelopment and laid down that only one-sixth of a force should be directed to a determined holding attack on an enemy's front, the remainder to be directed against one or both flanks. Sir John Slessor, The Great Deterrent (London: Cassell, 1957), p. 32.

10. Liddell Hart, "'Ten Commandments' of the Combat Unit," p. 293. Even the old German General of Cavalry, Friedrich von Bernhardi, while accepting that "the frontal battle has become the inevitable and characteristic feature of mass warfare," felt compelled to write in 1920 that "envelopment must also be the

final goal of the frontal attack. The enemy's line must be broken through at some point, in order that the portions of the line adjacent to the point of irruption . . . may be enveloped and the process of rolling up the rest of the front which is stationary may be begun." General Friedrich von Bernhardi, The War of the Future in the Light of the Lessons of the World War, trans. F. A. Holt (London: Hutchinson, 1920), pp. 25 and 27.

11. Balck referred to tactics as "psychology." Balck, Tactics, vol. 1, p. iv.

12. Fuller, "Strategical Paralysis as the Object of the Decisive Attack", On Future Warfare, p. 94.

13. Macksey, Tank Warfare, pp. 68-69.

14. Chief of the Truppenamt beginning in 1933.

15. Rosinski, The German Army, pp. 126-27 and 134-39; Dupuy, A Genius for War, p. 238; Goerlitz, German General Staff, p. 301.

16. Wilhelm Necker, Hitler's War Machine and the Invasion of Britain, trans. H. Leigh Farnell (London: Lindsay Drummond, circa 1940), pp. 154-7.

17. Gregory Blaxland, Destination Dunkirk: The Story of Gort's Army (London: William Kimber, 1973), p. 34; and Necker, Hitler's War Machine, pp. 154-7.

18. Major General H. Rowan-Robinson, "Lessons of a Blitzkrieg," Infantry Journal, 67 (1940): 222; Bidwell, Modern Warfare, pp. 52-53; and Necker, Hitler's War Machine, p. 31.

19. Dupuy, A Genius for War, p. 169; Liddell Hart, The Remaking of Modern Armies, pp. 223 and 225; and Thoughts on War, p. 211.

20. Messenger, The Art of Blitzkrieg, pp. 79-81, 99-101, 103-4, and 114; Liddell Hart, Thoughts on War, p. 254; Macksey, Tank Warfare, pp. 87, 93-94, and 96-98; and Goerlitz, German General Staff, p. 300. As a result of the Spanish Civil War the Soviet Union made a fundamental change of course with regard to the employment of tanks. Because attempts to use tanks on their own had proved to be disastrous, General Pavlov (the leader of the Russian forces in Spain) took the message back to Russia that tanks should only be used in the infantry close support role. Messenger, The Art of Blitzkrieg, p. 104.

21. Ernest R. Dupuy and Trevor N. Dupuy, Military Heritage of America (New York: McGraw-Hill, 1956), p. 438; and Horne, The Price of Glory, p. 336.

22. Major General Sir John Kennedy, The Business of War (London: Hutchinson, 1957), pp. 2 and 11.

23. General Charles de Gaulle, The Army of the Future, p. 93. The army proposed by de Gaulle was to move "entirely on caterpillar wheels." It was also to contain the latest equipment and be "independent in its movements." Speaking from a lofty height indeed, he vaguely described the operations of this mechanical army. The book is not as abstract or philosophical as The Edge of the Sword, trans. Gerard Hopkins (New York: Criterion, 1966); I can well understand why a lot of military thinkers would have ignored de Gaulle's writings. According to Theodore Ropp, de Gaulle is "highly overrated as a military thinker." Ropp, War in the Modern World, p. 303.

24. Blaxland, Destination Dunkirk, p. 27.

25. Marshall, Blitzkrieg, p. 79.

26. Deighton, Blitzkrieg, pp. 198-99. Unbelievably, in 1933, 1934, and 1935, an average of 47 percent of the total armaments credits available to the French army was left unspent by the military. Ibid.

27. Liddell Hart, The Remaking of Modern Armies, pp. 220 and 223; Hoffman Nickerson, Arms and Policy, 1939-1944 (New York: G. P. Putnam's Sons, 1945), p. 52; and Wynne, If Germany Attacks, p. 327.

28. Macksey, Tank Warfare, pp. 89 and 105.

29. Total divisional artillery, in two regiments, included 36 new, improved 75-millimeter guns, twelve 105-millimeter guns, and twelve 155-millimeter howitzers. Blaxland, Destination Dunkirk, pp. 27 and 405.

30. Liddell Hart, The Remaking of Modern Armies, pp. 224-5 and 247; and Blaxland, Destination Dunkirk, pp. 404-5. There were 16 machine guns in the close-support company and three 60-millimeter mortars and several antitank guns in the close-support section.

31. The French had stressed live-fire training in the Great War. Brigadier General Jack recorded: "Our Allies do not hesitate to use live ammunition in their exercises, and take risks that we consider too great for training." Terraine, General Jack's Diary, p. 193.

32. Liddell Hart, The Remaking of Modern Armies, pp. 21, 124, 243, 262, and 267.

33. Wynne, If Germany Attacks, p. 327.

34. The Czech military critic F. O. Miksche remarked sarcastically that French General Staff officers "trussed in telephone wire . . . became a real bureaucracy of the battlefield." Blitzkrieg, pp. 27 and 30. Fuller, in a related vein, stated that there was "nothing more dreadful" than watching a chain of command "in telephone boxes, improvised or actual, talking, talking, talking, in place of leading, leading, leading." Generalship, its Diseases, and Their Cure: A Study of the Personal Factor in Command (London: Faber and Faber, 1933). p. 55.

35. Liddell Hart, The Remaking of Modern Armies, pp. 222 and 244-5; Miksche, Atomic Weapons and Armies, pp. 58-59 and 90; and Miksche, Blitzkrieg, pp. 41-42 and 105.

36. Fuller, Armored Warfare, p. 102.

37. Liddell Hart, The Remaking of Modern Armies, pp. 222 and 244-5; Miksche, Atomic Weapons and Armies, pp. 58-59 and 90; and Miksche, Blitzkrieg, pp. 41 and 105.

38. Liddell Hart, The Remaking of Modern Armies, pp. 224 and 246-7; and Messenger, The Art of Blitzkrieg, p. 114.

39. Four light mechanized divisions (DLMs) were formed between October 1933 and May 1940. By 1938 there were three in existence. Each division on May 10, 1940, had approximately 240 tanks, mainly S.O.M.U.A., which were superior in speed and armor to most German tanks. Messenger, The Art of Blitzkrieg, pp. 88-89, 114-115, and 142; and R. H. S. Stolfi, "Equipment for Victory in France in 1940," History, vol. 55 (February 1970), pp. 11 and 13.

40. Miksche, Blitzkrieg, p. 33. Official military circles in France rejected the significance of the employment of the air arm in place of artillery during the Spanish Civil War; they argued that the Spanish experience was exceptional because the opposing forces were deficient in artillery. Ibid., pp. 62 and 83.

41. Liddell Hart, The British Way in Warfare, pp. 177 and 282.

42. Farrar-Hockley, Infantry Tactics, p. 36.

43. Liddell Hart, The British Way in Warfare, pp. 287-8. The Italians told their soldiers not to open fire until within 300 yards of the enemy, a poor dictum in Liddell Hart's opinion. Ibid., p. 283.

44. Liddell Hart, The British Way in Warfare, pp. 283, 285-6, and 291. Tanks were regarded as infantry close support weapons. Tank battalions comprised four companies, each of two four-tank platoons in peace and four in war. Interestingly, nearly all Italian generals at the time were from the artillery. Ibid., pp. 289 and 291.

45. Blaxland, Destination Dunkirk, p. 5.

46. "'Man-in-the-Dark' Theory," p. 22.

47. W. H. A. Groom, Poor Bloody Infantry (London: William Kimber, 1976), p. 68; Lieutenant General Sir G. M. Harper, Notes on Infantry Tactics and Training (London: Sifton Praed, 1921), pp. 24-5; and War Office, WO 1447, Infantry Training: Training and War (1937) (London: HM Stationery Office, 1937), p. 225. Interestingly, General Harper wrote that sections could be increased to ten men "for training purposes." He also stated that the platoon "is the largest unit whose action should be decided by verbal orders." Harper, Notes on Infantry, pp. 24-25.

48. Liddell Hart, The Remaking of Modern Armies, pp. 116-18.

49. Liddell Hart, The Remaking of Modern Armies, pp. 118-20. The Lewis gun weighed 29 pounds, fired a magazine of 47 rounds in three and a half seconds, and could incur stoppages due to overheating. On the whole, however, it was an effective weapon. Groom, Poor Bloody Infantry, p. 69.

50. Infantry Training (1937), pp. 1 and 21.

51. Liddell Hart, The British Way in Warfare, pp. 231, 233-4, and 271.

52. Liddell Hart, The British Way in Warfare, p. 273. To avoid learning false lessons, Liddell Hart urged that attacks not be allowed "to end with the assault, or be made against objectives arbitrarily selected as convenient finishing-points." The greatest weakness of the infantry he ascribed to the lack of section and individual training. Ibid., pp. 269 and 272.

53. Kennedy, The Business of War, p. 14.

84 A PERSPECTIVE ON INFANTRY

 54.. Not the 1916 "Defensive Battle" Die Führung der Abwehrschlacht but a 1917 companion manual on the layout and construction of field defenses entitled Allgemeines uber Stellenbau. The British, in fact, copied the wrong manual! Barnett, The Swordbearers, pp. 298-9.

 55. Tom Wintringham, Deadlock War, pp. 174-5 and 179. In the words of Wintringham: "The British Official History by Brigadier General Sir James Edmonds shows that the British Army early in 1918 took their whole defensive doctrine from a single German document, which was translated word for word and distributed to commanding officers. This document dealt only with the construction of strongpoints, pill-boxes, defensive outposts, and the larger defensive works in the battle zone. It did not deal with the organization and disposition of troops. It did not point out that each division should keep reserve battalions behind the battle zone for counter-attack. And it did not mention that the Germans had so far developed their theory as to hold, in addition, one complete division for counter-attack behind each pair of their ordinary divisions in the line. As a result the British dispositions in March 1918 were disposed in depth but were entirely inelastic. They had no arrangements for counter-attack and were therefore unsuccessful. This lack of success led to a refusal to accept the theory of defence by strong points arranged in depth." Deadlock War, pp. 176-7.

 56. Blaxland, Destination Dunkirk, pp. 6-7. Marching columns were also slimmed from four ranks to three. "Rifle" battalions during this transition period comprised four rifle companies of four platoons, each of three sections, and a headquarter company. "Mixed" infantry battalions comprised a headquarter company, a support company, and three companies each of four platoons of four sections. Infantry Training (1937), p. 2.

 57. The PSM was junior to the Company Sergeant Major (CSM) but senior to the Company Quartermaster-Sergeant (CQMS). The WO III practice was abandoned in 1940. David Erskine, The Scots Guards 1919-1955 (London: William Clowes, 1956), p. 7; and Blaxland, Destination Dunkirk, pp. 5-6.

 58. Major L. E. Ellis, The War in France and Flanders 1939-1940 (London: H. M. Stationery Office, 1953), p. 371; Blaxland, Destination Dunkirk, pp. 6-8; and Farrar-Hockley, Infantry Tactics, pp. 7-8. A new "battledress" was introduced in 1939, appearing very much like an afterthought. New pattern webbing had been

issued from 1937. There were no funds available for field firing exercises. Farrar-Hockley, Infantry Tactics, p. 7.

59. Infantry Training (1937), pp. 144-71, and 228-35. As an example of this, medium and light machine guns were normally sighted on the forward arcs on the assumption that the enemy would always be obliged to assault frontally. Farrar-Hockley, Infantry Tactics, p. 12. The standard "weapon pit" of the British army at the time measured six feet by three feet six inches wide at ground level by three feet deep. It accommodated two soldiers and could be dug by one in four hours in reasonable soil. Ibid, pp. 221 and 225.

60. Infantry-Training (1937), pp. 107, 116, 120, 126, and 153; and Farrar-Hockley, Infantry Tactics, pp. 7 and 10. The influence of Liddell Hart can be seen in Infantry Training: i.e., equipping riflemen "as lightly as possible" and dividing rifle companies into "forward and reserve" platoons, The sensible formation of "arrowhead" was not adopted as a field formation, however, losing out to "column, square, diamond, and figures 'T' and 'Y'." Infantry Training (1937), pp. 39, 112, 125 and 150.

61. There is no question but that these troops during the last two years of the war were the "shock" equivalent of the German Sturmtruppen. That they suffered fewer casualties than the average divisions of the British army while participating in equally hazardous operations is proof of this. They used the same tactical system as that of the British, yet every "honest observer" attested that they fought differently. Wintringham, Deadlock War, pp. 256-9. This was not due to superior courage, as some said, or because they got all the easy tasks, as still a few others mistakenly charged; it was simply because their commanders, Sir John Monash and Sir Arthur Currie, maintained open minds and devised new systems of tactics from the bottom up to suit the unfortunate debacle of the trenches. In the final battles of 1918, the Australians comprised only 9 percent of German prisoners, guns, and territories taken by all Allied forces in the offensives launched from April to October 1918 and more than the Americans. Beaumont, Military Elites, p. 28.

62. Fuller, Machine Warfare, p. 9; and Farrar-Hockley, Infantry Tactics, pp. 7-8 and 10.

63. Major General Sir Ernest D. Swinton (pseud. Ole-Luk-oie), The Green Curve and Other Stories (London: William Blackwood, 1915), p. 200.

4. Fair-Weather War

Of Infantry and Blitzkrieg

It is now generally recognized that the Germans in their 1940 blitzkrieg assault on France did not outnumber the Allies. The Anglo-French field armies were slightly smaller numerically, two million men organized in 112 divisions to 2.5 million Germans in 136 divisions; but counting 600,000 Belgians and 400,000 Dutch, the Allied total was the equivalent of 156 divisions. Also, contrary to popular belief, the Anglo-French alone outnumbered the Germans in tanks, roughly 3,600 to 2,574, with most of the Allied tanks being at least equal in quality and many superior in armor and armament.(1) Though the Germans compensated for this relative inferiority through a more efficient grouping of tanks, this single factor does not entirely account for their astonishing victory. A better explanation for their success might be found in the very Germanic philosophic conviction that "the Whole determines the Part," and not otherwise.(2) The panzers, though spectacular, and important, were but a "Part." The great mass of the field army was infantry, and its role was equally important to that of the ten German armored divisions. The real reasons for German success in the Battle of France appear, therefore, to lie less in the realm of material and organization than in the areas of doctrine, tactical development, and even psychology.

As mentioned earlier, the German army following the Great War was composed primarily of infantry. Imbued as it was, however, with Seeckt's doctrine of high mobility, with attack given precedence over defense, this particular arm was never regarded as anything less than an offensive means of effecting a decision on the battlefield.(3) The German infantry was never cast in a purely defensive role. From its beginning, it was organized as an army of attack, with quality stressed over numbers. Within its ranks, a professional military inquisitiveness and candor were widely encouraged, the old Prussian tradition of "absolute frankness, even towards the King,"(4) being strongly maintained. Renewed efforts were also made to revive Kameradschaft (comradeship) as "the god of the German Army."(5)

In rebuilding the Reichsheer from the ashes of the old Imperial Kaiserheer, military authorities were very much aware that an extremely serious gulf had developed between officers and men in the autumn of 1918. They were determined, therefore, without relaxing the outer forms of discipline, to put new relations between officers and men on a firmer footing. The bond established was to be based not merely on external authority but on an inner cohesion as well. From the beginning, the young officer was impressed with the necessity of gaining the confidence and comradeship of his men without forfeiting his authority. At the same time, conditions of service were made better for the soldier. Pay and barracks were improved, compulsory church and other parades terminated, and punishments eased insofar as they were left to the discretion of the individual officer, with emphasis laid upon appeal to the culprit's sense of honor. To provide additional pay and status incentive to the soldier, three new ranks were introduced between private and NCO, presumably Obergefreiter (corporal), Gefreiter (lance corporal), and Oberschütze (American private, first class). This extraordinary attention devoted to the physical and psychological needs of the men and much enhanced comradeship in arms was later to serve the Germans well.(6)

The new German army additionally stressed forward battle positions for commanders, although this was as much intended to facilitate clear passage of orders from rear to front as to cement leader-follower relationships through personal example and leadership. Here again, however, one can perceive the influence of Seeckt, whose dictum "The essential thing is action" permeated the German army. In the tradition of Auftragstaktik, which shall be explained shortly, German officers and soldiers were trained to exercise an exceedingly high degree of initiative in all circumstances. At stake, of course, was the entire tactical system of the German army. A major reason why the concepts of infiltration and defense in depth had been so strongly opposed in the Great War was the "scanty training of . . . [certain junior] officers, NCO's, and men." Ludendorff, in admitting that "it was a risky business"(7) to adopt infiltration techniques, probably recognized that he was taking quite a chance in trusting his junior leadership to the extent that he did.

After the war, especially strong emphasis was placed on the careful training of the commander of a section, "the largest unit that a leader can command by direct personal influence upon the individual soldier." It was generally recognized that it was upon the example of the section leader, upon his resolution and coolness in the face of difficulty and danger, that the performance of a section depended. The section commander, and not the company or platoon commander as in 1914, gave the order to fire. In an attack, it was the section commander who was the driving force behind the soldier's determination to get at the enemy. Simply stated, a good leader, an outstanding leader, meant a good section; a bad leader meant a bad section.(8) According to Shelford Bidwell, in rating the importance of the section leader so highly, the Germans were particularly perceptive as regards small group cohesion:

> The key men who occupy the place of father and uncle in this artificial family [of the close-knit platoon] are the corporals. The best indication of the spirit and stability of a fighting organization is their status and quality. It was not mere facetiousness which led to both Marlborough and Napoleon being nicknamed "corporal" by their troops. The corporal is the most important man in the private soldier's life.(9)

The training of the new German army prior to World War II was as intensive physically as it was intellectually. In spite of the importance attached to tanks and planes, the marching ability of infantry was not neglected. Many German units were capable of astonishing feats in this area, and 30 miles a day for several days on end appears to have been fairly common training practice. A good rate for a longer march was considered to be an impressive four to five kilometers an hour. Equally important, however, was that more than six kilometers was considered impossible. The Germans were not afraid to admit to human limitations.(10)

Small-group <u>Wehrsport</u> military competitions such as machine-gun crash-action drills and mortar shoots were also fostered to a great extent. Field camouflage from ground and air observation received considerable emphasis and was ultimately effected to a degree never equaled in other armies. Exercises were almost always conducted so as to adapt marching formations and actual tactics to probable interference from enemy aircraft.

The Germans also continued to train special assault squads, <u>Stosstruppen,</u> to overcome key strongpoints during infiltration. The tremendous initiative they instilled in all of their soldiers, plus their distinctive "plug in" concept of army organization--which stressed a balanced allocation of all arms down to the lowest levels--and well suited to the creation of such ad hoc groups, which remained a hallmark of the German army throughout the war. Unlike the British soldier, who generally displayed a rather narrow and clinging loyalty to his particular unit, the German appeared willing and able to fight under the direction of any officer.(11)

As previously adumbrated, standard infantry organization during the interwar period reflected both common sense and the general mobile "all arms" doctrine of the Germans, with special emphasis on communications. On the eve of the invasion of Poland, an infantry division comprised: three infantry regiments, a mixed artillery regiment of forty-eight 105-millimeter and 155-millimeter howitzers; a reconnaissance battalion; an antitank battalion with 37-millimeter guns; an engineer battalion; a signals battalion; and services. Each infantry regiment consisted of three battalions, a cannon company of six 75-millimeter and two 150-millimeter howitzers,(12) and an antitank company of twelve towed 37-millimeter guns. The battalions were four-company organizations, with the fourth, eighth, and twelfth (companies were numbered 1 through 14 in the regiment) filling the role of heavy-weapons companies. The rifle companies possessed a strength of 180 to 200 men, a total of nine light and two heavy machine guns, and three light (50-millimeter mortars each; the heavy-weapons companies, eight machine guns and six 81-millimeters mortars each.(13)

A German infantry platoon thus consisted of a platoon headquarters and three sections, each of ten men and a light machine gun. Platoon headquarters included a 50-millimeter mortar squad and an antitank squad of three 7.92-millimeter antitank rifles (later replaced by the <u>Panzerfaust</u>). During 1940, the 50-millimeter mortar squad was abolished and replaced with another light-machine-gun section (that there were six 81-millimeter mortars in the heavy-weapons company doubtless influenced this decision). Significantly, each German section comprised a leader, a five-man rifle group, and an MG 34 light machine-gun team of four, including two ammunition carriers. With a high cyclic rate of fire of 800 to 900 rounds per minute and a moderate weight of about 24 pounds, the MG 34 was probably the most advanced and effective machine-gun of its time. Putting forth the equivalent fire of 20

TABLE 4-1. GERMAN INFANTRY DIVISION ORGANIZATION
(1939)

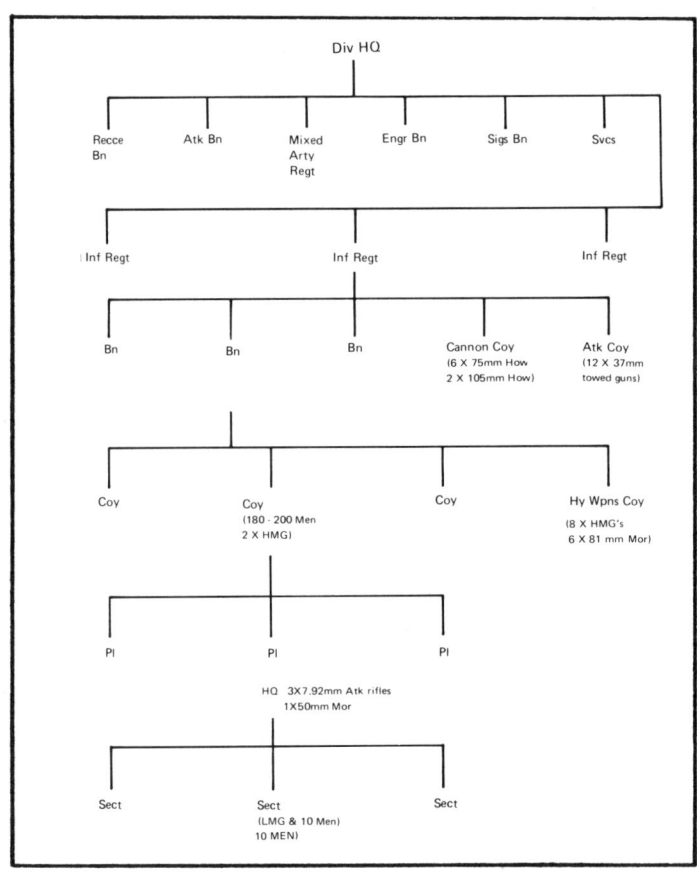

riflemen, it provided the German section with a
superior weapon of decision. On the Eastern Front it
was not uncommon to find sections consisting of one
NCO, three riflemen, two grenadiers, and two
machine-gun teams of two men each. By 1943, because of
manpower shortages, infantry section strength was
officially reduced to nine men.(14)

Motorized divisions were smaller than the standard infantry division by about 1,400 men. Each motorized infantry division comprised three infantry regiments and was organized much like a standard division except that all elements were transported by motor vehicle. Within this organization, each infantry battalion also had its own company of 37-millimeter antitank guns. (This weapon was eventually replaced within the German army by a 50-millimeter long-barreled version.) Motorized divisions were later upgraded to semiarmored status and renamed Panzergrenadiers, at which time an assault gun battalion replaced one infantry regiment while the other two were equipped with greater firepower. In addition to these divisions, the Germans also organized mountain, parachute, and Volksgrenadier divisions.(15)

The first three panzer divisions had been created in October 1935. As implied earlier, they were anything but "all-tank," each consisting of a motorized infantry brigade in addition to an armored brigade. The latter comprised 561 tanks organized into a headquarters and two tank regiments, each of two tank battalions of four light 32-tank companies. A panzer division infantry brigade consisted of one regiment of two motorized infantry battalions and one motorcycle battalion. Each motorized infantry battalion contained one motorcycle company, two motorized companies, one heavy machine-gun company, and one mixed company (engineers, anti-tank, and infantry gun platoons); the motorcycle battalion(16) consisted of three motorcycle companies and one mixed company. Each division included a motorized artillery regiments (of twenty-four 105-millimeter howitzers), an antitank battalion of towed 37-millimeter guns, a light engineer company, a reconnaissance battalion of armored cars and motorcyclists, and various signal land divisional service units. During 1938 and 1939, the motorized infantry regiment in each infantry brigade was increased from two to three motorized infantry battalions. The next three panzer divisions raised were established with two infantry regiments of two battalions each. On mobilization, tank dissipation further lowered the tank-infantry company ratio from 16 to 9 to 12 to 12.(17)

If balanced tank and infantry forces were pivotal to German army organization, then surprise was central to German tactics. Without doubt, the German army glorified surprise, a subject on which the German General Waldemar Erfurth had written at length.(18) The German infantry manual clearly stated that "every action ought to be based on surprise," the rule to be applied on the smallest scale: to machine-gun bursts and mortar fire, the making of a sound, and even the

TABLE 4-2. PANZER DIVISION ORGANIZATION (1938-39)

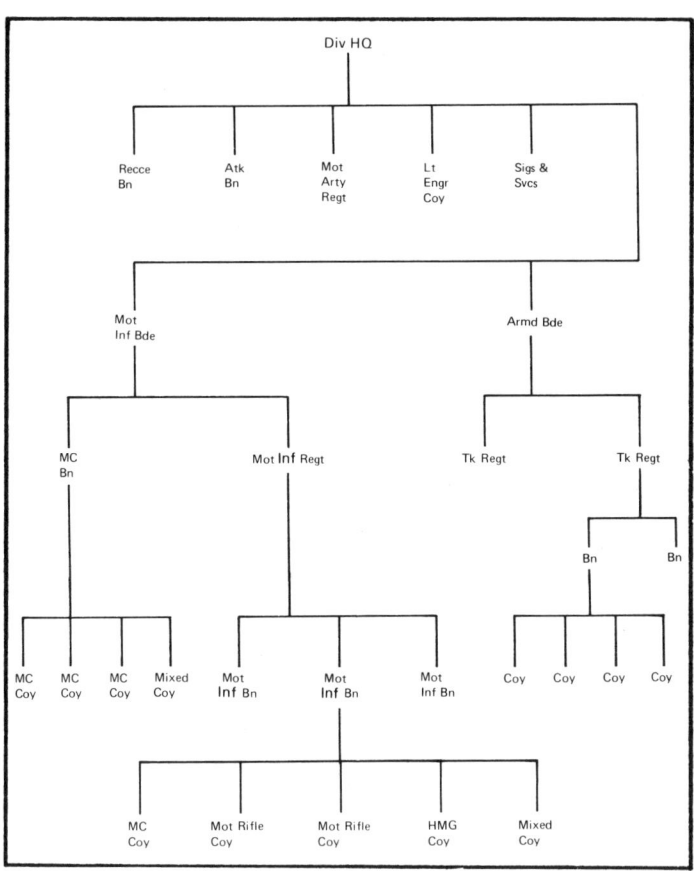

sudden shot of a rifleman. Positions of the smallest defending groups were to be changed at intervals so that the enemy was constantly faced with new problems. The German army was, in fact, an army saturated in the principle of surprise. Mobility and maneuver were but respective means to effect it in time and space.(19)

Unlike the French army, however, the German army exalted mobility and maneuver; even within the smallest units in large-scale attacks, outflanking maneuver was to be attempted. Having studied the application of firepower with a thoroughness superior to that of any other army, the Germans rejected as simplistic the basic French conception of two lines advancing on one another, the one side prevailing because of its superiority of fire. Instead, the Germans envisioned their future approach combats as merging into a series of local actions, followed by a steady progression through the enemy position by infiltration and outflanking of centers of resistance. In this scenario, the small-infantry section figured prominently, for the infantry attack was built around the light machine gun. Complementing this theory was the idea that every attack has a "centre of gravity."(20)

The latter idea, which requires expanding, is probably best explained by the two German words Schwerpunkt and Aufrollen. Together, they represent a single main pattern of German World War II tactics, a pattern into which most operations in the attack were fitted, whatever the size of the units involved. The word Schwerpunkt, first used by Clausewitz, is literally translated as "center of gravity"; however, this simplistic translation does not convey the exact sense of the German meaning into English. According to F. O. Miksche, a more militarily correct interpretation would be "thrust-point," to indicate the principal effort or concentration of force aimed at seeking out the weakest point of enemy resistance. In this regard, the concept was similar to the French effort principal, which was a course or direction of effort decided on before an attack and afterward seldom changed. The German Schwerpunkt principle differed dramatically, however, insofar as the weight or "thrust" of the effort was constantly altered during the attack to fit the circumstances of seeking a line of least resistance; it could mean the flexible switching away from the original direction of attack. Furthermore, whereas the French considered the principle as mainly applicable to grand tactics, the Germans regarded it as equally relevant to the movement of a section of infantry as to the maneuvering of an army corps. By such continual switching of the direction and place of thrust, the attention of the opposing defense was constantly confused as to real objectives and consequently tied down everywhere. A German attacking force, even in the smallest details of combat, maintained in this manner superiority, initiative, and surprise.(21) This concept of operations has been referred to by some commentators as the principle of the "unlimited objective,"(22) and it so permeated the

TABLE 4-3. SCHWERPUNKT AND AUFROLLEN

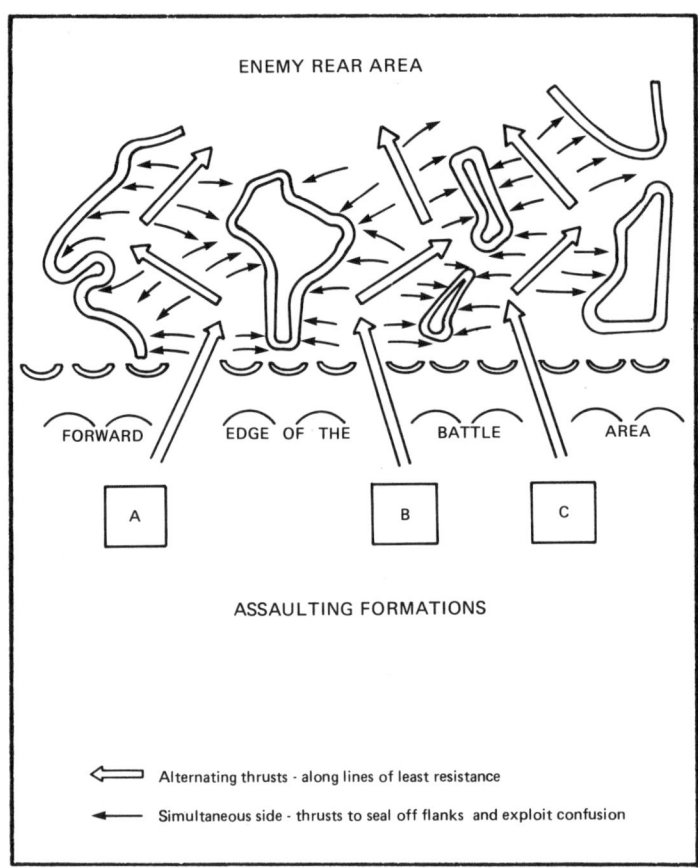

German army that Field Marshal Paul von Hindenberg's quip "An attack without a Schwerpunkt is like a man without character" was often quoted.(23)

To this Clausewitzian-inspired principle was grafted the Schlieffenian notion of Kesselschlacht, of striving constantly for a decision on the enemy flanks or rear, definitively described by the German word

Aufrollen. Translated in a military dictionary as "to clear or work along," the word appears to have been used in the Great War to portray the action of storm parties clearing trenches laterally. However, in the mobile scenarios painted by the post-Great War army, it came to be associated with a "rolling out" action accompanying a "thrust." Alternating with the movement of the Schwerpunkt, it was, in fact, the "immediate and methodical exploiting of each local thrust by side thrusts." The Aufrollen action thus sealed off and protected the flanks of the advancing units as well as widening the gap. At the same time, it dislocated enemy communications and disorganized his rear areas.(24) The normal German attack thus searched constantly for a weak point in the enemy's position and continually used this very point to secure ground from which flank attacks and attacks from the rear could be mounted against stronger points.(25) It bore a striking resemblance to the method of attack advocated by Liddell Hart in this theory of the expanding torrent. In fact, it established for the German army exactly what he had urged the British army to adopt: a framework for tactical action on which the most junior leader as well as the general could hang his hat.(26)

The two ideas of Schwerpunkt and Aufrollen, when put into practice, obviously meant that the ensuing battle would not be fought over a long linear front but rather over a wide area, characterized by a series of minor actions. The tactics required for such actions were described in German military literature as Flachen und Luckentaktik ("tactics of the space and gap"). In conducting a battle along these lines, attacking teams or Angriffesgruppe, had to be practically independent and tactically capable of fighting on their own. This, in turn, called for a larger degree of decentralization in both weapons allocation and decision-making authority. As we shall see, it also called for a greater role on the part of the infantry.(27)

The decentralization of tactics concept placed the Germans in diametric opposition to the classic school of French military thought, namely, the centralization of tactics. There was a thread of tradition in the German tendency, however, that dated as far back as the "Captain's War" against Austria in 1866. The concept of Auftragstaktik of "mission tactics" during that war made it the responsibility of each German officer and NCO, and even Moltke's "youngest soldier," to do without question or doubt whatever the situation required, as he personally saw it. Omission and inactivity were considered worse than a wrong choice of expedient. Even disobedience of orders was not inconsistent with this philosophy. A favorite story that Moltke liked to recount concerned a young major who, on receiving a reprimand from Prince Frederick Charles, offered the excuse that he was only obeying orders; the prince's prompt retort was "His Majesty made you a major because he believed you would know when not to obey orders." As far as the Germans were concerned, the first demand in war was decisive action."(28)

With the wedding of mechanization and motorization to the concepts of Schwerpunkt, Aufrollen, and Auftragstaktik the technique of blitzkrieg was formulated. Again, it was no great novelty but merely the adaptation of old German military principles to modern engines of war. When this technique was applied to the idea of infiltration, known to be the only effective method of piercing a defense in depth, it speeded up the process immeasurably. In Miksche's opinion, the faster or motorized form of infiltration was better described by the action work "irruption." The total process, therefore, was to concentrate a force on a narrow front, apply the principles of Schwerpunkt and Aufrollen to attain an irruption, and through such irruption create the necessary flanks to effect a Cannae.(29) As most centers of resistance to such a German attack were expected to be of an elastic nature, sited in depth with a capacity for counterattack, the role of the infantry was considered vital. Only that arm was considered capable of holding ground and providing the required flank protection for the exploitation of the irruption.(30)

Eight critical months elapsed between the outbreak of World War II and the German breakthrough at Sedan. The period of "Sitzkrieg" or "Bore War" on the Western front did little to dissuade the French from erroneously speculating that any German attack launched would take on the form of a Great War infiltration on a wide front. The French further prepared for a strategic repeat of the Schlieffen Plan by weighting their left flank with their best and most mobile troop formations. Unfortunately, as history records, they neglected their center just north of the Maginot Line. Nonetheless, the Germans in their original Fall Gelb, or "Case Yellow" plan, did come close to emulating their Great War strategy. This plan called for a Schwerpunkt on the right wing to be executed by General Fedor von Bock's Army Group "B," which was to include all ten German panzer divisions. A smaller Army Group "C" of 17 infantry divisions, under General Wilhelm von Leeb, was assigned the task of holding or fixing the French forces entrenched along the Maginot Line, while General Gerd von Rundstedt's Army Group "A" provided the connecting link and support to von Bock. All this was changed by the last-minute adoption of the Manstein Plan, which called for a surprise thrust through the

MAP 4-4. PLAN YELLOW

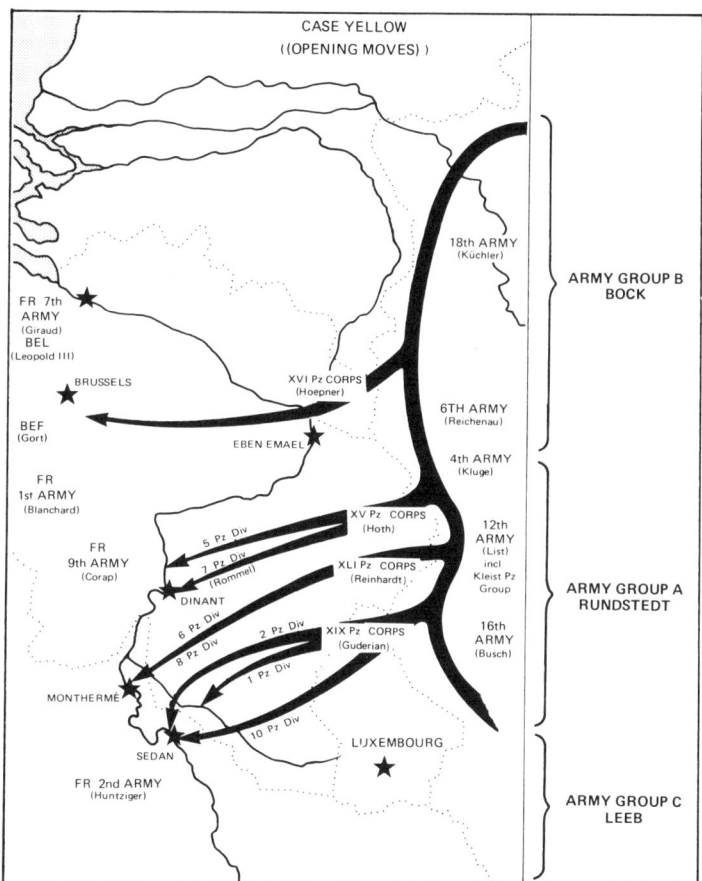

Ardennes between Namur and Sedan by a strengthened Army Group "A" comprising seven panzer divisions, three motorized and 34 infantry divisions. The new Schwerpunkt was aimed at the weak French center. Army Group "B" with three panzer divisions, one motorized and 24 infantry divisions, was to create a diversion by overrunning Holland and northern Belgium, using land and airborne forces.(31)

The German assault in the West commenced on May 10, 1940, with the simultaneous crossing of the frontiers of Belgium, Luxembourg, and Holland. By midday May 11, the strategically located Belgian fortress of Eben Emael, considered virtually impregnable by the Allies, along with two key bridges across the Albert Canal, fell to a small force of some 500 German glider and parachute troops.(32) This action virtually sealed the fate of Belgium and Holland. The latter nation surrendered on May 14, the former on May 28.

Although the Germans enjoyed the advantage of air superiority, no attempt was made to interfere with the flow of British and French formations into Belgium and southern Holland to meet the onslaught of Army Group "B." The Allied countermovement fitted in nicely with German plans, as the decisive stroke was to be delivered by Army Group "A," comprising the Fourth, Twelfth, and Sixteenth armies, and Panzergruppe Kleist.(33) The last group consisted of General Georg-Hans Reinhardt's Panzer Corps (Sixth and Eighth Panzer Divisions), Guderian's Panzer Corps (First, Second, and Tenth Panzer Divisions) and General Gustav von Wietersheim's motorized corps of five motorized divisions. General Herman Hoth's Panzer Corps (Fifth and Seventh Panzer Divisions) was under the command of the Fourth Army, which was assigned the task of crossing the Meuse River at Dinant. The main punch was to be delivered by Panzergruppe Kleist at Monthermé and Sedan, each the thrust point of a panzer corps. By May 12, Kleist's group had reached the Meuse; by May 16, it had broken clear through the French front west of the river and set out on its drive to the sea.(34)

The German breakthrough in the West was a paragon of blitzkrieg style. On each thrust point(35) were mustered concentrations of roughly two panzer divisions and two or three motorized divisions, all backed up by massed regular infantry divisions; the whole prepared to exploit success to the fullest if the opportunity presented itself. Accompanying the ground thrusts and absolutely essential to their success were the Sturzkampfflugzeuge or "Stukas," whose aim on establishment of air superiority was to provide local protection and direct support to the ground forces (Guderian used them as aerial artillery). On the advice of Germany army psychologists, sirens had been fitted to the Stukas to increase their terror effect. Against French artillery emplacements, they were deadly; on May 13, 1,000 aircraft in dense formations dealt the French artillery a blow from which it never recovered.(36)

The first phase of the blitzkrieg attack consisted

of reconnaissance accomplished through armored car probes, some acting as decoys, and swarms of more cautious motorcyclists (the action of the latter prompting some observers to remark that the Germans were waging motorized guerrilla warfare). The next phase, usually shielded by an army of refugees,(37) generally witnessed light tanks supported by mediums attacking along lines of least resistance attempting to take centers of enemy opposition in the flanks. Into the breach so effected would then flow a heavier combat echelon of more medium tanks, assisted as necessary by infantry to force it wider. Following the armored breakthrough, motorized infantry formations would hold and expand by infiltration the ground gained, protect the flanks of the continuing attack, and occupy dominating positions over enemy lines of communication. In some instances, motorized divisions were expected, after relief by regular infantry, to drive the main thrust point deep into hostile territory so as to force a decision in the flank or even rear of the enemy forces. In the pursuit, they often led. The task of the hard-marching foot-infantry divisions was to continue the process by mopping up, consolidating, and if still possible, exploiting the situation created, thereby making the success decisive. In this respect, the rigorous peacetime training of the German infantry paid off, as they normally had to march 30 miles a day and sometimes as many as 50. During the conquest of France, they exceeded even the best marching records of the old Imperial army.(38)

The importance of infantry to the blitzkrieg system was further illustrated by the May 13 action of the First Rifle Regiment, First Panzer Division, in crossing the Meuse River during the initial breakthrough by Guderian's forces at Sedan. With close indirect fire support provided by Stukas instead of artillery, this formation managed to cross the river in collapsible rubber boats and secure a bridgehead. Due to French inability to mount effective armored counterattacks, the First Rifle Regiment was also able to hold and expand the bridgehead. It had been a risky operation, however, as the regiment was without tanks, and three out of seven attempted crossings by the three panzer corps had failed. Nonetheless, it was such action on the part of infantry that eventually got Guderian's corps across the Meuse. This success seems to have strongly influenced Guderian, for he used the same method to cross the Ainse in June; von Kleist, on the other hand, vainly tried to throw armor against well-prepared defensive positions.(39)

The assault action of the First Rifle Regiment was considered normal for German motorized infantry, since they had been trained to mount quick attacks to secure bridgeheads across obstacles like rivers and mine fields, which either halted further advance by tanks or could not be bypassed. Old methods of operation such as extensive reconnaissance, detailed fire planning, and exhaustive orders were consequently discarded. Since 1937, the German army had stressed quick reaction, and it had accordingly trained its field commanders to arrive at tactical solutions--to problems posed on the ground--in a few minutes instead of hours, as was British custom. Whatever criticisms were levied in the process by instructors, none were ever made to discourage dash. A German motorized battalion with some regimental support was expected to execute an assault across an obstacle in about 40 minutes from the moment it encountered the obstacle. The assault was normally mounted on a 600-yard front; the basic method employed was infiltration and three phases were conceived. The first (niederhalten) involved pinning the enemy down and winning the fire fight (Feuerkampf), during which time brief reconnaissance and troop deployments could be carried out. The second (blinden) phase called for assault bodies to infiltrate forward, at which point the enemy defense would be smoked off and blinded in addition to being subjected to already initiated offensive fire. Finally, in the niederkämpfen phase, the enemy position would be saturated with intense fire while the assaulting troops destroyed it in detail.(40)

Although the German method of quick reaction attack involved the application of fire and maneuver, it differed substantially from British and French practice in one very important respect. In the first place, the German concept confined the attack to a very narrow frontage, thereby allowing the attacking commander to concentrate nearly all of his fire resources in this area. More important, he did not have to worry about much else since he expected as a matter of course that the flanks would be engaged by tanks or light armor and the enemy rear by artillery fire and Stukas action arranged by the regimental staff. This procedure dispensed with the delays involved in consultation and preparation of a comprehensive fire plan involving mortars, artillery, and aircraft. The time gained was subsequently translated into shock action that denied fire initiative to the enemy.(41)

Rommel perhaps best described the German approach when he said that in encounter battles the day normally went to the side that was the "first to plaster its opponents with fire."(42) The German system did, nonetheless, have the disadvantage that it risked misunderstanding and left much to chance in fire coordination, to the point where troops often shot one another. The Germans were generally convinced, however, that it reduced casualties overall. In any event, because of such methods, the French were caught off balance as much by German motorized infantry as by German tanks. Neither of the major Allies' infantry, with its rigid divisional organization and linear bias, could cope effectively with these German tactical methods.(43)

Most tragic of all, however, the French infantry during the period of the "phoney war" was allowed to stagnate. While the Germans were tearing the Polish army to bits, French soldiers in certain sectors of the Maginot Line were being ordered not to fire on German working parties. Apart from the abortive "Saar offensive" and some patrolling activity, French units generally settled down to a routine of guard duty. Although many infantrymen had still not fired a rifle and had never used their antitank and antiaircraft weapons, no compensatory training programs were instituted. Instead, a system of unit and divisional rotation was introduced so that troops could gain experience and familiarize themselves with activities and routine in the French defensive line. This policy contrasted sharply with that of the Germans, who kept troops designated for offensive operations away from such stagnating fronts. Their training was so demanding, in fact, that German infantry were ultimately glad for the relief provided by war.(44)

The situation within the French formations defending the area of the Ardennes was perhaps most critical of all, particularly in regard to training and tactical dispositions. The two armies stationed along the 95-mile Meuse-Chiers front, the Ninth and the Second, comprised but 12 divisions strung out in a long, thin line with few reserves. Artillery support, nevertheless, was reasonably strong. The main line of French resistance, which followed the river line, consisted of trenches and concrete pillboxes, each armed with an antitank gun and machine guns. The average density in the Sedan area was eight pillboxes and eight machine guns for every 200 yards of front. The Meuse itself was 60 yards wide and unfordable. Discipline within divisions had been eroded, however, by lack of leadership and challenge. Most of the troops in the vicinity of Sedan were described as "fat and flabby men in their thirties."(45)

When the Germans struck on May 13, the well-dug-in French artillery reacted vigorously. This response was relatively short-lived, however, due to the action of the German dive bombers; though they actually inflicted few casualties on the French gunners, the shock effect of the Stukas on such unseasoned and poorly led troops was enormous. Corps gunners, in fact, contributed to the general panic and disintegration of morale by spreading rumors of disaster and defeat over their telephone links. Indeed, the real collapse at Sedan began with the gunners and, according to the Commander of the Second Army, it quickly spread to the other arms:

> . . . the infantry cowered in their trenches, dazed by the crash of the bombs and the shriek of the dive-bombers; they had not developed the instinctive reaction of running to their anti-aircraft guns and firing back. They did not dare move. Five hours of this torture was enough to shatter their nerves. They became incapable of reacting to the approaching enemy infantry.

The fact nonetheless remains that on May 13 the German crossings succeeded at only one thrust point out of the three attempted. Such local checks demonstrated quite clearly that a general check could have been imposed by any adequately conducted defense. Instead of giving smaller groups their heads and manning the stop lines, however, the French staff repeatedly and vainly tried to mount massive centrally controlled counter-attacks. The panzer corps that managed to cross at Monthermé on May 14 did so via a deep gorge, "where one resolute and well-equipped battalion should have been able to hold up an army corps."(47) French failure to defend in depth, of course, allowed the German tanks to range far forward of their infantry, thereby opening an undefended and potentially dangerous breach. While the Germans doubtless realized that separating tanks from infantry was the first major step in countering blitzkrieg, the Allies were never able to capitalize on this desirable situation. French armor, though scoring some local successes, was generally ineffectual, as over one-third of it was used in infantry support. The three heavy armor breakthrough divisions that were organized in early 1940 were essentially all-tank and incapable of existing or fighting independently. The net result was that no forces flowed into the space behind the German tanks to cut off the spearheads. The wake of the panzers, devoid of nests or islands of resistance, thus remained a vacuum in which very little actual combat took place.(48)

The German records clearly indicate, nevertheless, that where French infantry formations offered determined resistance, German tank spearheads invariably got into serious difficulties. This was most evident when Panzergruppe Kleist tried without success to break through the Weygand Line at Amiens and Peronne in June. At this eleventh hour, however, it was too late for France, and von Kleist ultimately prevailed by altering his Schwerpunkt. It was the easiest-won victory in history, yet the outcome on the Meuse might have been different had only the French High Command trained its infantry by means of realistic field exercises and made them truly battle worthy.(49) It would take the Russian and German infantries to demonstrate what determined foot soldiers could accomplish against tanks.

In contrast with the French, the German infantry performed extremely well throughout the campaign. They demonstrated great skill in attacking pillboxes and reducing other centers of resistance. Except for the case of the S. S. division Totenkopf, which showed signs of panic under the onslaught of British tanks at Arras, the German infantry withstood the strain of enemy armored attacks quite well. On several occasions, in response to dangerous situations, they moved against French tanks using hollow charges to break tracks.(50) Such stalwartness and staying power would become even more important, indeed, absolutely indispensable to the success of German arms as the war progressed. Blitzkrieg, as it was executed in Poland and France, was essentially fair-weather war. Good flying weather and good roads had contributed to its success. It was highly unlikely that such conditions would always prevail.

NOTES

1. Dupuy, A Genius for War, p. 267; and Guderian, Panzer Leader, p. 472. Macksey gives totals as 2.35 million Germans with 3,379 tanks, mostly lightweight, against 2,862,000 Allies with 4,170 tanks. Kenneth Macksey, The Guinness History of Land Warfare, (Enfield: Guinness Superlatives, 1973), pp. 177-78.

2. Rosinski, The German Army, p. 188.

3. Seeckt, Thoughts of a Soldier, pp. 61-5; Rosinski, The German Army, pp. 129, and 134-35; and Necker, Hitler's War Machine, p. 23.

4. Liddell Hart, The Other Side of the Hill, p. 67.

5. Major General F. M. Richardson, Fighting Spirit; A Study of Psychological Factors in War (London: Leo Cooper, 1978), p. 20.

6. Rosinski, The German Army, pp. 125-6, and 139; and W. J. K. Davies, German Army Handbook, 1939-1945 (New York: Arco, 1974), p. 143. In the Great War an Unteroffizier commanded a Korporalschaft (section). The status of an Unteroffizier resembled that of a sergeant in the British army. It should be

mentioned at this point that the new German long-service volunteer army had to offer a somewhat more attractive life, even in a nation where the recruiting lines were long. Interestingly, the German army had no detention barracks during its build-up period from 1935 onward. John Laffin, Jackboot (London: Cassel, 1965), p. 161.

7. Ludendorff, My War-Memories, vol. 1, p. 387; and Necker, Hitler's War Machine, pp. 33-4.

8. Necker, Hitler's War Machine, pp. 33-4; and Laffin, Jackboot, p. 180.

9. Bidwell, Modern Warfare, p. 99. The corporal remains the section commander in the British system.

10. Nickerson, Arms and Policy, p. 69; and Colonel Hermann Foertsch, The Art of Modern Warfare, trans. Theodore W. Knauth (New York: Oskar Piest), p. 143. The marching ability of German soldiers in the Great War was also most impressive. In 27 consecutive days, the Thirty-eighth German Fusiliers marched 408 miles, an average of 15.1 miles a day. This period included at least ten battle days. On one occasion, this regiment marched 43.8 miles under the most difficult traffic conditions, with one three-hour halt. All marches were made under full pack. "Infantry in Battle," Canadian Defence Quarterly, 12 (1934):68-9.

11. Necker, Hitler's War Machine, p. 31; Liddell Hart, The Remaking of Modern Armies, p. 215; Davies, German Army Handbook, pp. 29 and 57-8; and Rosinski, The German Army, p. 131 and 139. The German Einheit concept of organization was that all larger military units would be made up of multiples of the simplest team which would perform basic missions. Jac Weller, Weapons and Tactics; Hastings to Berlin (London: Nicholas Vane, 1966), p. 142.

12. The gunners were infantrymen, not artillerymen, and the guns (range 5,000 meters) were normally used for direct (aim on line of sight) fire. Albert Seaton, The Russo-German War 1941-45 (New York: Praeger, 1970), pp. 74-5. The regimental gun companies were referred to as "shock artillery" by awestruck observers after German infantry in Greece overcame strongly fortified British positions without the support of tanks. Necker, Hitler's War Machine, p. 264.

13. Major Robert M. Kennedy, The German Campaign in Poland (1939) (Washington: Department of the Army Pamphlet No. 20-255, 1956), p. 30.

14. R. H. S. Stolfi, ed., Major L. O. Ratley, and Major J. F. O'Neill, German Disruption of Soviet Command, Control, and Communications in Barbarossa, 1941, A Study Prepared for Director, Net Assessment, Office of the U.S. Secretary of Defense (1980), pp. 62, 86-7, 94, and 113; Guy Sajer, The Forgotten Soldier, trans. Lily Emmet (London: Weidenfeld and Nicolson, 1967), p. 170; Weller, Weapons and Tactics, pp. 142-3; and Blaxland, Destination Dunkirk, pp. 34-5. In the German army the MG 34 served as both a light and heavy machine gun; with bipod mount, it was considered light, with tripod, heavy. It was the first general-purpose machine gun. It was later replaced by an improved version designated the MG 42. Davies, German Army Handbook, p. 137.

15. Kennedy, The German Campaign in Poland, p. 31; and Blaxland, Destination Dunkirk, p. 35. After 1943, all infantry units were theoretically termed grenadier. Davies, German Army Handbook, pp. 24, and 35-51.

16. These were abolished in 1942.

17. Ogorkiewicz, Armoured Forces, pp. 73-4 and 386; Guderian, Panzer Leader, Appendix XXIV; and Mellenthin, Panzer Battles, pp. 5 and 24. The dive-bomber ingredient of Blitzkrieg must not be forgotten as it was critical. Stukas had already been employed in Spain as aerial artillery in direct support of troops. The story of German entry into this field is interesting, however, as dive bombing had originally been discredited by the Air Ministry as too suicidal. The technique was nonetheless taken up seriously by the U.S. Navy. During one of its displays, a German ex-fighter ace turned stunt pilot and film star, Ernst Udet, was converted to the idea. Bribed into joining the German air force by Goering, who purchased two U.S. dive bombers, Udet spread the gospel throughout the Luftwaffe. The first Ju 87 was produced in 1935. The RAF went to war in 1939 without a dive bomber. General Hobart's pleas had fallen on deaf ears. Messenger, The Art of Blitzkrieg, pp. 75, 101-2 and 107; and Hans Herlin, Udet: A Man's Life, trans, Mervyn Savill (London: Macdonald, 1960), pp. 154-5, 165-5, 174-82, 197, and 203.

18. General Waldemar Erfurth, Surprise, trans. Dr. Stefan R. Possony and Daniel Vilfroy (Harrisburg: Military Service Publishing Company, 1943).

19. Liddell Hart, The Remaking of Modern Armies, pp. 217-18.

20. Liddell Hart, *The Remaking of Modern Armies*, pp. 222-4, and 229-39; and Captain Harlan N. Hartness, "Germany's Tactical Doctrine," *Infantry Journal* 46 (1939):250. The German infantryman at the time believed that even without tank support he could advance and defend against an enemy. Hartness, "Germany's Tactical Doctrine," p. 250.

21. Miksche, *Blitzkrieg*, pp. 17, 39 and 53-5; and *Atomic Weapons and Armies*, pp. 61-4. Rosinski claimed that Schwerpunkt had "annihilation seeking" overtones. *The German Army*, p. 188.

22. Palit, *War in the Deterrent Age*, pp. 93-4. According to Palit, as late as 1941 British army maneuvers "taught a stereotyped pattern for the offensive limited bounds and objectives, halting at intermediate stages for flank formations to catch up, and moving forward cautiously only after each line of objectives had been consolidated." To his mind, this was merely the old "advance-in-line habit of a bygone age." Ibid. Liddell Hart claimed it was folly to set separate, limited objectives for each attacking unit, as some would not reach them; a wiser course to his mind was to exploit the objectives attained. *Thoughts on War*, p. 307.

23. Rosinski, *The German Army*, pp. 187-8.

24. A major aim of blitzkrieg was to instil panic. In Fuller's words, "It was to employ mobility as a psychological weapon: not to kill but to move; to terrify, to bewilder, to perplex, to cause consternation, doubt and confusion in the rear of the enemy. In short, its aim was to paralyze not only the enemy's command but also his government." Fuller, *The Conduct of War*, pp. 256-57.

25. Miksche, *Blitzkrieg*, pp. 17-18, 39, and 51-5; and *Atomic Weapons and Armies*, pp. 61-4.

26. Liddell Hart, "The Soldier's Pillar of Fire by Night," pp. 618-22.

27. Miksche, *Blitzkrieg*, pp. 8 and 18-19; Liddell Hart, *The Remaking of Modern Armies*, pp. 229-30.

28. Miksche, *Blitzkrieg*, p. 8; Dupuy, *A Genius for War*, p. 116; and Hartness, "Germany's Tactical Doctrine," p. 250. Von Seeckt wrote: "Meetings, discussion, committees, councils of war are the enemies of vigorous and prompt decision and their danger increases with their size. They are mostly burdened

with doubts and petty responsibilities, and the man who pleads for action ill endures the endless hours of discussion." Thoughts of a Soldier, p. 127. However, he did end the dangerous practice of testing an officer's initiative by deliberately placing him in an exercise scenario, that demanded a disobedience of orders. This was customary German practice before 1914. Rosinski, The German Army, p. 196.

29. In the German conquest of Poland, all elements were evident. The German attack came as a tactical surprise to the Polish army despite the troop concentrations and worsening diplomatic situation. Attacking with two strong wings and practically no center, the Germans enveloped the Polish army, which was strongest in the center, and for all intents and purposes hopelessly compromised it in eight days. It was indeed a Polish Cannae. Throughout the campaign, panzer-led thrusts from but six armored divisions were selected and effected with regular precision. Meanwhile, the hard-marching German infantry (there were but 6 motorized out of 38 infantry divisions) were expected to cover 20 to 30 miles a day to keep up. Hoffman Nickerson, The Armed Horde (New York: G. P. Putnam's Sons, 1940), pp. 372-80; Messenger, The Art of Blitzkrieg, p. 131; and Major Kennedy, The German Campaign in Poland, pp. 78-130. As a result of the Polish experience, recommendations were made to decrease the German infantry soldier's load. General von Bock also criticized German infantry tactics, charging that they sacrificed too much to caution. He advocated placing greater emphasis on the mission and the aggressive action to accomplish it. The Poles appear to have impressed upon the Germans the value of the night attack. Kennedy, The German Campaign in Poland, pp. 134-5.

30. Miksche, Blitzkrieg, pp. 50-3.

31. Guderian, Panzer Leader, pp. 89-98; Mellenthin, Panzer Battles, pp. 10-12; and Messenger, The Art of Blitzkrieg, p. 139.

32. William L. Shirer, The Rise and Fall of the Third Reich (New York: Simon and Schuster, 1960), pp. 720-9; and Messenger, The Art of Blitzkrieg, p. 145. Eben Emael, garrisoned by about 1,200 Belgian troops who subsequently surrendered, fell to 85 assault engineers in nine gliders that landed on top of the fortress. Macksey, The Guinness History of Land Warfare, p. 177; and Alexander McKee, The Race for the Rhine Bridges (New York: Stein and Day, 1971), pp. 29-30 and 62.

108 A PERSPECTIVE ON INFANTRY

33. General Ewald von Kleist

34. Mellenthin, Panzer Battles, pp. 11-16.

35. Two or three miles wide.

36. Messenger, The Art of Blitzkrieg, pp. 127, 137, 145; Miksche, Blitzkrieg, pp. 68, 81, and 116; Laffin, Jackboot, p. 176; and Mellenthin, Panzer Battles, p. 13.

37. Marshall, Blitzkrieg, p. 112. The consequent clogging of transport and communication facilities in Allied areas contributed in great measure to the inability of Allied forces to maneuver against the German armies and so to their inability to stop them. Deitchman, Limited War and American Defense Policy, p. 87.

38. Miksche, Blitzkrieg, pp. 113-14 and 117-19; Messenger, The Art of Blitzkrieg, pp. 80-1; Necker, Hitler's War Machine, p. 216; Nickerson, Arms and Policy, p. 69; Rosinski, The German Army, p. 139; and Macksey, Tank Warfare, pp. 120-22.

39. Guderian, Panzer Leader, pp. 102-6 and 123; Mellenthin, Panzer Battles, pp. 13-16 and 20; and Messenger, The Art of Blitzkrieg, p. 147.

40. Mellenthin, Panzer Battles, pp. 13-15; and Farrar-Hockley, Infantry Tactics, pp. 14-15 and 15-17. Patton would later write that "it takes at least two hours to prepare an infantry battalion to execute a properly coordinated attack." General George S. Patton, War As I Knew It (New York: Pyramid, 1970), p. 350.

41. Farrar-Hockley, Infantry Tactics, pp. 17-18.

42. B. H. Liddell Hart (ed.), The Rommel Papers, trans. Paul Findlay (London: Collins, 1953), pp. 7 and 75.

43. Farrar-Hockley, Infantry Tactics, pp. 17-18.

44. Colonel A. Goutard, The Battle of France, 1940, trans. Captain A. R. P. Burgess (London: Frederick Muller, 1958), pp. 68-70 and 79-81.

45. Goutard, The Battle of France, pp. 88-89 and 126-29; and Messenger, The Art of Blitzkrieg, pp. 143-4.

46. Goutard, The Battle of France, pp. 132-6 and 141; and Alistair Horne, To Lose a Battle (London: Penguin, 1979), pp. 331 and 346-51.

47. B. H. Liddell Hart, Defence of the West (London: Cassell, 1950), pp. 9-10.

48. Dupuy, A Genius for War, p. 269; Miksche, Atomic Weapons and Armies, pp. 88-99; Bidwell, Modern Warfare, p. 70; Goutard, The Battle for France, pp. 30-31; and Messenger, The Art of Blitzkrieg, 142. It was not until December 1938, well after Munich, that the French took the decision to form two armored divisions, both to be ready by 1941. Two were raised in January and another in April of 1940; a fourth, eventually commanded by de Gaulle, was in the process of formation when the Germans struck. Interestingly, the French army managed to place all its armored and motorized divisions in the path of the German advance. The problem was that the French leadership, though it realized the importance of the German attacks near Sedan and Dinant on May 13, 1940, did not allow the situation which was still murky to clarify itself. Instead of determining just how serious the situation was around Sedan and coordinating attacks from both north and south against the German bridgeheads, the French High Command dissipated its armored divisions in several hasty and ill-coordinated attacks. Stolfi, "Equipment for Victory," pp. 11, 15-18, and 20; Messenger, The Art of Blitzkrieg, p. 115, 142, 147-8; and Horne, To Lose a Battle, pp. 219, 371-2, and 657-8.

49. Miksche, Atomic Weapons and Armies, p. 76; and Mellenthin, Panzer Battles, p. 20.

50. Deighton, Blitzkrieg, p. 255.

5. The Backbone of Land Forces

Infantry of the Eastern Front

The German armies that attacked the Soviet Union at first light on the historic Sunday morning of June 22, 1941 were again destined to fight outnumbered. The actual strength of the Red Army as of that moment was approximately 5 million men, comprising 303 field divisions, 24,000 tanks, and 7,000 aircraft. In May, the Germans had credited that same army with having only the equivalent of 121 rifle divisions and 21 cavalry divisions available in the first instance to withstand a German attack. About armor, they were less specific, believing the Russians to have no more than 10,000 tanks total, with about five tank divisions and 33 mechanized brigades in the west. The Marcks plan assumed that 55 Soviet infantry divisions, nine cavalry divisions, and 10 armored brigades were located at the frontier, and that a total of 96 infantry divisions, 23 cavalry divisions, and 28 armored brigades would be thrown into the fight against Germany in the short term. German intelligence was generally convinced, however, that the Red Army "was not fit for modern war and could not match a boldly led and modern enemy."(1)

According to Marshal Georgy Zhukov, 149 Red Army divisions--including 36 armored, 18 motorized rifle, and 8 cavalry--were stationed in the Soviet western frontier districts when the Germans attacked.(2) The Soviet high command was apparently convinced that this force could hold a German attack at the frontier. As near as can be determined, since force structure interpretations vary, the Germans hurled against this disposition and the nation behind it 108 infantry, 19 panzer (now comprising 150 to 200 tanks each), and 14 motorized infantry divisions. Better than 3 million soldiers, 2,000 aircraft, and 3,350 tanks were considered sufficient to knock out a world power of some 170 million, just as roughly two-thirds that number had humiliated a France of 40-odd million. That the Soviet Union had the capacity to produce 12,000 tanks and 21,000 aircraft a year appears to have been largely overlooked, as German armored production in 1941 was but 2,800 tanks. However, no German general except the prescient von Bock put forward any objections on the grounds that German resources were inadequate for the task.(3)

Flushed with their recent victories in the West and casting such fears as they had aside, the Germans struck. In an almost classic page from Erfurth, their shock assault achieved total tactical surprise, the effect of which was gauged even more devastating, as no Soviet political or military leader believed that the Russian armed forces could be taken so off guard. In a very short time, three German army groups were thrusting into Russia: Army Group North, under Field Marshal von Leeb, aiming at Leningrad; Army Group Center, under Field Marshal von Bock, striking toward Moscow; and Army Group South, under Field Marshal von Rundstedt, cutting into the Ukraine.(4)

Sixteen hours after the opening of Operation Barbarossa the German army in the east had virtually unhinged two Soviet army "fronts" or army groups. By June 28, the Soviet western front had in effect been broken up and smashed. In the battle of the Bialystok-Minsk pocket, the Germans claimed the destruction of 2,585 Russian tanks and the capture of 290,000 prisoners. A further 350,000 prisoners and 3,000 tanks fell into German hands with the closure of the Smolensk pocket in August. The German encirclement tactics of <u>Keil und Kessel</u> (wedge and cauldron) yielded an even greater victory in the area of Kiev at the end of September with the capture of about 450,000 Russian prisoners. The encirclement of Leningrad was also completed in the meantime, though Hitler forbade direct attack on that city. The truly super-Cannae came in October in the area of Vyazma-Bryansk, which pockets yielded, according to the Germans, a staggering 650,000 prisoners. In the major encirclement battles before November 1, 1941, the Germans claimed to have taken a grand total of 2,053,000 prisoners; 17,000 tanks were also supposedly destroyed by German action. At this juncture, it seemed obvious that Hitler would succeed in his aim to seal off Asiatic Russia along a general line of Archangel-Volga-Astrakhan. He joyfully announced to the world that the Red Army had been annihilated.(5)

Why the Soviets were caught so completely off balance can be attributed to a number of reasons, not the least of which was the seriously damaging and lengthy Stalinist purge of the military in 1937-38. In that period, the head literally was chopped off the Red Army. Marshal Mikail Tukhachevskii, reputedly the most gifted general officer in the Soviet hierarchy, was executed along with about 60 percent of all Soviet officers from divisional commander upward; some 30-40 percent of officers from colonel downward, to and including company commander, were either liquidated or imprisoned. The ever-suspicious Stalin had quite clearly judged Tukhachevskii and his progressive military clique to be a potential political threat. The political commissar system, which had fallen into disuse, was consequently reinstituted in 1937.(6)

The decapitation of the Red Army left the Soviet land forces weak in basic leadership and without the dynamic doctrine postulated by Tukhachevskii. Under the influence of this officer, Chief of Staff since 1926, the Soviet army had become a pace setter in its own right, surging ahead of all other countries in the development of mechanized forces.(7) Arguing that the Red Army should be a highly mobile professional regular force, Tukhachevskii stressed encirclement and maneuver. In accord with M. V. Frunze, Soviet War Commissar from 1924 to 1925, he further advocated committing the Red Army to the offensive rather than the defensive posture proposed by Leon Trotsky. New infantry regulations, published in 1927 to replace the provisional Frunze manual, consequently stated that firepower and mobility were the cardinal principles for success in infantry combat, the aim of all maneuver being to envelop, or turn, one or both enemy flanks.(8)

Though the Russians were aware of the writings of Liddell Hart, de Gaulle, and Guderian, they were apparently more directly influenced by the theories of General Giulio Douhet and Fuller. Over 100,000 copies of a Russian edition of On Future Warfare were sold in the Soviet Union, strongly indicating a Red Army preference for Fuller's relatively more specific ideas on tank warfare as a subject for military discussion. Lectures on F.S.R. II was also supposedly made a military "table book" around the same time. It should be noted, however, that many Soviet officers regarded both Douhet and Fuller as much too extreme; they therefore continued to exercise their own distinct brand of combined arms doctrine. The 1936 Field Service Regulations, which reflected Tukhachevskii's tactical ideas, saw a ground offensive developing in four stages: first, an attack by assault groups (of infantry, tanks, and artillery) on the weakest point in the enemy's line; second, where one or more groups succeeded, a push through by support groups backed by additional artillery fire; third, a decisive breakthrough by tanks, cavalry, and motorized infantry splitting up enemy units to form local encirclements that, in turn, would be attacked by mass infantry, artillery, and air; and, finally, the pursuit. The Russian concept differed sharply from the German in that the breach tended to be effected at the slower pace of the infantry; in short, the Soviets advocated using "mass" first rather than in the follow-up role, as did the Germans (but then Soviet doctrine also presupposed that static conditions of warfare would prevail initially).(9) Another fundamental difference between the two military concepts was that the Soviets, from the beginning, were convinced of the necessity for attacking on as broad a front as possible, with strong artillery fire support throughout its breadth.(10)

Though Tukhachevskii's ideas were written into the 1936 Field Service Regulations, there were indications that further refinements were necessary; infantry tactical methods, for example, left room for improvement, as did tank-infantry cooperation. It was also arguable whether the progressive doctrine contained in the regulations had filtered down to all levels of the Red Army establishment, certain members of which had in 1935 scorned theories of blitzkrieg warfare. Emphasis still tended to be placed on exact implementation of plan, which had to be rigidly specific to compensate for deficiencies in tactical abilities and training at the lower levels. Tukhachevskii had, nonetheless, personally stressed and encouraged maximum initiative and "nerve" at all levels.(11) What the Red Army really needed at this critical juncture was a lengthy period of tactical training rather than political indoctrination. But this was not to be.

The experience of the Spanish Civil War led the Soviet military to make a fundamental change of course in regard to the employment of tanks. Because attempts to use them on their own had proved disastrous, General D. G. Pavlov, the commander of Russian forces in Spain, took back the message that tanks should be used only in an infantry close-support role. Stalin agreed, and the mechanized formations were subsequently disbanded, the tanks dispersed in support of infantry formations. Ironically, Tukhachevskii was blamed for advocating the employment of tanks on their own!(12) An eventual result of the purge of the "Tukhachevskii clique" and the concurrent, more politically palatable recommendations of Pavlov was that the Red Army returned to its older proletarian tradition of revolutionary mass, minus technique but with more powerful commissars.(13) Although its training and indoctrination remained primarily offensive in outlook, many modernization programs were now handed on to much younger and inexperienced commanders for direction:

114 A PERSPECTIVE ON INFANTRY

> The men who followed Tukhachevskii lacked . . .
> that insight into the probable forms of modern
> mobile warfare which had so preoccupied the purged
> commanders; they lacked any intellectual curiosity
> simply because they disposed no intellect, either
> singly or as a group. They mouthed slogans but
> understood nothing of principles, they paraded
> statistics about firepower without grasping any of
> the implications of the new weapons their own
> designers were developing, they were martial in a
> swaggering sense without the least grasp of the
> professionalism necessary to the military.(14)

The Red Army, now under the firm control of Stalin thus continued to veer toward French doctrine developed from the experience of the Spanish Civil War. Fighting on the Mongolian-Manchurian border against the Japanese in 1938-39 nonetheless confirmed the basic wisdom of the 1936 Field Service Regulations. Though Tukhachevskii's mobile theories were by no means faultless--and Soviet infantry paid heavily for deficiencies in platoon and company training in the 1938 engagements with the Japanese around Lake Khasan--Soviet forces under General Zhukov applied them with brilliant success in the 1939 Battle of Khalkhin Gol. Attacking on a broad front of some 48 miles with infantry supported by tanks and aircraft, Zhukov completed the encirclement of the Japanese with a maneuver group of tanks and motorized infantry. Over 18,000 Japanese were killed and 25,000 wounded; total Russian casualties were but 10,000. Interestingly, this creditable performance attracted minimal attention in the West, where other war clouds were gathering. For the Russians, unfortunately, it provided an illusion of security, to the point where they began to believe their own propaganda that German armies could never pierce the western frontier.(15)

From most indications, German blitzkrieg methods in Poland attracted substantial attention within Soviet military circles, but it really took the Finnish experience to confirm Red Army inadequacies. Driven by an almost paranoiac concern for the security of Leningrad following the German defeat of Poland, Stalin demanded the readjustment of the border around and in the Gulf of Finland (with Finland to be compensated by turning over parts of Soviet Eastern Karelia). The Finns refused Stalin's demand and on November 30, 1939, the Red Army invaded with a massive air and ground attack mounted by four Soviet army groups deployed for operations along the border from Leningrad to the Arctic Ocean. The main thrust was launched by the 350,000-strong Seventh Army Group against the Mannerheim Line, a fortified zone of fire points, "dragons teeth" tank traps, and trenches running diagonally across the Karelian Isthmus. Frontal attacks were the order of the day: first, with tanks preceding infantry; then, when the light and medium tanks fell easy prey to the Finns, by infantry in mass hoping to create breaches for massed tanks held in reserve in the rear. These attacks were made more frequently at night, but the enterprising Finns merely switched on searchlights and poured merciless machine-gun fire on to the Russian masses.(16) In the other army group sectors, the action took on a more mobile flavor.

Unlike the Soviets, the Finnish ground forces were almost entirely infantry. They were formed into ten divisions (at least six of which defended the Mannerheim Line), each consisting of about 15,000 men organized into three infantry regiments and a battery of 36 guns. The specialty of the Finnish army was motti fighting, a form of fast-moving fluid warfare in which ski troops operating in small units encircled and isolated small pockets of enemy, mainly by night. The standard of individual training was extremely high among the Finnish infantry,(17) especially in such skills as camouflage and the use of ground. Marksmanship was also highly developed, and according to Russian Colonel G. I. Antonov, Finnish snipers were able to get their man at distances of 800 to 1,000 meters. These skills were complemented by an ebullient Finnish morale and an army junior leadership abounding in initiative.(18)

In the area of Suomussalmi on the waist of Finland, the Finnish army scored its greatest success, wiping out at least one Soviet division and badly mauling another. However, on February 2, 1940, a new offensive under the direction of Marshal S. K. Timoshenko was renewed against the Mannerheim Line. Although Soviet troops had been put through a winter indoctrination course and an intensive training program in the storming of fixed fortifications, the Finns were, on the whole, bludgeoned into submission mainly by mass and the use of more modern equipment such as the heavy KV-1 tank and armored sledges. All in all, it was a humiliating performance for the Russians, particularly when contrasted with that of the Germans in Poland. Almost 1.2 million Russian soldiers had to be deployed to defeat a nation of some 3.7 million. The Soviets admitted casualties of 68,000 killed and missing and 130,000 wounded.(19)

Many shortcomings of the Red Army came to light in Finland. Divisional organization was considered too rigid and the army generally far too road bound. Though the Russian soldier's doggedness and ability to withstand severe hardships had shown through, there was no denying that his standard of training was low and his marksmanship particularly poor. Soviet infantry were not properly trained for close-quarter fighting, and all arms cooperation was lacking. The fashion of group tactics, using small detachments of 8 to 12 men acting independently to exploit the fire of light machine guns, had never been favorably greeted in the Red Army, as many officers were concerned about the adverse effect this might have on command and control. The result was that in Finland battalions initially attacked in tightly packed formations. Later assaults on the Mannerheim Line did, of course, prove group tactics to be invaluable.(20)

Reforms were not long in coming, however, as the man of the hour, Timoshenko, was made Defense Commissar. Infantry manuals again were amended, and although the mass infantry attack was still recommended as the ultimate method of annihilating the enemy, infantry tactics were made more flexible with greater initiative allowed to junior leaders. Winter equipment was improved and production of the T-34 tank was given the go ahead. No longer were light automatic weapons dismissed as "police weapons." Combat training was intensified and made more realistic, with special emphasis placed on fighting under "difficult conditions." Programs were set in motion to exercise troops by day and night in all weathers, with rigorous physical exertions, so that sections, units, and formations could maneuver on any terrain. The infantryman was additionally taught how to dig in quickly and how to deal with surprise attacks. Equally important, a new disciplinary code was introduced: whereas before troops had been exempted from fulfilling "criminal" (i.e., anti-Soviet) orders open to wide interpretation, they were henceforth to obey all orders unconditionally. Commander's prerogatives were also buttressed in other areas, though the word "officer" was still banned as being essentially repugnant to a revolutionary socialistic army. Significantly, in August 1940, the powers of political commissars were restricted to advising on political matters and troop indoctrination. All military and political training became the responsibility of the commander.(21)

Such wide-sweeping reforms obviously did not have time to permeate the Red Army before the Germans struck. Furthermore, although Timoshenko's reforms basically represented a return to the system that prevailed before 1937-38, in themselves they contributed to more turbulence. Between 1938 and 1941, Soviet military manuals underwent no less than a staggering 26 major modifications or radical amendments. The dramatic collapse of France added to the confusion, as Soviet planners rushed to rethink their doctrine of tanks in the infantry close-support role. A major reorganization of armored and mechanized forces was therefore underway when the Germans attacked. To make matters worse, despite all reforms instituted to revitalize the army, many commanders on the night of June 21 lacked even the initiative to keep their units in a state of readiness. The basically linear deployment of the Soviet forces, which left a number of vulnerable salients, also did not help matters. There is some evidence, however, that the Soviets were attempting to create a strong mobile counterattack force based on Tukhachevskii's original plan for a defense in depth against German attack. In any case, it was not completed in time.(22)

Thus, it was that on the anniversary of Napoleon's invasion of Russia in 1812 the Soviet army was caught totally by surprise. The effects of the purge, the misreading of combat experiences in Spain and the Far East, and the eleventh hour reorganizations were all to take their toll. Yet, unlike the French, the Soviets were not beaten before they started. Geography had blessed them with broad reaches, great forests, and a savage climate. By themselves, however, these had not been enough to save Russian masses in the Great War from crushing defeat by superior German military technique. That this did not occur in 1941 must be attributed, in large part, to the stiff resistance of the offensive-minded Red Army and, in particular, to the defensive skill of the Russian infantry that made up its bulk. In July 1941, it was not the T-34 tank that was the Soviet superior weapon but rather the dogged, determined Russian infantryman, who remained a quality unto himself. What he lacked in tactical or technical ability, he made up for in sheer ruggedness. Not stupid, he was educated on the battlefield where he took the measure of the German panzers. Incapable of expelling the enemy from his beloved country, he nonetheless saved her from her erstwhile follies by blunting the German onslaught. By the time the Japanese attacked Pearl Harbor, the military art form known as blitzkrieg had met its match in the Red Army.(23)

Though many Soviet formations disintegrated and thousands of infantry surrendered at the first impact of battle, many other Red Army soldiers bitterly contested the German advance, hanging on tenaciously in pockets of resistance. German howitzers and antitank weapons were directed at many Russian bunkers to no avail; only demolition teams with shaped charges and flamethrowers could take them out. Not surprisingly, the Germans soon began to realize that the Russian campaign was not to be a "manoeuvre with ammunition" as had been the campaign in France. Furious local counterattacks launched by the Red Army as early as the first days of August confirmed this conjecture. The casualties began to mount. From a shocking low of but 102,000 German casualties for the entire Eastern Theater on July 16 they steadily began to rise: by the end of August, casualties amounted to 440,000, of which 94,000 had been killed; by September 26, losses had risen to 534,000, which, although only 15 percent of the total manpower on the Ostfront, represented over 30 percent of German infantry strength.(24) By the end of November, total casualties had reached 743,000, of which 200,000--including 8,000 officers--were dead or missing. By comparison, total German casualties in Poland and France had been 44,000 and 156,000 killed, wounded, and missing, respectively.(24)

The main reason for these substantially higher casualty rates must largely be ascribed to the fighting qualities of the Russian troops. The combat characteristics and military effectiveness of Soviet units did not, of course, adhere to any common pattern, as the Red Army was essentially a multinational and somewhat polyglot organization. However, it was noticeable that in broken or close country such as mountain waste, forest, or marsh, where panzers were less effective, the Red Army soldier generally fought better than in the open. Though the fighting performance of Soviet formations and units in the Ukraine often varied from high to low, the standard of resistance in the wooded swamps near Leningrad was uniformly bitter. Around Moscow, there were occasional variations, a case in point being the Fifty-seventh panzer corps rounding up the drunken, insensible soldiery of the Forty-third (Red) Army while nearby the Ninety-eighth (German) Division fought for its life against the elite Fifth Soviet Airborne Corps.(25) In this respect, the Russo-German War showed time and time again that a handful of good men at a certain spot at a given time often saved the day for all.(26)

Obviously, geography exerted a tremendous influence on the character of the Russo-German conflict. To be sure, European Russia is in many ways a natural fortress. Though its southern portion of arid steppes and sand flats is practically devoid of woods, the foothills of the Caucasus contain impenetrable thickets. The northern portion around Leningrad is essentially a woodland interspersed with swamps. In the central area, the broad marshes and forests of the Pripet region adjoin the huge tracts of forest around the centers of Gomel, Minsk, Bryansk, Borisov, Orsha, and Vyazma. A large triangle of pine- and fir-heavy forest is thus composed, with corners at Leningrad, Ufa (on the southern Urals), and Lvov. In 1941, the roads leading to Moscow from the north and west were for the most part narrow forest corridors, ideally suited for

MAP 5-1. WESTERN RUSSIA

defense against a road-bound enemy. To reach Moscow, the Germans had to engage the Russians in terrain where maneuver off the highways was difficult at best. Less forestation was found on the southern approach to Moscow, but low, marshy lands around the city of Tula proved an ideal defense against tanks. To make matters worse for a road-bound army, twice in the course of every year, prior to the onset of winter and again in early spring, the soil of Russia is softened by rain and thaw. It is the period of the rasputitsa, when the roads become bottomless and the countryside turns into a morass. In 1941, this phenomenon took the Germans completely by surprise. In October, on the Smolensk-Vyazma highway, 6,000 Wehrmacht supply trucks carrying ammunition, rations, and fuel bogged down.(27)

In the forests and swamps of European Russia, the infantry arm assumed a particular importance and position of dominance. Panzers, and even artillery, lost much of their effectiveness in forested areas. In such circumstances, the advantage accrued to the Russian infantryman not only because Timoshenko's reforms had stressed fighting under special and difficult conditions, but because the Soviet citizen generally was more adept than the German at coping in the wilderness. Furthermore, the Russian army again began to emphasize close-quarter combat. The ghost of old General Dragomirov still haunted the Russian infantry; bayonet training, of which he was a stern advocate, regained its former vigor and appeal. For these and other reasons, the Russian infantryman was well suited for close combat. His German counterpart, on the other hand, though expert in river crossings and open warfare, had not had the opportunity to practice combat in deep forests and swamplands. In fact, lack of suitable training areas, the secret nature of Barbarossa, and involvements elsewhere, all contributed to most German units arriving on the Eastern Front without such special training. The Germans consequently tended to avoid combat in close country, their most typical forest operation being limited to mopping up dispersed Russian forces. The Russians, however, deliberately chose for their most determined efforts swampy, forested terrain where superiority in material was least effective.(28)

The Germans were, of course, ultimately forced to meet the Russian infantry on ground of its own choosing. As the advancing panzers outflanked and pierced Russian positions, skirting swamps and bypassing large forested areas in their rush to seek decision in the open, Red Army units retired laterally into the depths of the bypassed areas. In the early stages of Barbarossa, the Germans considered this to be a desirable result, and there were numerous occasions when the Wehrmacht deliberately drove the Russians into swamps. It was thought that in this way they would be either neutralized or induced to surrender through starvation and the general hopelessness of their situation. This, of course, proved to be a fatal error, for the Russians did not always surrender. Thus, while the immediate vicinities of trails and improved roads were usually cleared of hostile forces to German satisfaction, the panzers no sooner surged past than the Russians once more emerged from the forests and resumed fighting. German infantry divisions, which sometimes followed a considerable distance behind the panzers, were consequently often faced by the same, sometimes sizable, Soviet forces. Many a pitched and bloody battle was therefore forced on the German infantry, which, like the panzers, was goaded on by an ever increasingly anxious high command. By the fall of 1941, the resurgence of dismembered Russian forces in forests and swamps behind the German front had assumed serious proportions.(29)

The effect of such Russian tactics was to turn the Kessels created by blitzkrieg into fortresses or islandlike strongpoints in depth. Overall, it was a basically infantry defense that, in Kenneth Macksey's opinion, came to "baffle" the Germans.(30) The formidable nature of the defending Russian infantryman accounted for most of this bafflement. According to Colonel Joachim Peiper, a German officer who fought for three years on the Eastern Front, the Russian infantryman surpassed all others in defense.(31) Expert in camouflage, deception, and the ways of the woods, he chose his ground carefully and well. Particularly adept at preparing and concealing an individual trench, he doubtless rendered Patton's description of the German as the "champion digger" inaccurate.(32)

Even poorly trained and indifferent Russian soldiers, given spades, could melt into the earth. A great believer in overhead protection, the average Russian infantryman would often lay logs as thick as telegraph poles, five layers deep, over his trench so that it could withstand any direct hit. To ensure that his position remained undetected for as long as possible, he normally held his fire until the enemy closed to the very shortest of ranges. When he did engage by fire, it was often delivered as part of a volley to gain maximum psychological effect. Sharp-shooting, a consequence of the Finnish War, was nonetheless stressed to a great extent within the Red Army; interunit competitions were commonplace, and snipers were treated with great respect. Interestingly, the Russians were never satisfied with being able merely to dominate an area by fire; it had to be occupied by infantry. Like the French before them in the Great War, they never abandoned ground they had gained in an attack.(33)

When fully dug in, the Russian infantryman was particularly dangerous to tanks, often allowing them to pass over his almost invisible trench so that he could take them in the rear. This tactic also aimed at separating enemy tanks from their infantry, in which regard it proved exceedingly effective. It is, of course, worth noting that the first effective man-portable infantry antitank weapon, the American "bazooka," was only introduced to Russia in 1942.(34) The Russian infantryman was thus forced to improvise imaginative antitank measures from the beginning of the war. In this, he succeeded admirably, his German enemy even paying him the compliment of copying many of his methods and naming one of his improvised weapons the "Molotov cocktail." (35)

    The areas in which Soviet infantrymen best applied their defensive skills, however, were the swamps and forests. Unlike the German infantry, Russian foot soldiers did not take up positions in front of a wood, or on a wood's edge, but invariably moved right inside it in a defensive circle, preferably behind swampy ground. A particular feature of these wooded defensive positions were well dug in infantry trenches with fields of fire mainly to the rear, sited primarily to pick off the enemy from behind after he had punched through. Whereas the Germans tended to clear lanes of fire by considerable felling of trees, leaving themselves open to detection from the air, the Russian infantryman cut down undergrowth only up to waist height, creating tunnels of fire both rearward and to the sides. This gave him cover from view from air, a definite frontal invisibility, and a clear field of fire all at the same time. Behind these all-round positions would be husbanded Russian tactical reserves. This method of wood defense had a nerve-wracking effect on German clearing parties.(36)

    Lack of experience in fighting in forests was a great disadvantage to the Wehrmacht in the Russian campaign. German infantrymen, trained to depend on air support(37) and the massed fire of all the combat arms, had to adapt themselves to terrain where infantry alone had to carry the main burden. Even then, the infantry arm was restricted to using the rifle, machine pistol or submachine gun, grenade, mortar, and light antitank weapon. The machine gun generally had limited effect in dense woods, and suitable observation positions for artillery were rare. In fact, from repeated experience, the Germans came to the conclusion that except in unusual circumstances artillery in thick forests was just an impediment and should be left behind. They preferred instead heavy mortars such as the Russians used in large quantities and with great effect. In the main, combat in Russian forests, whatever the size of the forces originally committed, eventually assumed the character of a series of small-unit actions. Well-trained men were therefore more important than firepower.(38) Until the Germans mastered this type of fighting, however, they had to pay a price exacted in terms of casualties and morale. In the din of battle in the forest, with its ricocheting bullets, the close-combat cunning and loud shouts of the Russian infantryman predominated. Moreover, he thought he had discovered the chink in the German infantry's armor: close-range fighting.(39)

In general, the Russian infantry soldier was extraordinarily tough; so tough, in fact, that from the outset of the war German overall tactical superiority was partly compensated for by the greater physical fitness of Russian officers and men. Seemingly inured to the effects of unfavorable weather conditions--snow, rain, extreme cold, and ice--the Russian soldier appeared to have a kinship with nature. Endowed with a good sense of direction, he moved freely by night or in fog, through woods and across swamps. The average rifleman was able to hold out for days without hot food, prepared rations, bread, or tobacco. At such times, he subsisted on wild berries or the bark of trees. During the winter campaign of 1941, a Russian regiment was surrounded in a forest near Volkov. The Germans, too weak to attack, decided to starve them out. After one week, Russian resistance had not subsided; after another week, only a few prisoners were taken, the majority having fought their way through the German encirclement. According to prisoners taken by the Wehrmacht, the Russians had subsisted during those weeks on a few pieces of frozen bread, leaves, and pine needles. It had never occurred to any of them to surrender, although the temperature dropped to 30 below zero Fahrenheit.(40) It would be silly to say, of course, that the Russians were unaffected by such extremes of weather and deprivation. They were simply more accustomed to their own country's climate than the Germans and, most significant, much better equipped and clothed to exist in it.

The hardiness of the Soviet infantryman was reflected in his weapons and equipment. Though his winter uniform and "varlenki" felt boots were generally of excellent quality (thanks to the Finnish experience), his personal kit was meager: a small haversack, rolled great coat, and occasionally one blanket, which had to suffice even in severe weather. He was nonetheless capable of marching long distances carrying a heavy load, a major reason, no doubt, why Russian infantry were able to escape in large numbers from many German encirclements. (Some divisions marched 30 miles a day for 10 days on end to escape.) It is a mistake to think, however, that because the Russian infantry man packed many of the stores commonly held by Western armies in advanced depots, that it was a heavily overburdened infantry. On the contrary, the Red

Army traveled lighter than any army of modern times, and its infantrymen had great battlefield mobility. They went into combat with the bare essentials for survival. One indication of this was that the Russian foot soldier was never issued with a scabbard for his bayonet, which was consequently carried, almost invariably, permanently affixed to his rifle.(41)

From the outbreak of war, the Russian infantry was reasonably well equipped with small arms. The Degtyarev section light machine gun proved sturdy and effective, its basic simplicity reflected in other Soviet small-arms designs. Though the semiautomatic Tokarev rifle was startlingly in advance of similar European national weapons in 1941, it was soon replaced in quantity by the less sophisticated, more reliable third variant of the bolt-action 7.62 Moisin-Nagant rifle, first adopted by the Russian army in 1891. The Germans preferred to use the Soviet PPSh 1941 submachine gun until an improved version was brought out for their own MP 38/40 model, which tended to jam at low temperatures. The simplicity of Soviet small arms was indeed remarkable; the PPSh 1941, for example, had but 83 metal parts compared to over 200 for the popular American Thompson.(42)

Ironically, though perhaps not surprisingly, the infantry of the Red Army was referred to as the tsarita of the battlefield. The 1936 and 1940 Field Regulations, the 1942 Theses on Offensive Combat, and the 1942-1945 Infantry Combat Regulations repeatedly stated that "combined action of all types of troops is organized in the interests of the infantry, who fulfill the main role in combat." This statement did not really represent much of a change from Tukhachevskii's idea that infantry, supported by tanks, artillery, and air power, would decide the land battle. His advocacy of the exercise of initiative and "nerve" by junior commanders was, however, somewhat overshadowed by emphasis on the use of mass.(43) Interestingly, when General Martel visited the Red Army in 1943, he found the officers in almost every rank he met "capable and confident," the junior officers possessing the trust of their men and the men proud to serve their officers. This contrasted sharply with his 1936 impression that Russian junior officers were tactically deficient, lacking in confidence, and hesitant to take charge. In 1943, Martel had "no doubt" that the Russian had a "natural flair for war."(44)

The basic unit in the Soviet infantry was the section of nine men. A platoon comprised four sections, two of them armed with two Degtyarev light machine guns and two with one. Platoon headquarters included one or two trained snipers, and the platoon was commanded by a junior lieutenant. A rifle company, usually commanded

THE BACKBONE OF LAND FORCES 125

TABLE 5-2. RED ARMY INFANTRY REGIMENTAL ORGANIZATION
(WORLD WAR II)

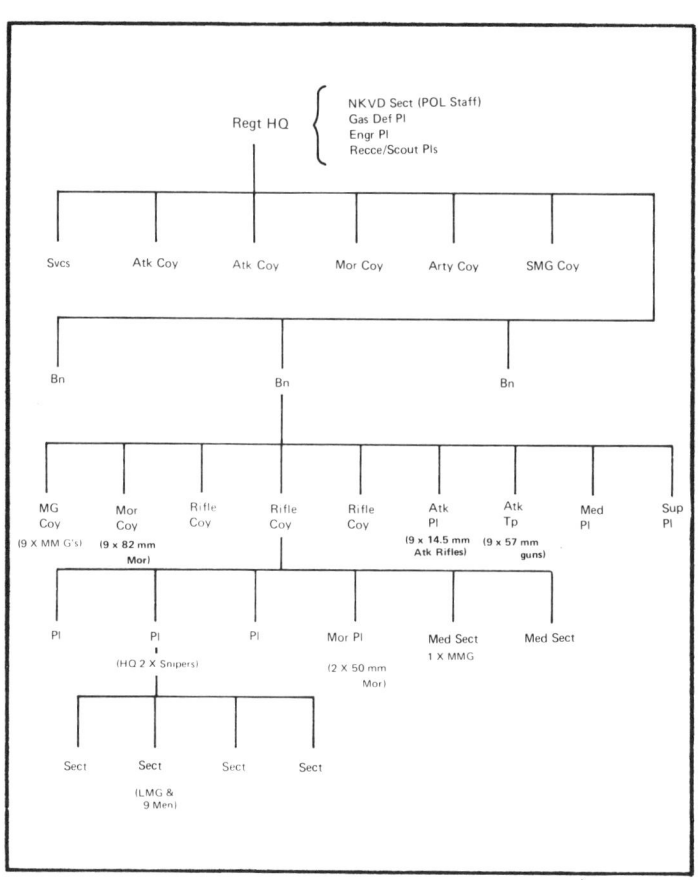

by a senior lieutenant or captain, consisted of three
platoons. It also had a mortar platoon of two
50-millimeter mortars, a machine-gun section of one
medium machine-gun (two in guards units), and a medical
section. Three companies made up a battalion, commanded
by a major or lieutenant colonel. The battalion also

had a machine-gun company with nine medium machine guns; a mortar company with nine 82-millimeter mortars; an antitank platoon with nine 14.5-millimeter antitank rifles; an antitank troop with two 57-millimeter guns; a medical platoon of one doctor, three orderlies, and a horse-drawn ambulance; and a supply platoon with horse-drawn ammunition vehicles and field kitchens. Three battalions made up a regiment, normally commanded by a lieutenant colonel or a colonel. At regimental headquarters, there were mounted reconnaissance and scout platoons, a mine laying and clearing engineer platoon of about 20 men, a gas defense platoon, an NKVD section, and a political staff headed by the deputy regimental commander, a major. A regiment's supporting units consisted of a submachine gun company of 90 men tasked to do reconnaissance and to spearhead the regiment in street fighting; an artillery company with four horse-drawn 76-millimeter howitzers; a mortar company with six motor-drawn 120-millimeter mortars; two antitank companies; and supply, medical, and veterinary services. From 1941 onward, the title "guards" was allotted to regiments and larger units that distinguished themselves in battle. Guards soldiers received double pay, and guards units were somewhat stronger in firepower.(45)

Theoretically, Red Army small-unit tactics in the offense throughout the war were based on taking successive strongpoints within an enemy position by double envelopment. Units were to penetrate the weaker portions of the enemy defense on each side of the objective and overrun it from the rear and flanks simultaneously. For this reason, many assault units tended to fight with subordinate elements abreast. (In some cases battalions actually placed all rifle platoons in line.) The idea was generally presumed to be practical because of the tremendous firepower organic to rifle units and available from supporting tanks and artillery. To lend "weight" to an attack, frontages were often narrowly restricted: 200 to 300 yards for a company, and 300 to 700 for a battalion. Normally, an attacking battalion was organized into two echelons, with at least one-third of its strength in the second echelon. No doubt reflecting a specific state of training, rifle companies in the assault were permitted to use only three types of formation: line, wedge, or reverse wedge.(46) A natural consequence of such antiquated and unimaginative tactical approaches were actions such as that reported near Zelva in June 1941 and described in lucid detail by German observers:

> Again and again they swept up against the German positions with their unnerving cries of "Urra!"--companies, battalions, regiments. The

picture was one that made the German troops' imagination boggle. The Russians were charging on a broad front, in an almost endless-seeming solid line, their arms linked. Behind them a second, a third, and a fourth line abreast.(47)

Although the foregoing Great War-style tactics were not infrequently employed by Russian infantry units on the Ostfront during the Great Patriotic War, they were nonetheless definitely more aberrant than the rule. According to General Fuller, Soviet minor tactics throughout that war "appear to have been of a high order."(48) In fact, the principal method of Russian infantry attack, at least when the Soviets were on the defensive, was the short-distance raid. Where possible, the men crawled silently, and without firing to a starting position or "jumping-off" spot. (For more deliberate attacks, troops in contact would dig to within 200 yards of the enemy and prepare jumping-off positions for assault units in depth.) There they would wait, often in the cold, until the hour of attack, which would normally be launched under cover of darkness. At other times, they would assault in a driving snowstorm or under cover of a howling wind; the Russians were masters at penetrating German positions without visible preparation or major fire support. Moving by night and disappearing by day, hiding in villages and woods, the Russian infantryman could be expected to strike only at unexpected times under the most adverse conditions. The Germans paid a rather backhanded compliment to the stalking ability of Russian assault infantry generally by adopting the policy of holding their fire until the last minute; they knew that the sooner they opened fire, the sooner the Russians would go to ground and merely creep up under cover. Red Army infantry had to be crushed decisively by the first blow.(49)

Perhaps the most impressive characteristic of Russian infantry in the offense was its unmatched ability to infiltrate enemy positions. According to General von Mellenthin, practically every Russian attack was preceded by large-scale infiltrations of small units and individual men. The Russian methods, described by Liddell Hart as an "ant strategy," usually followed the same pattern. During the first night, a few men would infiltrate German positions and vanish in the forest. During the second night, reinforcements would bring the force up to platoon strength. In this manner, provided no countermeasures were taken, a whole battalion group could be lodged in the rear of German lines within one week. The remedy--strongly manned lines, well organized in depth and continuously patrolled by men wide awake and alert, all backed up by local reserves--was considerably troop intensive. Such tactics forced the Germans in north and central Russia to attempt to form continuous fronts, with all the wearisome patrolling, fighting, and expenditure of reserves that these entailed. At the same time, Russian skill in speedily exploiting even the smallest infantry penetration made it a matter of utmost urgency for the Germans to eliminate immediately the infiltrating group, however small their own counterattack force.(50) Although Russian infiltration tactics and methods proved most effective in winter when they were practically invited by the German strongpoint system of defense, they were also used with similar disrupting effect in river and stream crossings during all seasons. That these "bridgehead tactics" were taken seriously by the Germans is clearly illustrated in an appreciation by the Chief of Staff of the Forty-eighth Panzer Corps:

> Bridgeheads in the hands of the Russians are a grave danger indeed. It is quite wrong not to worry about bridgeheads and to postpone their elimination. Russian bridgeheads, however small and harmless they may appear, are bound to grow into formidable danger-points in a very brief time and soon become insuperable strongpoints. A Russian bridgehead, occupied by a company in the evening, is sure to be occupied by at least a regiment on the following morning and during the night will become a formidable fortress, well equipped with heavy weapons and everything necessary to make it impregnable. No artillery fire, however violent and well concentrated, will wipe out a Russian bridgehead which has grown overnight. Nothing less than a well-planned attack will avail. The Russian principle of "bridgeheads everywhere" constitutes a most serious danger, and cannot be over-rated. There is again only one sure remedy which must become a principle: If a bridgehead is forming, or an advanced position is being established by the Russians, attack, attack at once, attack strongly. Hesitation will always be fatal. A delay of an hour may mean frustration, a delay of a few hours does mean frustration, a delay of a day may mean a major catastrophe. Even if there is no more than one infantry platoon and a single tank available, attack! Attack when the Russians are still above ground, when they can be seen and tackled, when they have had no time to organize their defense, when there are no heavy weapons available. A few hours will be too late. Delay means disaster. . . . (51)

In a military struggle that generally alternated between the maneuver of open warfare and vicious in-fighting at close quarters, the dogged defense of the Russians gradually sapped the Germans of their combat strength. The addition of "fortified" cities to the equation of "islands of resistance" in depth plus constant counterattack at all levels witnessed the inadvertent Russian formulation of a "web"-style defensive doctrine. Advocated by Miksche as the modern shield to the modern sword of the blitz attack, it could have been employed equally well by the French in 1940 as by the Russians in 1941. French failure to fortify any villages or towns in 1940 showed a gross misreading of the lessons of Spain and Poland; Warsaw alone, for example, held out for 12 days against German panzers. Fortunately for the Russians, they possessed an infantry that by its very refusal to surrender in many cases was able to wear down the German onslaught from isolated islands of resistance.(52)

Also, fortunately for the Russians, the season of the rasputitsa neutralized the German advantage in mechanical mobility, causing blitzkrieg to founder for the first time. By the end of October 1941, only the German infantry was capable of continuing the advance, and then only at a rate of eight miles a day. The panje wagon replaced the motorized infantry lorry, becoming the mainstay of the German army. Still the Germans pressed on, to paraphrase Zhukov, "gnawing their way forward." The encirclement of Leningrad having turned into a seige, the Germans now depended for success in their Moscow thrust on the efforts of a number of ill-clothed, unsupported, and extremely tired spearhead infantry battalions. The rest of the mighty Wehrmacht stood powerless and idle, immobilized in the mud. To make matters worse, some infantry companies at the time could muster but 20 to 30 soldiers; Guderian's companies, clad only in denims in subzero weather, were reduced to an average of about 50 men. Frostbite had become a serious problem, and weapons malfunctioned because of the cold. The German war machine was slowing down.(53)

The three weeks from November 24 to December 15, 1941, marked the real turning point in the Russian campaign. The German offensive was beginning to grind to a halt along the entire front. At the same time, the Russians began to seize the initiative. In the north, counterattacks were mounted to relieve Leningrad, the garrison itself striking out even as it was beginning to die for lack of food. In the center, in the battle for Moscow, the Russians loosed furious local spoiling attacks against Army Group Center, which launched its final drive on Moscow on November 15. In this titanic struggle, Zhukov's Western Front--initially inferior in numbers, guns, and tanks--managed to hold on. Formations like General I. V. Panfilov's 316th (later Eighth Guard) Division, covering the Volokolamsk highway, were ground to pieces fighting panzers but not before they took their toll. On November 16, 28 men from Panfilov's division knocked out more than 14 panzers. Although most of them were killed, they held out for four critical hours, buying valuable time for the tactical deployment of fresh regular troops from the Far Eastern Front that had just begun to be transferred to the Moscow sector.(54)

On November 17, when Guderian's ragged, shivering infantry of the 112th Division were bumped by well-equipped Siberian riflemen outfitted in quilted winter uniforms with white camouflage smocks, the German soldiers broke and ran for the first time in the campaign. To Guderian, the warning was clear: the combat ability of the German infantry was at an end, and they could no longer be expected to perform difficult tasks. This first German rebuff was shortly followed by another more serious reverse; on November 28, the Soviets attacked and captured Rostov on the Don, the first positive defeat inflicted on the Germans in any theater of the war up to that time. In 161 days after opening hostilities against the Soviet Union, the Wehrmacht was forced to retire. On December 5, Army Group Center ceased its attack. On December 6, Zhukov, believing the Germans to have been bled white, commenced a major counteroffensive in the Moscow area. It was a remarkable feat of arms. Army Group Center, stranded in heavy snow in an extended, vulnerable, almost desperate position, was forced over onto the defensive.(55)

The Wehrmacht never really recovered from the defeat suffered in front of Moscow. There its strength was broken, the striking power of its divisions blunted forever. By the end of November, German ground forces in the Russian theater were short 340,000 replacements. On the Moscow front alone, from November 16 to December 4, the Germans lost 777 tanks, 55,000 soldiers were killed, and another 100,000 were wounded or disabled from frostbite.(56) It was such losses rather than the danger and hardship that caused a subsequent deterioration in German morale. By the time the Soviet offensive bogged down, around the beginning of March, the Germans had been driven back 50 miles. The blitzkrieg technique that had prevailed in Poland, France, and the Balkans had been soundly defeated.(57)

The Wehrmacht nevertheless remained a formidable force with which to be reckoned. The realization that the Germans were retreating, while having a salutary effect on Red Army morale, also gave rise to a dangerous tactical recklessness. Casualties rose alarmingly as Russian units began to attack without regard to losses. It was not long before Zhukov felt compelled to intervene. Lashing out at costly and uninspired frontal-attack methods, he directed a change in tactics, ordering his troops to bypass all major centers of enemy resistance. Soviet troops, though well equipped for winter warfare and consequently more mobile than their enemy, were still not skilled enough in offensive techniques to tactically master the Germans.(58)

After Moscow, the Germans were drawn further into a battle of attrition deep within Russia. A major success near Kharkov in May 1942 was followed by a rapid advance to the lower Volga in late August. By mid-September, encircling German armies had reached the suburbs of the city of Stalingrad, population 600,000. Into this incipient "Verdun on the Volga" were initially drawn three panzer, two motorized, and six infantry divisions. Unfortunately for the Germans, this was insufficient force to overcome the savage Soviet resistance that shortly developed within the city. By the end of September, German infantry companies were reduced to strengths of roughly 60 men each, while the Panzers, largely unsuited for fighting in city streets, were being slowly incinerated in the fires of close combat. At the height of the battle for Stalingrad, 22 German divisions, comprising almost 300,000 men, were committed to what was essentially a "dead Schwerpunkt."(59)

In the early stages of the battle for Stalingrad, the Germans enjoyed a superiority of three to one in men and six to one in tanks. The Luftwaffe also possessed complete air supremacy. However, in the rubble of the city, the Germans forfeited all their advantages in mobile tactics, while the less well supported but extremely tenacious Russian infantry came into their own. Horrifyingly savage battles were fought for cellars, rooms, and sewer drains as the Germans smashed their way, usually by day, building by building to the Volga. Panzer divisions were split up into company packets of from three to four tanks to support the German infantry. In assaulting a building, the infantry normally were sent into draw fire and hopefully locate enemy positions; once these were pinpointed, the tanks would cover each other while they battered away at pointblank range until the building fell down. What the Germans gained by day with the aid of the Luftwaffe, however, they had to be prepared to defend around the clock. This onerous task, which fell primarily to the German infantryman, was an unenviable one indeed since the Russians often counterattacked by night, feet bound in sacking to deaden their footfalls.(60) The Russian defense of Stalingrad was a truly active defense.

The Stalingrad defense was based on the "centre of resistance" comprising a number of "strongpoints." A "strongpoint" usually consisted of a building or a group of buildings, especially those constructed of good stone or brick. Every strongpoint, prepared for defensive purposes, depending on its size and importance, was occupied by a section, a platoon, a company, or sometimes even a battalion. Strongpoints were adapted to permit all-round defense and could fight independently for several days. Normally, they were linked with other buildings or groups of buildings by means of trenches, the spaces between them strengthened by obstacles and covered by fire. A group of strongpoints, with a common firing network, under a single administration and also equipped for all-round defense constituted a "center of resistance." Superimposed upon this defensive web, which essentially left the Germans in the streets and the Russians hidden in fortifications, was a rather novel system of active defense based on small-unit counterattack teams composed mainly of soldiers on foot. According to the Soviet defender of Stalingrad, General Vasili I. Chuikov,(61) commander of the Sixty-second Army, the Russians were content to let the Germans penetrate their defenses, as surprise counterattacks launched against the latter's flanks and rear invariably separated tanks from infantry and destroyed both piecemeal.

By Chuikov's reasoning, the success of German tactical methods stemmed from a superior coordination of all arms, none of them in themselves of outstanding quality. He noted that in most engagements German tanks would not attack without Luftwaffe support and that until the tanks reached their objectives, the infantry normally were not prepared to attack. Noting also a reluctance on the part of the German infantry for close combat, Chuikov decided that the most effective way to break down the Wehrmacht system was to keep as close to the Germans as possible; this tactic not only negated the striking power of the Luftwaffe but further served to separate tanks from infantry. In applying such tactical measures, the Russians were greatly assisted by the particularly special nature of the urban environment, which caused both sides to expose their flanks to a high density of fire when attacking and counterattacking. According to Chuikov:

> City fighting is a special kind of fighting. The buildings in a city are like breakwaters. They broke up the advancing enemy formations and made

their forces go along the streets. Our commanders and men learned to crawl right up to enemy positions during enemy bombardments and bombing, and by doing so avoid being killed. German airmen and artillerymen would not risk attacking our units for fear or hitting their own troops. . . . [Russian] troops defending the city learned to allow German tanks to come right on top of them--under the guns of our anti-tank artillery and anti-tank riflemen; in this way they invariably cut off the infantry from the tanks and destroyed the enemy's organized battle formation.(62)

Boldly rejecting tactical methods and organizations that basically were unsuited to urban conditions, Chuikov and his officers decided to "break down the formations that existed in the Army." Recognizing the futility of counterattacks by whole units and even sections of units (that is, battalions and companies) against strongpoints and centers of resistance captured and fortified by the Germans, they turned instead to violent flank and rear surprise attacks by small units infiltrating between enemy-occupied positions as the most effective means of combating the German onslaught. The aim of such active defense was to give the enemy no respite, to make every German soldier feel he was perpetually under the muzzle of a Russian gun. The "centre of the stage," to paraphrase Chuikov, was consequently "taken by small infantry groups" in which the individual "soldier . . . was on occasion his own general."(63)

At Chuikov's personal urging, the main battle unit for the Russian active defense of Stalingrad became the independently operating "storm group." Tailored to suit their assigned mission, "storm groups" were specifically developed for offensive and defensive operations in built-up areas. Essentially based on reinforced infantry sections that often had artillery and tanks attached to them, they represented a radical departure from various contemporary methods of waging war. According to Chuikov, storm groups were not designed in accordance with any tactical blueprint but were rather perfected in the course of battle for the precise conditions of city fighting. The value of allowing for the "broad use of weapons for short-range firing"(64) was well recognized by the Russians, who also became "convinced with practice that it was essential to make up the storm groups out of the personnel of one small unit. There was no question of constructing them at company strength. Every platoon, every squad, every soldier, had to be able to carry out an assault."(65)

The Russian storm group consisted of three distinct elements: assault, reinforcement, and reserve. Together, they constituted one whole designed to carry out a single task. The strength and constitution of each storm group depended on the objective it was to attack. The special features of the operations of each group were worked out on the basis of reconnaissance information about the nature of the objective and the size of its garrison. These special features were naturally crucial, for without clarifying them, it was normally impossible to come to grips with the tactics of the battle for a fortified building. In an actual engagement, the spearhead of the storm group was the assault element, composed of several subgroups of between six to eight men in each. The task of this element, under one commander, was to break in swiftly to a building and wage battle independently inside it. Assault-element members were lightly armed with submachine guns, grenades, daggers, and spades (the edges sharply honed for hand-to-hand fighting).(66)

The reinforcement element was usually divided into separate parties that entered the building simultaneously from different directions once the assault element fired off signal rockets or flares indicating a successful break-in. Armed with heavier weapons (heavy machine guns, mortars, antitank rifles and guns, picks, crowbars, and explosive charges), the reinforcement element included sappers, snipers, and soldiers of various trades. Commanded by the storm-group commander, its task was to quickly seize firing positions to create a defensive network against the enemy to prevent the latter from coming to the aid of his beleaguered garrison. The reserve element supplemented and strengthened the assault element, securing the flanks as necessary or taking up blocking positions.(67)

The offensive tactics of the storm group were based on rapid action and boldness on the part of every soldier. Since city fighting generally consisted of an endless series of assaults on well-fortified houses, buildings, and other objects, attacks by storm groups had to be short and sharp, their execution marked by swiftness and daring. Characterized by a high degree of decentralization, storm-group tactics also called for maximum flexibility and an unbridled sense of initiative. Timing and surprise were the two most important ingredients required for success. The timing of an attack was usually fixed in accordance with the enemy's behavior, his sleeping and eating habits as well as relief times receiving constant and particularly detailed scrutiny. Launched almost always by night or under cover of smoke screens, storm-group attacks were frequently made without any preliminary artillery bombardment. In fact, it became a rule that when enemy weapons were concentrated solely inside a building or other object transformed into a strongpoint, an attack would be carried out without prior artillery preparation, relying for its success on the surprise factor alone. Experience, recorded Chuikov, had taught the Russian that both timing and surprise were always attainable.

> Get close up to the enemy's position; move on all fours making use of craters and ruins; dig your trenches by night, camouflage them by day; make your build-up for the attack stealthily, without any noise; carry your tommy-gun on your shoulder; take ten to twelve grenades. Timing and surprise will then be on your side.(68)

The storm group also played a key role in the occupation of Russian strongpoints and centers of resistance during the battle for Stalingrad. A classic example of the former occupation was the defense of "Pavlov's house," a four-storied edifice into which Sergeant Jacob Pavlov of the Thirteenth Guards Division managed to cram 60 men, mortars, heavy machine guns, antitank weapons, and a full complement of skilled snipers. The antitank weapons were deployed on the ground floor, machine guns on the higher stories, and infantry at all levels, including the basement. This virtual fortress covered the approaches to a square in which Pavlov had skillfully mined all the open ground leading to his "house." From the third story, his observers detected Germany ground movement, while tank attacks came to grief on mines. Bombing and artillery and mortar fire finally wrecked most of "Pavlov's house," but for 58 days it beat off every German assault.(69)

There can be no question but that in the special kind of fighting that was Stalingrad, the Russian infantry excelled. Snipers and sharpshooters were undoubtedly among the most valuable soldiers, and here again the Soviets appear to have had the edge. Their ascendancy, in fact, became so marked that the head of the German sniper school at Zossen was reportedly sent to Stalingrad to help redress the balance. (He was subsequently shot and killed by the famous Soviet sniper Vasili Zaitsev.)(70) In night fighting and foul-weather war generally, the Russians also appear to have held the advantage. In the main, however, it was the novel Chuikovian approach to the defense of an urban area that blunted German blitz tactics. Active counterattacks by Red Army storm groups were the main factor in the Russian defense that kept the enemy in a constant state of tension. Day and night, the storm groups attacked and counterattacked, penetrating behind enemy lines and "firing at point-blank range at anyone who tried to lift a finger above the ground." Ultimately, under the attack of these groups, the Germans had to abandon not only buildings but strongpoints as well. The enlightened Chuikov would later write, with some elation, that the "most important thing that I learned on the banks of the Volga was to be impatient of tactical blueprints"; yet, without denying him due credit for certain originality of tactical thought, one could well argue that he merely, even if inadvertently, applied Great War storm tactics to an urban setting:

> Experience showed that the storm groups and the strongpoints were the most important facets of our defence. The Army beat off enemy attacks, itself attacked, made bold sallies, and took the initiative out of the enemy's hands. The power of our troops lay in the fact that, while defending themselves, they attacked the whole time. In conclusion I would note that modern city warfare is not street fighting in the literal sense of the word. In the city battle raging in Stalingrad the streets and squares were empty.(71)

After Stalingrad, "in a sense the battle of the war," according to John Keegan,(72) the Wehrmacht was placed almost entirely on the defensive in Russia. By spring 1943, four armies--Italian, Rumanian, Hungarian, and one German--had been lost and 50 divisions (half of them German) annihilated. Not surprisingly, most of the German generals by this time favored a shift to an elastic defense based on a series of strongly defended urban "hedgehogs" backed up by large, mobile counterattack forces. Hitler would have none of this, however, and ordered instead Army Groups Center and South to liquidate the "Kursk salient." No less than 12 panzer and 6 Panzergrenadier divisions were committed to this operation, given the code name Citadel.(73)

By 1943, the Red Army had reached its full strength of roughly 500 divisions, and in comparison with the army that had started the war, it was a completely new organization. More military than revolutionary, it was in every way a modern battle-hardened force that had the measure of its enemy. The morale of its soldiers was high and its leadership experienced and capable.(74) It was this army that fought the Battle of Kursk, which saw the last violent heave of the German army before its death agony.

In contrast to earlier German offensives, the Red Army was well informed of the German plans for Citadel.

Accordingly, it constructed in the Kursk area a layered defensive sector comprising no fewer than eight defensive belts extending over a depth of between 120 and 180 miles. The forward zones consisted of a tight web of strongpoints, each of which disposed three to five 76-millimeters antitank guns, about five antitank rifles, up to five mortars, and one section each of sappers and infantry armed with submachine guns. Groups of these strongpoints, under one commander and fire system, formed an "antitank area." The Russian method of controlling antitank-area gunfire was reputedly copied from the Germans and refined. Known to the Germans as a Pakfront, it was based on the use of groups of up to 10 well-camouflaged antitank guns under a single commander, who was responsible for concentrating their fire on a single target at a time in broadsides. The idea was to draw attacking armor into a web of enfilade fire, which was held until the last possible moment. In addition to such antitank centers of resistance, mines were also laid at an average density of 2,200 antitank and 2,500 antipersonnel per mile of front (six times the density in the Battle of Moscow and four times that used at Stalingrad). During the waiting period, the Russian soldiers were toughened up by a rigorous program of physical training, forced marches, and political indoctrination.(75)

When the Germans attacked on July 5, 1943, they did not do so in classic blitzkrieg fashion. Instead, they advanced in a succession of armored wedges, known as Panzerkeil, heavy Tiger tanks leading with Panthers and Pzkw IVs fanning out behind. It was a tactic that had been forced on the Germans by the tenacity of the Red Army in holding close to the sides of a breach and bringing to bear a heavy weight of effective antitank firepower. The action of the new German tactic, however, was further akin to the swing of a broadax than to the thrust of a sword.(76)

In the Battle of Kursk, Russian infantrymen proved their worth once more not just as defenders of terrain but as tank killers in their own right. The "Porsche" Tiger tank, through a design error, had been produced without a coaxially mounted turret machine gun. This fault was soon discovered by the Russian infantry, which immediately began to hunt the Tiger down as the helpless brute it really was. Devastating against Russian tanks at a distance, it had no means of protecting itself against close-in infantry action. The modified version, the Ferdinand self-propelled gun, was taken on in similar fashion by Russian tank killers emerging from the ground. As for the splendid Panther tank, the aggressive Russian infantry found that it burned even more spectacularly than other German tanks.(77) On such a note, the Red Army went over to the offensive, the qualities of its infantry now becoming more obscure as fascination with the maneuver of mass tank armies increased.

There is no question but that the German soldier was shocked by the conditions of the Eastern Front and the fighting capabilities of the Russian infantryman. Described alternately as "soullessly indifferent" and "unpredictable," the latter was nonetheless recognized from the beginning as a first-rate fighter and one who would become more effective with experience. There were, supposedly, two sounds a German serving on the Ostfront never forgot: one, the terrifying ripping noise of the Katyusha rockets; the other, the haunting battle cry of the Russian infantry. According to the veteran Colonel Wolfgang Mueller, however, the German soldier did not fear the Katyushas ("Stalin organs" as they were called), no more than he dreaded the T-34 tank; but he did retain a lasting trepidation for the Russian assault infantry with its chilling battle cry of "Urra!" Even on quiet fronts, the German initially suffered such heavy casualties at their hands that infantry manpower remained critically lacking for the rest of the war.(78)

To reinforce this point, there were precious few World War II German field commanders who were not aware that the difference between the war in the East and the war in the West was the difference between day and night.(79) The fighting in the forests around Moscow in 1941 was a far cry from the conflict in Flanders and France in 1940. The following catechism given to German infantrymen proceeding to the Ostfront reveals just how much:

> The soldier in Russia must be a hunter. The Bolshevist's greatest advantage over the German is his highly developed instinct and his lack of sensitivity to the weather and to the terrain. One must be able to stalk and creep like a huntsman.
>
> The soldier in Russia must be able to improvise. The Bolshevist is a master of improvisation.
>
> The soldier in Russia must be constantly on the move. Hardly a day passes on which the Russians, however weak they may be, do not attempt to push against our lines. Day after day they work to improve their positions.
>
> The soldier in Russia must be suspicious.
>
> The soldier in Russia must be wide awake. The Russian practically always attacks during the night and in foggy weather. In the front line there is nothing to be done but to remain awake at

night and to rest during the day. But in Russia there is no front line or hinterland in the military sense of the word. Anyone who lays down his arms east of the old Reich frontier may greatly regret this a moment later.

The soldier in Russia must reconnoiter. Reconnoitering is the main component of all fighting in Russia.

The soldier in Russia must be hard. Real men are needed to make war in forty degrees of frost or in great heat, in knee-deep mud or in thick dust. The victims of the Bolshevist mass attacks often present a sight against which the young soldier must harden his heart. He must reckon with the possibility of losing his life. Only men who not lose their nerve when death threatens them are fit to be fighters against Bolshevism. Weak characters must realize that the leadership is sufficiently hard to punish cowardice by death.(80)

Although his march performance in Russia was from the beginning even more unbelievable than that in France (normally 25 miles per day over the most atrocious roads, all good ones being reserved for motorized movement), the German infantryman, in comparison with his Russian counterpart, was much too spoiled. He had become accustomed to barracks with central heating and running water, to beds with sheets and mattresses; it was not, therefore, easy for him to adapt to the extremely primitive conditions of the Soviet Union. As a first adjustment to local conditions, the Wehrmacht revised the standards for selecting small-unit commanders; their average age was lowered, and the physical-fitness requirements were raised. It was said that the studious kind of officer who relied chiefly on maps was out of place in Russia. Excess personal baggage and staff cars were also left behind and impedimenta generally reduced to a minimum. One measure specifically introduced for warfare in the East was the organization of light-infantry divisions. This was done after the Germans discovered that their mountain divisions were the most effective type of unit for sustained combat in forests and swamps.(81)

As the very vastness of Russia and the peculiarities of the war there led often to the repeated isolation of units, great stress was placed on all-round defense and security measures. In winter, in order to survive the cold let alone the enemy, German companies normally formed a defensive center around a town or other inhabited point. In summer, since the Germans were not strong enough to hold a continuous line, the same system of bastion towns came to be used;

trenches would be dug around a village and a large portion of automatic weapons placed along the perimeter. Stealing a page from the book of Marshal Maurice de Saxe, however, only a few troops were used to man these lines, the bulk of the soldiery being held back centrally as a counterattack force in a mobile defense. When mixed forces were involved, as they often were, tanks would form an inner ring capable of firing over the heads of dug-in infantry comprising an outer ring, from which would also issue security patrols and outposts. This form of tactics, frequently not mentioned in standard field service regulations, was adopted by most units and referred to as "hedgehog" defense.(82)

The Germans also learned early in the war not to site their positions on forward slopes where they could be engaged by long-range enemy fire and attacked by tanks. The reverse slope position, protected by wire and mines and backed up by mortars and heavy weapons, was much more likely to be chosen by German infantry units. A counterattack element, normally armor hidden in hollows, was usually retained to protect the reverse slope from being overrun. The art of field camouflage was also developed to an advanced degree, particularly since the Germans lost their air superiority during the battle for Moscow. In this regard, dummy and alternate positions were used to a great extent, and the improvement of natural obstacles to an enemy advance was brought to a high pitch. Interestingly, when defending along river lines, the Germans favored placing their positions on the enemy side, thus making the river an integral part of their antitank defense.(83)

The infantryman's natural fear of tanks was heightened acutely on the Eastern Front where Russian armor, particularly the dangerous T-34, became more and more active as the war went on. To counter "tank terror," intensive training with armored and assault-gun units was instituted for the German infantry.(84) The Germans early established special antitank teams composed of infantry and combat engineers and attached them to leading spearhead elements. Training was thus as offensive oriented as defensive. The remarks of a member of the <u>Grossdeutschland</u> division who underwent such <u>training</u> in Russia bear this out:

> As we had already been taught to dig foxhole in record time, we had no trouble opening a trench 150 yards long, 20 inches wide, and a yard deep. We were ordered into the trench in close ranks, and forbidden to leave it, no matter what happened. Then four or five Mark-3s rolled forward

at right angles to us, and crossed the trench at different speeds. The weight of these machines alone made them sink four or five inches into the crumbling ground. When their monstrous treads ploughed into the rim of the trench only a few inches from our heads, cries of terror broke from almost all of us. . . . We were also taught how to handle the dangerous <u>Panzerfaust,</u> and how to attack tanks with magnetic mines. One had to hide in a hole and wait until the tank came close enough. Then one ran, and dropped an explosive device--unprimed during practice--between the body and the turret of the machine. We weren't allowed to leave our holes until the tank was within five yards of us. Then, with the speed of desperation, we had to run straight at the terrifying monster, grab the tow hook and pull ourselves onto the hood, place the mine at the joint of the body and turret, and drop off the tank to the right, with a decisive rolling motion.(85)

The Germans in Russia approached training for mine-field breaching in much the same realistic fashion. As mine fields were normally covered by fire, they found the procedure of using heavily protected engineer detachments to clear lanes and gaps too cumbersome. They also considered the more specialized training of infantry as unsatisfactory. The expedient finally adopted as the preferred method for crossing mine fields was to thoroughly instruct all infantrymen in mine-laying techniques and in spotting mines by using captured enemy mine fields as training grounds. They were then run through such mine fields in rear areas. German troops thus trained were able to breach mine fields on foot in battle in reasonable safety. To give the men further confidence in their ability to negotiate mine fields safely, German commanders at all levels would often hold their orders groups in mine fields.(86)

Another impressive feature of the German army in Russia was its capacity for employing its rear-echelon troops in the front line with effect. Large frontages and heavy casualties left the Wehrmacht critically short of infantrymen by the end of 1941. Battalions, combat teams, and even divisions were therefore improvised and thrown into battle. Though this practice was ultimately wasteful, as it had the effect of leaving many essential logistic areas untended, there was no questioning its necessity during the Russian counteroffensive in early 1942. In many cases, "emergency alert units" were formed to garrison and defend strongpoints or directly reinforce forward units.(87) In other cases, however, these units actually took the offensive. The manner in which they did so remains a tribute to German military skill, organization, and understanding of the human psyche.

In forming fighting units from rear-echelon troops, the Germans, where possible, left them with their own commanders. Once they had proved themselves capable of holding a defensive line or position, they were often given small objectives to seize. However, the Germans selected only the most vulnerable, and hence attainable, objectives for emergency units to attack. The psychological consequence of such discerning tasking was that these ad hoc units rapidly gained confidence and ultimately performed the exceptional levels. The "snail offensive," mounted by the Sixth Panzer Division on a 40-mile front in February 1942 by soldiers armed with but rifles and one or two machine guns on average per company, netted in one month approximately 80 small villages. German emphasis and insistence on training all Wehrmacht personnel for combat, particularly in the use of antitank weapons, no doubt contributed to this military versatility. Given such a training system, it comes as no surprise that a divisional bakery company was able to force Russian tanks to retreat in an action in 1943.(88)

It must be stated finally, however, that the German soldier in general found Russia a great, foreboding place. The psychological effect of that country on many infantrymen was shattering, while to still others it was "a perpetual shivering fit."(89) It was indeed a wry soldierly humor that referred to the Eastern Medal struck for the Battle of Moscow as the "Order of the Frozen Meat." In his tight jackboots and summer uniform, with only a greatcoat and blanket, the German soldier suffered perhaps more than most.(90) After Kursk, it was his destiny to suffer even more, though this never compromised his exceptionally high standard of performance.

The Battle of Kursk was the swan song of German armor. From this point, the Red Army maintained a relentless strategic pressure that ultimately carried it to Berlin. Unlike the Wehrmacht, it relied more on mass than finesse in its offensive tactics. Although it attempted double envelopments at every opportunity, they were in most instances effected only when favorable ratios, sometimes as high as seven to one, predominated. With its hardy infantry clustered on tanks, the Red Army nonetheless constituted a formidable force. When it struck the Mannerheim Line once again in 1944, it accomplished in ten days what it had taken three months to do in 1939-40.(91)

The Wehrmacht, now decidedly on the receiving end, managed to prolong its defeat only through sheer tactical brilliance. The magnificent German army, though badly mauled at Moscow, Stalingrad, and Kursk, remained a viable force to the end; it was an army that refused to die. Buoyed up primarily by the superior training of its junior commanders(92) and the capacity of its smaller units--down to and including section level--to act independently, it still retained the grudging respect of its eastern enemy. That the German army did not lose its cohesion and hence its effectiveness as a fighting force has been attributed largely to the "high degree of primary group(93) integrity" it managed to maintain within its ranks.

The German soldier's ability to resist appears most definitely to have been a function of the "capacity of his immediate primary group (his squad or section) to avoid social disintegration." Where primary groups developed a high degree of cohesion, morale was likewise high and resistance effective, or at least very determined. For the ordinary conscripted German soldier, the decisive fact was that he was a member of a squad or section that maintained its structural integrity and satisfied some of his major primary needs. For the German army, the decisive fact was that leadership responsibility in battle devolved principally upon small unit officers and NCOs. So long as the small group possessed leadership with which the German soldier could identify himself, and so long as he gave affection to and received affection from its other members, he tended to go on fighting.(94) Whether intentional or not, the traditional emphasis of the German army on low-level leadersip and Kameradschaft(95) may well have been the most critical element in its battlefield excellence. "We are beginning to understand," concluded the perceptive Lord Moran, "that the secret of the awful power of the German army is not in tanks and aircraft, but in a certain attitude of mind of her manhood."(96)

Significantly, nearly 70 percent of the German fighting strength that advanced into Russia in 1941 did so on its feet. This proportion was never reduced. While the German army originally entered Russia with 3,350 tanks, there were, by January 23, 1943, only 495 fit for battle over the entire Eastern Front. The plain truth was that the mighty German army, once the pride of the Reich, had actually become one of the poorer armies in the world.(97) Yet this same field force was able to exit Russia under pressure of tremendous odds in almost as classic a military fashion as it had entered. With their panzer forces reduced to minimally effective levels, three German armies were able to hold off for nine months the repeated attacks of 15 Soviet armies. While their ability to do so can be explained partly in terms of the growing ascendancy of the

144 A PERSPECTIVE ON INFANTRY

defense, much of the credit must accrue to the German soldier--in particular, to the indefatigable German infantryman around whose perseverance and holding power the mobile "hedgehog" defense was constructed. Like the Red Army infantry that proved the salvation of Russia, the German infantry sustained the Wehrmacht throughout its withdrawal even unto the Götterdämmerung that eventually consumed it. Both of these splendid infantry arms, for better or worse, remained the backbone of their respective forces.

NOTES

1. Seaton, The Russo-German War, pp. 17-18, 46, and 60-62; and Addington, The Blitzkrieg Era, p. 183.

2. Marshal Georgy Konstantinovich Zhukov, Memoirs (London: Jonathan Cape, 1971), p. 250.

3. Albert Seaton, The Battle of Moscow (London: Rupert Hart-Davis 1971), pp. 30 and 38-39; and The Russo-German War, pp. 61-62, 74-75 and 95; Gen. Wadyslaw Anders, Hitler's Defeat in Russia (Chicago: Henry Regnery, 1953), p. 18; John Erickson, The Road to Stalingrad (London: Weidenfeld and Nicolson, 1975), pp. 98; Addington, The Blitzkrieg Era, p. 187, and Liddell Hart, The Other Side of the Hill, pp. 255-7. Under Speer's direction, German tank production increased to 18,000 a year by 1944. British aircraft production had already outstripped that of Germany in 1941. Seaton, The Battle for Moscow, p. 30.

4. Seaton, The Russo-German War, p. 61.

5. Malcolm Mackintosh, Juggernaut; A History of the Soviet Armed Forces (New York: Macmillan, 1967), pp. 133, 143-6, and 156; Davies, The German Army Handbook, pp. 54-55; John Erickson, The Soviet High Command (London: Macmillan, 1962), p. 625; and The Road to Stalingrad, pp. 210 and 219; Seaton, The Russo-German War, pp. 130 and 184; and The Battle for Moscow, pp. 45, 47-48, 65, 74, and 83; and Anders, Hitler's Defeat in Russia, p. 58. German claims are probably largely exaggerated as units often competed with one another in stating them. Alan Clark, Barbarossa: The Russian-German Conflict, 1941-1945 (London: Hutchinson, 1965), p. 137.

6. Erickson, The Soviet High Command, pp. 463-4 and 506.

7. Liddell Hart, Europe in Arms (London: Faber and Faber, 1937), pp. 31-32.

8. Messenger, The Art of Blitzkrieg, pp. 60-63 and 83-87; Mackintosh, Juggernaut, pp. 80-81; and Erickson, The Soviet High Command, p. 317.

9. Messenger, The Art of Blitzkrieg, pp. 60-63 and 83-87; Erickson, The Soviet High Command, p. 317; Mackintosh, Juggernaut, pp. 80-81; Trythall, "Boney" Fuller, p. 175; and Luvaas, Education of an Army, p. 374.

10. Farrar-Hockley, Infantry Tactics, p. 45; and Miksche, Atomic Weapons and Armies, pp. 89-90.

11. Erickson, The Soviet High Command, pp. 317, 444, 507, 537, and 567.

12. Edgar O'Ballance, The Red Army (London: Faber and Faber, 1964), p. 135; Mackintosh, Juggernaut, pp. 80-82, and 95-98; Messenger, The Art of Blitzkrieg, pp. 104 and 105; and Erickson, The Soviet High Command, pp. 444 and 537.

13. D. Fedotoff White, The Growth of the Red Army (Princeton: University Press, 1944), pp. 393-7; and Erickson, The Soviet High Command, p. 479.

14. Erickson, The Road of Stalingrad, p. 7.

15. Erickson, The Soviet High Command, pp. 499, 502, 518-22, and 532-7; and Mackintosh, Juggernaut, pp. 108-10.

16. Erickson, The Soviet High Command, pp. 541-4.

17. According to Seaton, the Finn was "the finest warrior on the side of the Axis and for winter and forest warfare he was inferior to none." Seaton, The Russo-German War, p. 94.

18. O'Ballance, The Red Army, pp. 144-5 and 149-52; B. H. Liddell Hart, ed., The Soviet Army (London: Weidenfeld and Nicolson, 1956), pp. 83-85 and 89-91; and Mackintosh. Juggernaut, pp. 119-23.

19. Liddell Hart, The Soviet Army, pp. 83-85 and 89-91; O'Ballance, The Red Army, pp. 144-52; and Mackintosh, Juggernaut, pp. 119-23.

20. Erickson, The Soviet High Command, pp. 206 and 552-3; and The Road to Stalingrad, p. 30; Mackintosh; Juggernaut, pp. 122 and 124; and O'Ballance, The Red Army, p. 154.

146   A PERSPECTIVE ON INFANTRY

21. Erickson, The Soviet High Command, pp. 554-5; O'Ballance, The Red Army, pp. 154-5; Mackintosh, Juggernaut, p. 127; and Liddell Hart, The Soviet Army, p. 86.

22. Erickson, The Soviet High Command, pp. 576 and 583; O'Ballance, The Red Army, pp. 154-5; Zhukov, Memoirs, pp. 250-1; and Mackintosh, Juggernaut, pp. 131-6.

23. Seymour Freidin and William Richardson, eds., The Fatal Decisions, trans. Constantine Fitzgibbon (London: Michael Joseph, 1956), p. 31.

24. Seaton, The Russo-German War, pp. 171 and 175; and The Battle For Moscow, pp. 65, 89 and 91; Small Unit Actions During the German Campaign in Russia (Washington: Department of the Army Pamphlet No. 20-269, 1953), p. 13; and George E. Blau, The German Campaign in Russia: Planning and Operations (1940-1942) (Washington: Department of the Army Pamphlet No. 20-261a, 1955), pp. 65 and 88.

25. Seaton, The Battle for Moscow, pp. 83-84 and 124. It should also be remembered that the power of the political commissar had by this time been increased to a position of parity with the military commander to give additional fiber to the Soviet resistance. Furthermore, Stalin's criminal law, which held hostage the relatives of soldiers taken prisoner, presumably prevented some troops from deserting or surrendering. The Russian soldier was therefore motivated by more than just love of country. Ibid., p. 84.

26. Although Seaton claims that "the part played by the Red Army in 1941 in halting the enemy advance has been exaggerated by Soviet historians," he does attribute "the mounting German casualty rate" to the great "determination and obstinacy" of "the larger part of the Soviet troops [who] fought on." Seaton, The Russo-German War, p. 221.

27. Terrain Factors in the Russian Campaign (Washington: Department of the Army Pamphlet No. 20-290, 1951), pp. 4, 5, 9, 28 and 30; and Addington, The Blitzkrieg Era, pp. 208-9.

28. Terrain Factors in the Russian Campaign, pp. 13-15 and 37-39.

29. Seaton, The Battle for Moscow, p. 109; Terrain Factors in the Russian Campaign, p. 31 and 37; Rear Area Security in Russia: The Soviet Second Front

Behind the German Lines (Washington: Department of the Army Pamphlet No. 20-240, 1951), p. 23; and Combat in Russian Forests and Swamps (Washington: Department of the Army Pamphlet No. 20-231, 1951), pp. v and 24.

30. Macksey, Tank Warfare, p. 161; and Palit, War in the Deterrent Age, p. 96.

31. Marshall, The Soldier's Load, p. 62; Mellenthin, Panzer Battles, p. 287; and James Lucas, War on the Eastern Front: The German Soldier in Russia (London: Jane's, 1979), pp. 53-54.

32. Patton, War As I Knew It, p. 349. His reference to the Russian as "an all out son of a bitch, a barbarian, and a chronic drunk" leaves out mention of the latter's more sterling qualities. Martin Blumenson, ed., The Patton Papers (Boston: Houghton Mifflin, 1974), vol. 2, p. 734.

33. Small Unit Actions During the German Campaign in Russia, pp. 4, 33, 42, and 60; Alexander Bek, Volokolamsk Highway (Moscow: Foreign Language Publishing House, 1944), pp. 41-42, 154, 252, 302, 313, and 325-6; Seaton, The Battle For Moscow, p. 280; and Walter Kerr, The Russian Army: Its Men, Its Leaders, and Its Battles (London: Victor Gollanz, 1944), p. 61.

34. The Germans copied it in their 1943 Racketenpanzerbüshe. John Weeks, Men Against Tanks (New York: Mason Charter, 1975), p. 65. The Panzerfaust also entered service in 1942.

35. Paul Carell, Hitler Moves East, 1941-1943, trans. Ewald Osers (Boston: Little, Brown and Co., 1964), p. 80; Kerr, The Russian Army, p. 57; and Clark, Barbarossa, pp. 198-9. The "Molotov cocktail" was actually first used in Spain by Republican troops along with the Austurian miners' satchel charge. Weeks, Men Against Tanks, pp. 30-31.

36. Carell, Hitler Moves East, pp. 69, 105, and 243.

37. According to Guy Sajer, who fought on the Ostfront, the "confidence the infantry placed in the Luftwaffe was absolute." Sajer, The Forgotten Soldier, p. 72.

38. Seaton, The Battle for Moscow, p. 226; and Small Unit Actions During the German Campaign in Russia, pp. 236 and 240.

148 A PERSPECTIVE ON INFANTRY

39. Strategy and Tactics of the Soviet-German War by Officers of the Red Army and Soviet War Correspondents (London: Hutchinson, circa 1942), pp. 57-58, 66, and 91. Loud shouting was considered to be effective in forest fighting.

40. Small Unit Actions During the German Campaign in Russia, pp. 2-3 and Martin Caidin, The Tigers Are Burning (New York: Hawthorn 1974), pp. 124-7; and Freidin, The Fatal Decisions, p. 38. The Russians claimed "very few cases of frostbite"; they at least knew better than the German how to avoid it. Canadian Army Training Memorandum, no. 16 (July 1942), p. 57; and Carell, Hitler Moves East, p. 176.

41. Lucas, War on the Eastern Front, p. 56; Marshall, The Soldier's Load, pp. 60-63 and 87; A. J. Barker and John Walter, Russian Infantry Weapons of World War II (New York: Arco, 1971), p. 25; Martel, An Outspoken Soldier, pp. 218-19 and 224-55; Freidin, The Fatal Decisions, p. 48; and Bek, Volokolamsk Highway, pp. 85 and 92. Gas masks were also carried initially. The Soviets seemed so desperately anxious not to allow gas warfare to begin, if they could help it, that they avoided the use of smoke during the opening months of the war. O'Ballance, The Red Army, p. 175. The Germans shared this fear; the "possibility of gas warfare weighed heavily upon us" recorded Kurt Emmrich under the pseudonym Peter Bamm in his book, The Invisible Flag, trans. Frank Herrman (London: Faber and Faber, 1957), p. 128.

42. Lt. Col. N. Yelshin, "Soviet Small Arms," Soviet Military Review (February 1977), pp. 15-17; Weller, Weapons and Tactics, pp. 150-52; Barker, Russian Infantry Weapons, p. 21; Effects of Climate on Combat in European Russia (Washington: Department of the Army Pamphlet No. 20-291, 1952), p. 69; and Kerr, The Russian Army, p. 31.

43. Raymond L. Garthoff, Soviet Military Doctrine (Glencoe: The Free Press, 1953), pp. 212 and 299; and Malcolm Mackintosh, "The Development of Soviet Military Doctrine Since 1918" in The Theory and Practice of War, ed. Michael Howard, p. 254.

44. Martel, An Outspoken Soldier, pp. 141-43, 224, and 227. One of the most severely criticized faults a Russian officer could display was that of being on too familiar terms with his men. Such officers were regarded as inefficient and could be reduced to the ranks for this shortcoming alone. Mackintosh, Juggernaut, p. 223.

45. Mackintosh, Juggernaut, pp. 222-24. Barker claims that a rifle company consisted of 110 men at full strength. Three sections of nine men made up a platoon, three platoons a company. The company also had a machine-gun platoon of 16 men divided into three sections with one medium machine gun (MMG) each. Barker, Russian Infantry Weapons of World War II, p. 8. According to Weller, a regiment in 1945 comprised 2,474 men. Each regiment had three rifle battalions and a headquarters that included a mortar company of seven 120-millimeter mortars, an antitank battery of six 45-millimeter guns, an antitank company with twenty-seven 14.5 antitank rifles, and a submachine-gun (SMG) company with 100 men, each armed with an SMG. Each rifle battalion consisted of about 600 men, divided into three rifle companies (143 men each), a machine-gun company (58) with nine .30 caliber MMGS, a mortar company (61) with nine 82-millimeter mortars, an antitank platoon (23) with nine 14.5 millimeter antitank rifles. Each rifle company had three platoons and a weapons platoon; these had together two MMGs, 18 light machine guns, and two 50-millimeter mortars. Weller, Weapons and Tactics, p. 151.

46. Col. Louis B. Ely, The Red Army Today, (Harrisburg: The Military Service Publishing Company, 1949), pp. 24-26; and Farrar-Hockley, Infantry Tactics, p. 45.

47. Carell, Hitler Moves East, pp. 49-50.

48. Fuller, Machine Warfare, p. 173.

49. Small Unit Actions During the German Campaign in Russia, pp. 6, 15, and 32; German Defence Tactics Against Russian Break-throughs (Washington: Department of the Army Pamphlet No. 20-233, 1951), pp. 32 and 37; Liddell Hart, The Soviet Army, p. 34; and Carell, Hitler Moves East, p. 95. In the forests around Moscow, darkness sometimes came as early as 1500 hours.

50. Small Unit Actions During the German Campaign in Russia, pp. 4 and 248; German Defence Tactics Against Russian Break-throughs, p. 32; Effects of Climate on Combat in European Russia, p. 16; Mellenthin, Panzer Battles, pp. 180-81; and Liddell Hart, The Soviet Army, p. 130.

51. Clark, Barbarossa, pp. 231-32; and Mellenthin, Panzer Battles, p. 181.

52. Clark, Barbarossa, p. 194; and Miksche, Blitzkrieg, pp. 41-42; 45, 181, 183, 189, 204, 212, and

227. I can find no evidence that this was the formal defensive doctrine of the Red Army before the war, although there is some reason to believe that certain ideas of Wilhelm von Leeb were incorporated in the 1936 Soviet Field Service Regulations. Palit, War in the Deterrent Age, p. 96; and Field Marshal Gen. Ritter Von Leeb; Defense, trans. Dr. Stefan T. Possony and Daniel Vilfroy (Harrisburg: The Military Service Publishing Company, 1943), pp. iii, vii, 28, 94-95, 97-99, 103-11, and 118-21.

53. Seaton, The Russo-German War, pp. 172-74 and 186-91; and The Battle for Moscow, pp. 78-79, 111-12, 114, 116-17, and 126-28; Zhukov, Memoirs, pp. 334-39 and 343-45; and Guderian, Panzer Leader, pp. 247-48. In many German divisions the frostbite incidence was as high as 40 percent. Carell, Hitler Moves East, p. 176.

54. Blau, The German Campaign in Russia, pp. 86-87; Erickson, The Road to Stalingrad, pp. 257-58, 262, 266-67, 269-70, and 273; Clark, Barbarossa, pp. 150-52, V. Ryabov, The Soviet Armed Forces Yesterday and Today (Moscow: Progress, 1976), pp. 70-71; and Bek, Volokolamsk Highway, p. 15. This last book, a novel narrating actions in the Battle of Moscow, was supposedly widely read by Israeli army officers. See Yeshayahu Ben-Porat, Hezi Carmel et al., Kippur, trans. Louis Williams (Tel Aviv: Special Edition, 1973), p. 304.

55. Guderian, Panzer Leader, pp. 248-49 and 258-60; Clark, Barbarossa, p. 157; Seaton, The Battle for Moscow, pp. 158-60, 287 and 294; Erickson, The Road to Stalingrad, pp. 265-66 and 273; and Zhukov, Memoirs, pp. 348-51.

56. By March, 1942, the Germans had lost 108,000 men killed, 268,000 wounded, 228,000 victims of frostbite, and 250,000 more suffering from other diseases, exhaustion, and exposure. Seaton, The Russo-German War, p. 228; and The Battle for Moscow, p. 288.

57. Carell, Hitler Moves East, pp. 191-92; Addington, The Blitzkrieg Era, pp. 212-13; and Freidin, The Fatal Decisions, pp. 31 and 61. At the end of March 1942, the 16 panzer divisions on the Ostfront had a total of but 140 serviceable tanks. Blau, The German Campaign in Russia, pp. 80 and 120.

58. Seaton, The Battle for Moscow, p. 228; Erickson, The Road to Stalingrad, p. 291; and The Soviet High Command, p. 652.

THE BACKBONE OF LAND FORCES   151

59. The War in Eastern Europe (June 1941 to May 1945) (West Point: U.S.M.A. Department of Military Art and Engineering, 1949), pp. 69-72 and 77-9; Erickson, The Road to Stalingrad, pp. 412, 416-17, and 421; and Clark, Barbarossa, pp. 205-6.

60. Clark, Barbarossa, pp. 198-99, 209, and 217; Mellenthin, Panzer Battles, p. 155; Erickson, The Road to Stalingrad, pp. 392-93, 402-4, 409, 414, 441, and 461; and Mackintosh, Juggernaut, pp. 184-85.

61. Marshall Vasili I. Chuikov, The Beginning of the Road, trans. Harold Silver (London: MacGibbon and Kee, 1963), pp. 284, 288, and 290.

62. Chuikov, The Beginning of the Road, pp. 146-7, 283, and 291.

63. Chuikov, The Beginning of the Road, pp. 109, 146, 150, 288, and 291-3.

64. In the fighting for the Krasny Oktyabr factory, even 203-millimeter guns were used for direct support, firing at ranges of 200-300 yards. Chuikov, The Beginning of the Road, p. 302.

65. Chuikov, The Beginning of the Road, pp. 150, 286, 291, and 205.

66. Chuikov, The Beginning of the Road, p. 294.

67. Ibid.; Erickson, The Road to Stalingrad, pp. 442-3; and Clark, Barbarossa, p. 212.

68. Chuikov, The Beginning of the Road, pp. 292 and 295-8. According to Chuikov, the soldier's "irreplaceable weapon" in the assault was the grenade.

69. Erickson, The Road to Stalingrad, p. 441.

70. Erickson, The Road to Stalingrad, p. 444; Mackintosh, Juggernaut, p. 185; and Clark, Barbarossa, pp. 215-17. For a detailed account of the duel between Maj. Konings, the head of the German sniper school, and Zaitsev, see Chuikov, The Beginning of the Road, pp. 142-5.

71. Chuikov, The Beginning of the Road, pp. 147, 283-4, and 292-3.

72. Keegan, The Face of Battle, p. 286.

73. Anders, Hitler's Defeat in Russia, pp. 149,

154, and 155-6; Freidin, The Fatal Decisions, p. 166; and Messenger, The Art of Blitzkrieg, p. 191.

74. The War in Eastern Europe, p. 90.

75. Clark, Barbarossa, pp. 289-90 and 295; Mellenthin, Panzer Battles, pp. 225-6; and The War in Eastern Europe, pp. 95-96. The Soviet 76-millimeter antitank gun was judged to be "the most efficient tank killer of its time." Lucas, War on the Eastern Front, p. 47.

76. Clark, Barbarossa, pp. 294-6; and Mellenthin, Panzer Battles, pp. 226-7.

77. Liddell Hart, The Other Side of the Hill, pp. 329 and 336; and Gen. Frido von Senger und Etterlin, Neither Fear Nor Hope, trans. George Malcolm (London: Macdonald, 1960), p. 123.

78. Liddell Hart, The Soviet Army, p. 326; Garthoff, Soviet Military Doctrine, p. 226; and Caidin, The Tigers are Burning, pp. 120-22.

79. Military Improvisations During the Russian Campaign (Washington: Department of the Army Pamphlet No. 20-201, 1951), p. 13.

80. Ely, The Red Army Today, pp. 20-21.

81. Freidin, Fatal Decisions, p. 47; Blau, The German Campaign in Russia, p. 47; Small Unit Actions During the German Campaign in Russia, pp. 3-4; Combat in Russian Forests and Swamps, p. 9; and Military Improvisations During the Russian Campaign, p. 17.

82. Fall, Ordeal by Battle, pp. 108-9; Strategy and Tactics of the Soviet-German War, p. 23; Military Improvisations During the Russian Campaign, p. 22; and The War in Eastern Europe, p. 57. De Saxe, noting that infantry in entrenchments often fled when the entrenchment was breached, recommended holding back the mass, which could always be counted upon to counterattack with zeal when the enemy was most disoriented and confused. This was how "a trifle changes everything in war and how human weaknesses cannot be managed except by allowing for them." Marshal Maurice de Saxe, Reveries on the Art of War, trans./ed. Brig. Gen. Thomas R. Phillipp (Harrisburg: The Military Service Publishing Company, 1944), p. 100.

83. German Defence Tactics Against Russian Break-throughs, p. 28, 30 and 68; Davies, German Army Handbook, p. 58; Mellenthin, Panzer Battles, p. 160; and Seaton, The Battle for Moscow, p. 197.

84. Blau, The German Campaign in Russia, p. 139; German Defence Tactics Against Russian Break-throughs, p. 69; and Combat in Russian Forests and Swamps, p. 22. The Russians did not employ armor in mass until May 1942. Blau, The German Campaign in Russia, p. 56.

85. Sajer, The Forgotten Soldier, p. 166. On Eastern Front tank fighting techniques and antitank weapons. See Lucas, The War on the Eastern Front, pp. 155-60.

86. Military Improvisations During the Russian Campaign, pp. 17-19.

87. German Defense Tactics Against Russian Break-throughs, pp. 21 and 34. The critical shortage of men was reflected in German infantry organization. Squads (sections) contained ten men originally but were reduced to nine about 1943. The fourth squad in each platoon, a mortar squad, had disappeared earlier. Weller, Weapons and Tactics, p. 143.

88. German Defense Tactics Against Russian Break-throughs, p. 34; and Military Improvisations During the Russian Campaign, pp. 4-6, 9, 13, and 77. Divisional strength in the Sixth Panzer Division had dropped to an unbelievable 57 riflemen, 20 engineers, and three guns in January, 1942.

89. Sajer, The Forgotten Soldier, p. 30.

90. Freidin, Fatal Decisions, pp. 38 and 69; and Clark, Barbarossa, p. 159. German casualties in the field from June 22, 1941 to March 31, 1945 were 1 million dead, 3,966,000 wounded, and 1,288,000 missing. Seaton, The Russo-German War, p. 586.

91. Mellenthin, Panzer Battles, p. 225; Ely, The Red Army Today, p. 13; and The War in Eastern Europe, pp. 114-15 and 121.

92. Mellenthin, Panzer Battles, pp. 205 and 264.

93. A primary group is one based on "intimate face-to-face association and cooperation." The primary group par excellence is, of course, the enduring family unit; in combat, however, it has been found to be the infantry squad or section, or its equivalent. Donald Light, Jr., and Suzanne Keller, Sociology (New York: Alfred A. Knopf, 1979), pp. 191-2; and John T. Doby, Alvin Boskoff, and William W. Pendleton, Sociology: the Study of Man in Adaptation (Lexington: D. C. Heath, 1973), pp. 141-2. Gabriel and Savage erred in

describing the German infantry company a primary group. Gabriel and Savage, *Crisis in Command*, p. 36. The company unit is, in fact, a "secondary group" with "limited face-to-face interaction." Light, *Sociology*, p. 192.

94. Morris Janowitz and Edward A. Shils, "Cohesion and Disintegration in the Wehrmacht in World War II," in *Military Conflict: Essays in the Institutional Analysis of War and Peace*, ed. Morris Janowitz (Beverly Hills: Sage, 1975), pp. 178-82, 197, 216, and 218. This finding is supported by the comprehensive study, Samuel A. Stouffer et al., *The American Soldier: Combat and its Aftermath* (Princeton: University Press, 1949), pp. 135-6.

95. Richardson, *Fighting Spirit*, p. 20. Richardson, too, agrees with Janowitz and Shils that "in the last ditch . . . the soldier will be thinking more of his comrades in his section or platoon than of 'The Cause', Democracy, Queen and Country, or even--dare I suggest it?--of the Regiment." Ibid., p. 12.

96. Lord Moran, *The Anatomy of Courage* (London: Constable, 1967), pp. 171-2.

97. Seaton, *The Russo-German War*, pp. 219 and 352.

6. Cutting Edge of Battle

Second-Front Infantry

The Anglo-American infantry, which was projected over the seas during World War II to fight the Axis, did so under an expanding and finally unequaled mantle of air power. This state of affairs reflected to a large extent a traditional Anglo-American military approach in which, historically, well-supported expeditionary forces figured more prominently than mass armies. Incorporating the democratic model of the citizen-soldier, the Anglo-Americans further inclined toward the idea that technological solutions could be found to most military problems. Such thinking was certainly manifest in the military organization and tactical doctrines of the American and British land forces, both of which suffered from acute shortages of infantrymen yet at the same time maintained the highest "tail to teeth" ratios of all belligerent armies. There were, of course, important differences between the two English-speaking forces, the Americans generally adhering more to Continental-Clausewitzian doctrine than the British,(1) whose Great War experiences left them strongly reluctant to engage in costly battles of attrition. Still, the Anglo-Americans were bound, much more than separated, by a common language, history, and military tradition. These commonalities, as much as differences, were evident to some degree in the performances of their infantry arms.

Though the British Expeditionary Force (B.E.F.) was badly humiliated at Dunkirk, its infantry during the Battle of France and Flanders gave a reasonable, though not spectacular, account of itself. According to a report prepared by the German IV Corps, which had fought the B.E.F. from the Dyle River to the coast, "The English soldier was in excellent physical

condition . . . [and in] battle he was tough and dogged
. . . a fighter of high value . . . In defence . . .
[he] took any punishment that came his way."(2)
Coordination between arms was abysmally poor however,
the only successful check imposed on the Germans--the
British tank attack at Arras--ultimately failing for
lack of infantry support.(3) Important lessons were
nonetheless learned from the fighting in France,
notably the manner by which the "magnificently drilled
[German] Infantry, supported by weapons subordinated to
a single purpose, went forward in a resistless
combination of fire-power and rapid movement, infiltrating every position, making full use of stratagem
and cover, and constantly seeking to close and
destroy."(4)

After the evacuation, a serious attempt was made
to inculcate the rapidly expanding British army(5) with
new tactical ideas. The appointment of General Sir Alan
Brooke as Commander-in-Chief Home Forces, with General
Sir Bernard Paget his Chief-of-Staff did much to
accelerate change. Assisted by veteran generals like
Sir Harold Alexander and Bernard Law Montgomery,(6)
they began the process of regeneration. For the British
infantry, it represented an attempt to restore to that
arm a truly offensive capability, only this time one
based on small-unit fire and movement. A director of
infantry was accordingly appointed at the War Office
with a staff in every theater of war to study the
special requirements of infantry and ensure it proper
weapons. Divisional and general headquarters "battle
schools" were also established to set the new standard
of infantry training. The infantry, to quote General
Paget, was to become "the cutting-edge of battle."(7)

Though the 1936 reorganization of companies and
platoons was accepted as a reasonably solid starting
point, the revitalized tactical training of the British
army was presented as a totally "new look." The basic
lessons of the new tactical methods were taught as
"battle drills" at the special battle schools that had
been established. Designed to promote instinctive
teamwork and combat effectiveness, "battle drills" were
to have universal application; they were, in fact, an
expedient for creating a workable tactical framework,
the actual formulation of which Liddell Hart had urged
years previously. The requirement to have common
tactics for the infantry as a whole was not, of course,
intended to stifle individual initiative but to infuse
soldiers at all levels, and junior officers and NCOs in
particular, with the will to win.(8)

A cardinal assumption of the British "new look"
was that battles were won by the actions of small units
that continued fighting even when cut off from supply
or communication with higher headquarters. The first

stage of infantry instruction accordingly involved convincing trainees of this fact. Sections and platoons were further instructed that in both defense and offense, battle drills offered the surest hope of survival and success, instinctive reaction compensating for momentary bewilderment in the heat of battle. Thus, while individual fieldcraft skills continued to be stressed, every basic military action that could be was taught as a drill. There were drills for advancing, reacting to enemy fire, launching assaults, and breaching obstacles as well as drills for giving orders, cooperating with tanks, and organizing a defense or patrol.(9) Each drill aimed at saving time, streamlining battle procedures, and subsequently speeding up actions.

Significantly, all British infantry battle drills began with the section. Though it was emphasized that a section could rarely expect to operate by itself, the idea was nonetheless fostered that if a section happened to be left on its own, it should continue to fight independently. This, of course, smacked of Liddell Hart theory since he had always maintained that in British organization the platoon, not the section, was the basic maneuver unit:

> The fact that the section is the unit of command pre-supposes it to represent the largest number of men who can be controlled in action by a single leader. This in turn means that it is incapable of tactical sub-division and therefore is limited to frontal action.(10)

While the platoon was considered to be the maneuver unit, the rifle company was judged to be the smallest unit capable of conducting independent operations for short periods of time. At this level, and that of the battalion to which it belonged, the contradictory demands for speed of reaction and time for reconnaissance, planning, orders, coordination of fire support, movement, and assault were felt acutely. Company commanders had to learn, therefore, how to make judgments on the move just as their platoon commanders were obliged to do. Again, drills were stressed and a "battle procedure" laid down; time was saved, for example, by company commanders taking their artillery and mortar advisers with them when they went forward to reconnoiter and plan. Such new speedier methods were not always readily acceptable, however, as a conservative element with a decided bias toward deliberate methods of tactical preparation remained strong and influential within the British army. More haste and less speed became its countercry.

For whatever reason--failure to embrace the new

methods or a too rigid adherence to the drills taught--British tactical performance on the battlefield was often assessed in surprisingly uncomplimentary terms by the German enemy. Many of the reasons for later British failures in the offense they attributed to an "unwieldy and rigidly methodical technique of command . . . [and an] over-systematic issuing of orders, down to the last detail, leaving little latitude to the junior commander."(12) While praising the courage and toughness of the individual British soldier, one official German manual stated that:

> The British . . . [had] not yet succeeded in casting off . . . [their] congenitally schematic methods of working, and the clumsiness thereby entailed. This clumsiness continues down the scale of command by reason of the method of issuing orders, which go into the smallest detail.(13)

If basic British tactics were sometimes lackluster and uninspired despite emphasis on lower-level battle drill training, tank-infantry cooperation was to remain one of the Commonwealth forces' major coordination problems for much of the war. It was so serious, in fact, that there developed at one stage "throughout the Eighth Army . . . a most intense distrust, almost hatred, of . . . [friendly] armour."(14) "Second echelon" drills with armor were accordingly introduced at section and platoon level, their aim to teach the infantry how to take advantage of tank firepower and mobility and, in turn, how to best support tank troops. The inability of the British army to equip more than the motorized battalions of armored brigades with adequate vehicular lift, however, relegated the speed of most tanks in action to the pace of marching "lorried" infantry.(15) This naturally tended to focus attention on bridging the gap in mechanical mobility rather than on the essential requirements of tank-infantry cooperation.

There thus grew up in British circles the specious belief that once sufficient motorized transport could be made available to the infantry, most of the conditions of modern war could be met. Forgotten or ignored were the far more germane requirements for a common doctrine, a sound system of communication, and the absolute necessity of training together. In this respect, the war in the Western Desert was most misleading; resembling as it did a war at sea,(16) it reinforced the post-blitzkrieg British conclusion that the key to modern land operations must be large-scale tank-to-tank battles.(17) The place of the infantry was largely ignored except insofar as it required motorization. That motorization even in this area was only a small part of the British problem, however, was indicated by the apposite German comment that "British commanders were certainly quick to realize what was wrong, but motorization alone, however excellent, could not put the trouble right." What was required was a complete revamping of the system, including the retraining of officers and commanders and the reorganization of the command machinery.(18)

Despite this crucial weakness, the British army that assumed the offensive in Africa performed in superlative fashion against the "new Romans" of Il Duce. Many British shortcomings were obfuscated, of course, by the flash and heady flush of victory over what was unquestionably an inferior enemy. The Italian army, though it had trained its infantry through strenuous physical courses and even experimented with the formation of tough parachute units,(19) had neglected its military essence. Thousands of "human panthers" were consequently taken captive by a slightly more mobile British force. In the Desert Campaign of 1940, General Richard N. O'Connor's Western Desert Force of two divisions destroyed in 62 days ten Italian divisions, collaterally capturing a total of 130,000 prisoners. British casualties amounted to roughly 3,000, including 500 killed.(20) This dismal Libyan performance was slightly bettered by the Italian forces in Eritrea and Abyssinia.

According to German observers, the Italian army suffered from more than just inferior armament and equipment; the training and effectiveness of junior commanders was limited, and the Italian infantry, in addition to being basically nonmotorized, was poorly trained in the tactics of modern war. Field Marshal Albert Kesselring, while refusing to condemn the Italian soldier as naturally poor military material, nonetheless remained convinced that the entire "Italian armed forces were trained more for display than for action." The most soul-destroying aspect of the Italian army, however, and the one that most adversely affected morale was the all-pervading differentiation that existed between officers and men:(21)

> An Italian officer led a segregated life; having no perception of the needs of his men, he was unable to meet them as occasion required, and so in critical situations he lost control. The Italian private, even in the field, received quite different rations from the officers. The amount multiplied in ratio to rank. . . . The officers ate separately and were very often unaware of how much or what their men got. This undermined the sense of comradeship which should prevail between men who live and die together.(22)

The British were soon faced by a qualitatively different enemy, however, with the arrival of General Rommel and his Afrika Korps. In very short order, Rommel's aggressive panzer forces recovered all of Cyrenaica and beseiged the British-Australian garrison at Tobruk. The spirited resistance of this garrison, which tied up one German and four Italian divisions from April to November 1941, nonetheless demonstrated the formidable holding power of a well-prepared infantry defense. When the Axis attacked on May 6, 1941, they suffered about 1,200 men killed. In Rommel's opinion, this engagement made it "only too evident that the training of . . . [the German] infantry in position warfare was nowhere near up to the standard of the British and Australians." Being a firm believer that even "in the smallest action there are tactical tricks which can be used to save casualties" and that these "must be made known to the men," he instituted a crash training program in positional warfare for German infantry. In the small-scale infantry tactics that were subsequently prescribed, Rommel stressed "a maximum of caution, combined with supreme dash at the right moment."(23)

Obviously, even in North Africa where the tank ruled as "Queen of the Desert,"(24) there was a place for infantry skills. As Rommel saw it, the primary role of infantry was "to occupy and hold positions designed either to prevent the enemy from particular operations, or to force him into other ones." It was essential, however, that such infantry elements be mobile enough to get away and not become hostage to any particular location.(25) An imaginative method of employing standard infantry within a desert field force was, of course, amply demonstrated during Rommel's assault on the heavily mined Gazala Line in May 1942. While his unmotorized infantry divisions fixed the opposing British forces, his panzer elements did an end run. British attacks against his bridgehead, called the Cauldron, failed due to a lack of infantry and the startling effectiveness of German antitank guns, particularly "88s." In the ensuing German attack on Tobruk the infantry of "Group Menny," supported by artillery and Stukas, successfully assaulted through minefield gaps prepared by engineers to effect a breakthrough.(26)

In every battle from mid-1941 onward the British outnumbered the Axis both on the ground and in the air; at Alam Halfa and El Alamein, this favorable disparity was even greater. In these battles, also, the British command was more in its element--that of static warfare and deliberate tactical methodology. The final breakthrough by General Montgomery was from the beginning spearheaded by infantry; its aim was to effect the methodical destruction of the enemy infantry sustaining the Axis defensive system, in which circumstance Axis tanks could not stand idly by watching. Rommel was nonetheless allowed to escape and run. The British method of waging war, recorded General Mellenthin, was "slow, rigid and methodical"; in fact, he did not "think that the British ever solved the problem of mobile warfare in open desert." Afrika Korps victories, on the other hand, he attributed largely to the principle of "Co-operation-of Arms" and to the German panzer division being a "highly flexible formation of all arms."(27)

To be sure, Rommel made masterly use of his antitank weapons during his Libyan campaign of May-June 1942. His tanks were followed into battle by his antitank guns, so that if the former encountered strong resistance, they could retire under cover of the guns. In turn, if the guns were threatened, the tanks could support them by advancing against the flanks of the enemy. Cooperation between arms was therefore intimate; the balance, optimum. Interestingly, the principal cause of disproportionately heavy losses of tanks on the British side during the Desert War was the German 50-millimeter antitank gun, which had replaced the 37-millimeter. Comparatively small and handy, it was employed aggressively in both attack and defense. The British for the longest time did not know that they were being shot up by 50-millimeter guns, ascribing their effect, in fact, to the more visible tank.(28)

Significantly, the British in 1942 reorganized their armored divisions to include more infantry. Formerly comprising two armored brigades, each with an integral rifle battalion, they were redesigned to compose but one armored brigade and an infantry brigade of three battalions. By this action, infantry battalions in a British armored division now outnumbered armored regiments (battalions) four to three, the total number of tanks dropping from 368 to 188.(29) The reason for the increase in infantry strength was partly inspired by the example of the German panzer operations and partly in anticipation that the undulating countryside of Tunisia would warrant a higher ratio of infantry.(30)

The bitter slugging match in Tunisia presaged the more difficult combat up the rugged boot of Italy and the fighting in the bocage country of France. Infantry independent action and prowess suddenly became much more important, and many of the methods of the desert had to be discarded. Antitank defense, of course, remained a most vital concern; whereas positions in 1940 had been sited around machine guns, battalion commanders now superimposed on an antitank plan-- usually established for them--all of the defensive means at their disposal. As forces were rarely isolated to the extent they were on the Eastern Front, even brigade commanders were commonly obliged to follow the same pattern of adherence to a higher-level antitank plan. Unit and small-unit battle options thus became more and more limited.(31) Strictly speaking, this process tended to suit the British method of employing infantry, which was to have it specialize in direct aimed fire at company level and below. This was reflected in British infantry company organization that, with a strength of roughly 125 and no major organic support weapons, was one of the smallest in the Second World War.(32)

In Tunisia, where hill and valley tactics supplanted those of the desert, the British infantry had to learn the hard way that training in the minor details of war at the lowest levels is important. During this campaign, many platoon commanders were unnecessarily made casualties at first by being too eager and moving right up with their leading sections. It was eventually determined that a platoon (and company) could be fought much better, once the source of enemy fire had been located, by the commander remaining in a position from which the maneuver of reserve sections (and platoons) could be controlled. There was at first also a tendency for platoons to rush into an attack, though this was not always the platoon commander's fault. Such action was soon discovered to be casualty intensive, and "really thorough" reconnaissance was recommended as the only solution: it was absolutely vital to know the exact location of the cunning and often invisible enemy. In carrying out an attack, the "pepper pot" method of moving forward by small group fire and maneuver was advocated. It was only suggested, however, because German positions were invariably mutually supporting and therefore difficult to outflank.(33)

A coincident British comment on defensive tactics was that owing to emphasis on the offense, they were not covered in enough detail in battle schools. Consequently, the Germans "were [considered] much cleverer at it" than the British. Deep and thorough digging (down to five feet) was accordingly advocated and practical tactics urged as follows:

> However huge an area of country you are given, in placing your troops imagine you have only three-quarters of your platoon. Put your spare quarter aside as a mobile reserve; then forget all the books and put the rest wherever your common sense and your knowledge of Boche habits tells you. Whenever possible, you want to be on reverse slopes--any movement on forward slopes brings the

shells down, and it is not easy to stay still all
day. If the ground forces you to take up forward
slope positions, keep the absolute number at
battle posts to observe, and the rest in cover
until you are attacked. It is then that your fire
control comes in. The first time, unless you drum
it in daily, everyone will blaze off at any range
at the first Boche to appear, giving all your
positions away. It is much more satisfying to let
the Jerries come up a bit and catch them in
numbers on some open patch. If by chance they
knock out one of your posts and start getting in
among you then you thank God for that quarter you
kept in reserve and nip in the counter-attack
straight away. If you have got a counter-attack
properly rehearsed with supporting fire, etc., for
each of your posts, you should be able to get it
in almost as soon as they arrive, or better still,
get them in a flank as they advance.(34)

The foregoing pattern of small-unit defense is striking in its similarity to that recommended in theory by Liddell Hart in 1920 and, of course, in its essential features by the Marshal de Saxe many years before. It may also have been copied from the contemporary German enemy.

Interestingly, platoon commanders were further advised to familiarize themselves with antitank mines, which were plentiful in North Africa, and to train their men to lay them. Waiting for the Royal Engineers to place them was considered most imprudent. Small-unit defense was additionally tailored to counter the peculiar night-fighting methods of the enemy. German patrols had a habit of lashing positions with indiscriminate automatic fire, hoping to draw British return fire. The correct training response was to condition the soldier to hold off until a positive target could be identified and then aim to kill. It was totally foolish to pursue German patrols, as they normally retired under cover of mortars specifically registered to engage pursuers. Mortars with their uncanny accuracy and killing power were the most feared of Axis weapons.(35)

The American experience in North Africa against the common enemy was much the same as that of the British, though U.S. fighting methods and organizations, being different in many respects, were dissimilarly affected. Because the Americans have not been discussed in any detail to this point, however, it is considered advisable to digress briefly to examine them in a depth equivalent to that accorded the British, German, and Russian infantries. To be sure, American arms were distinctly developed and conditioned in their own particular schools of battle experience, starting with the Revolution. Though springing from basically Anglo-Saxon military roots, they were heavily influenced by Continental military doctrines. The French-Jominian model, especially, served both sides in the American Civil War, an epic conflict that had a singular, unique effect on the subsequent exercise of American arms.(36)

Battle experiences as a regimental commander in the Civil War were responsible for prompting General Upton to produce, in 1867, a new system of loose-order infantry tactics based on movement and organization by two ranks of fours. With the introduction of this "double rank" formation, he, in effect, created the distinctly American eight-man infantry squad, although it was not so called at first. The trenches of the Great War with their demand for specialist combat skills, however, forced the American army to depart from its traditional concept of open warfare. The 1917 basic infantry "squad" of one corporal and seven riflemen was thus replaced by the more tactically relevant and specialized "section." Again, the French army served as a model.(37) By a table or organization in February 1918, the strength of a platoon was established at 53, divided into four sections of respectively, 12, 9, 17, and 15 specialists (that is, hand bombers, rifle grenadiers, riflemen, and automatic riflemen). In other variations, and there were many ad hoc arrangements, two squads made up a section, or "half platoon." Although the squad was submerged tactically within the section in any case, it appears to have been retained for administrative and movement purposes. The 1919 regulations sought to rationalize the Great War loss of the identity of the infantry squad in the following manner:

> The section leader guides his unit. He looks at it only when his exercise of control demands it--his eyes should be fastened upon the enemy. The section must be bound to its leader, who, under all circumstances, is the rallying point. The squad leaders maintain the positions assigned to them and see that the platoon and section leaders' orders are executed. They transmit the commands and signals when necessary, observe the conduct of their squads, and assist in enforcing fire discipline. When the ability of platoon and section leaders to control the actions of their units ceases, squad leaders lead their squads on their own initiative, lending each other mutual support.(38)

Following the Great War, the American infantry

battalion was "triangularized"; one rifle company was eliminated and replaced by a machine-gun company, the remaining rifle companies being reduced from four to three platoons. This restructuring as well as traditional preferences ensured a return to a squad-based organization. From 1920 to 1932, the American army infantry squad largely consisted of a corporal commander, a two-man automatic rifle team, a rifle grenadier, and four riflemen. Shortly after this period, another change was effected that served to highlight the difference between the American and European approach. With the introduction of the semiautomatic M-1 Garand rifle from 1934 onward, it was thought that the infantryman's dependence for advancing on the covering or supporting fire of automatic rifle teams would be reduced. By 1938, the superior rate of fire of the Garand (from 10 or 15 aimed shots to 20 or 30 a minute) had resulted in the elimination of the automatic rifle from the infantry squad. An automatic rifle section was instead created within each platoon. Thus, whereas in Europe infantry firepower was increased principally by augmenting the number of light machine guns, in America the increase tended to come from the faster-shooting shoulder arm of the individual rifleman.(39)

It was generally recognized, however, that in comparison with foreign armies the American infantry around this time was in "an obsolete condition in almost every particular: armament, organization and tactics." In subsequent review, beginning with the fundamental study of the infantry squad (thanks to the tactical good sense of the Chief of Army Ground Forces, Lieutenant-General Lesley J. McNair), an entire divisional organization was put together piece by piece. It was the most thorough examination the American infantry had ever received. Sections and brigades were in the process eliminated as "triangularization" was applied as a principle of organization at battalion, regimental, and divisional levels.(40)

Under this 1939-40 reorganization, which recognized that the eight-man squad had not been large enough to absorb casualties and continue action in the Great War, the American infantry squad was given a war establishment of 12. The automatic rifle, the Browning 1918 A2 (BAR), was by 1942 again included within the rifle squad and the automatic rifle squad disbanded. This was the basic organization with which the American army went to war in North Africa. The squad consisted of eight M-1 riflemen (including a more experienced sergeant squad leader and corporal assistant squad leader from late 1940 onward and two scouts), a three-man BAR team (two with M-1 rifles), and a sniper

166 A PERSPECTIVE ON INFANTRY

TABLE 6-1. AMERICAN INFANTRY SQUAD ORGANIZATION

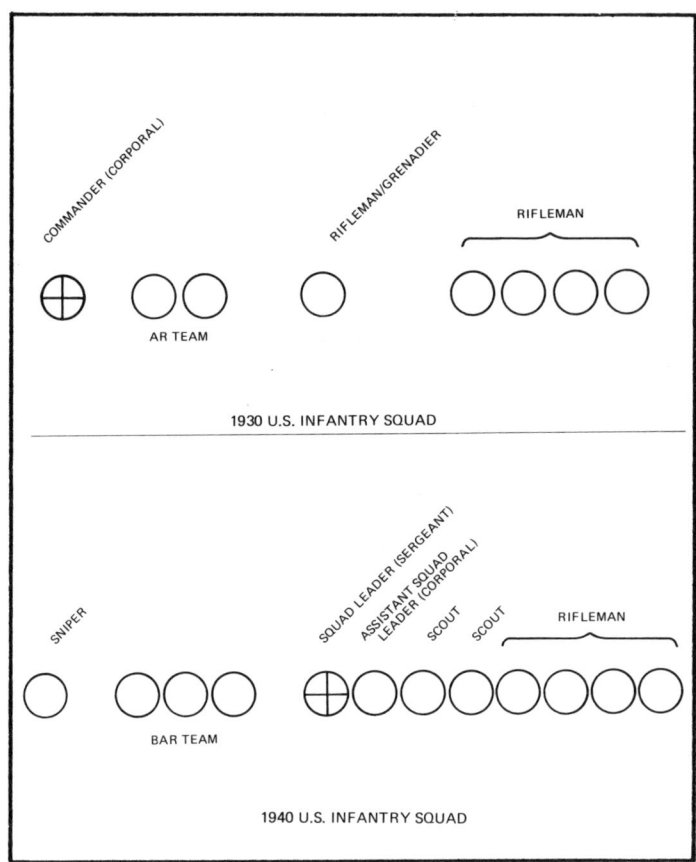

equipped with an M1903 Springfield. Three such squads formed a platoon, three of which plus a weapons platoon, in turn, formed a rifle company of about 198 all ranks. The weapons platoon of a rifle company contained two .30-caliber air-cooled "light" machine guns and three 60-millimeter mortars; in 1943, it lost two automatic rifles but gained three 2.36-inch

TABLE 6-2. AMERICAN INFANTRY REGIMENTAL
ORGANIZATION (1942)

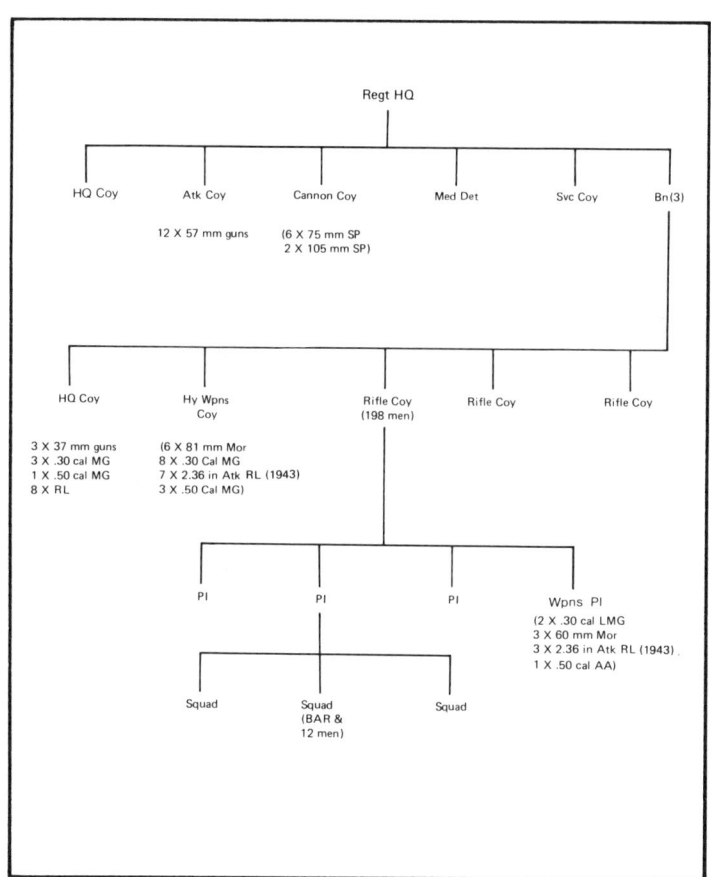

"bazooka" antitank rocket launchers and a .50-caliber machine gun for antiaircraft defense.(41)

Three rifle companies were included in an infantry battalion along with a heavy weapons company that contained six 81-millimeter mortars, eight .30-caliber water-cooled machine guns, and, by 1943, seven antitank rocket launchers and three .50-caliber machine guns.

Three battalions made up an infantry regiment, which also comprised an antitank company of twelve 37-millimeter (later 57-millimeter) antitank guns, a cannon company of six self-propelled 75-millimeter howitzers and two self-propelled 105-millimeter howitzers (in 1943 it was abolished and replaced by three cannon platoons of six short-barreled, towed 105-millimeter howitzers), and a service company. Three regiments plus supporting arms made up an infantry division, the triangular concept having been accepted throughout the American army.(42)

At this point, it should be mentioned that infantry serving with American armored formations was eventually organized somewhat differently. Unlike foot infantry, not organically motorized, and motorized infantry equipped to move in trucks, "armoured infantry" moved in self-contained fashion by White M-3 half-track carrier. At first, the armored infantry squad was organized (by July 1940 establishment) as a standard peacetime infantry squad of one corporal commander and seven riflemen, the BAR having been eliminated by this time at rifle-squad level. The March 1942 organization of the armored division nonetheless called for an armored infantry squad war establishment of 12 men, to be commanded by a sergeant, with a corporal his assistant squad leader. The automatic rifleman and his assistant were dropped, however, as a .30-caliber detachable machine gun was mounted on the carrier. With the addition of a driver, the armored infantry squad thus comprised 11 all ranks, eight of whom were riflemen armed with the M-1 (A 2.36-inch rocket launcher was later added in April 1943.) The armored infantry normally fought dismounted, sometimes supported by their vehicular firepower. Otherwise, there was little difference in their tactical approach from that of the standard infantry.(43)

The expansion of the American military was indeed a monumental feat of ingenuity and organizational efficiency. Eighty-nine relatively large army divisions were sustained in the field, as well as six marine divisions, a gigantic strategic air force, and a huge two-ocean navy.(44) Starting from an initial strength of 456,000, the combined services of the American army, navy, and marine corps between June 30, 1940 and June 30, 1945 grew to an overwhelming total strength of 11,857,000--68 percent of which was army.(45) The American infantry from 1941 to 1943 increased by 600 percent.(46) To German officers, such as Rommel, it was a miraculous mobilization and an everlasting tribute to the American "extraordinary sense of the practical and material and . . . complete lack of regard for tradition and worthless theories."(47)

Notwithstanding these adulatory remarks, however,

many serious problems were encountered in raising a force of such magnitude in so short a time frame. Although certain progressive methods were employed in organization and training, overall quality did begin to suffer as the system came to resemble a gigantic "sausage machine," the very situation that General Beck had worked so hard to avoid in the expansion of the Reichswehr. Individual attention was at a premium as infantry range practices were conducted on a continuous basis, with group upon group passing through the hands of dedicated cadres. Huge classes, for weapons training and other subjects, were necessitated because of a shortage of qualified instructors. Without question, this pointed to the most critical problem experienced by the American army at this time, namely, the lack of physically fit, dynamic, trained leaders, particularly at the junior level.(48) In 1941, General George C. Marshall reported that junior officers lacked experience, had little confidence in themselves, and hence failed to assume or discharge their responsibilities. Not surprisingly, field exercises and maneuvers continued to show glaring weaknesses in basic and small-unit training.(49) Although a return to fundamentals was urged, these weaknesses could not be kept from the battlefield. As late as November 1944, General Patton would rage, "Our chief trouble in this war is the lack of efficiency and lack of sense of responsibility on the part of company officers."(50)

It is perhaps for these reasons that American infantry manuals on minor tactics were published in such exhaustive detail. Though not, in principle, dissimilar from battle-drill publications of the British infantry, they offered many more variations of situation and operational responses.(51) Although the influence of Liddell Hart's infantry theories could be perceived,(52) references to old-style infantry tactical methods still appeared: "the reserves of small units . . . [being] called supports," for example, those of a larger force "the general reserve." A rifle company, ordinarily employed as part of its battalion, might participate in a "decisive" or holding attack, but the normal company attack was considered to be "frontal, except as interior or local flanks of the enemy result from the penetration of 'soft spots'."(53)

The American small-unit infantry attack was divided into three principal phases: the advance in route column, the deployment (including the approach march), and the attack proper. Infantry attack formations were many and varied, any rigidity being proscribed. In crossing the "line of departure," signifying the commencement of the attack phase, all units in the front line were to deploy "as skirmishers," a loose, irregular, squad-based formation, and open fire. During the deployment phase, squads had the option, depending on the terrain, of adopting a column formation, column opened out, or diamond formation. Within infantry platoons, commanders had a choice of: "column of threes"; "column of twos"; "platoon column"; "line of squads"; or "line of squads echeloned." In the attack itself, platoons were usually allocated "zones of action" within a company frontage of from 200 to 500 yards. In the attack, the company commander was to post himself where he could best control the battle, if possible from an observation post; once fire began, his influence was thought to be largely limited to the employment of support fire from the weapons platoon, the three 60-millimeter mortars of which normally remained under his control.(54)

The value of fire and maneuver in conducting an attack was constantly stressed, although in the assault these tactics were recommended "only as necessary to advance," in which case the squads moved forward by rushes. In theory, each squad had a two-man scout (Able) team, a four-man BAR (Baker) team, and a five-man maneuver and assault (Charlie) team. The squad leader was normally with Able until the enemy was located and fixed; after making his plan, he then signaled Bravo to provide fire support while he led Charlie in an attack by short rushes. The problem with this system was that the squad leader was too often pinned down with Able, while two or three casualties within the squad rendered the entire arrangement impractical. Many other solutions were proposed and tried, with and without the sanction of higher authority.(55) Some, like Colonel J. C. Fry's 350th Infantry of the Seventy-seventh Division in Italy, employed an assault battle drill for squads (divided simply into fire teams and maneuver teams) and platoons; the fire and maneuver stressed, however, had a definitely frontal bias up to and including platoon level.(56) Others, like General Patton, advocated a system of "marching fire" in which infantry, supported by every larger weapon available, went forward in a skirmish line firing at anything that appeared to be capable of containing enemy. Patton thought it invited disaster to employ a unit smaller than a battalion against an active enemy.(57) He was much opposed to "rushing" tactics:

> Today, when the chief small-arms fire on the battlefield and the majority of the neutralizing fire is delivered by machine guns, mortars, and artillery, there is no advantage in advancing by rushes, because, until you get within three hundred yards, small-arms fire has very little effect, whereas when you lie down between rushes,

you expose yourself to the effect of shrapnel.
When you get to three hundred yards, your own
small-arms fire, which is superior to anything now
existing [the M-1 rifle] . . . , will neutralize
that of the enemy small-arms fire, so that you do
not have to advance by rushes. I say this very
feelingly because I have seen, on many occasions
in maneuvers and in battle, troops advancing by
rushes when they were defiladed behind hills and
could have gone forward in limousines, had they
been available, with perfect impunity. The proper
way to advance, particularly for troops armed with
that magnificant weapon, the M-1 rifle, is to
utilize marching fire and keep moving. This fire
can be delivered from the shoulder, but it can be
just as effective if delivered with the butt of
the rifle halfway between the belt and the armpit.
One round should be fired every two or three
paces. The whistle of the bullets, the scream of
the ricochet, and the dust, twigs, and branches
which are knocked from the ground and the trees
have such an effect on the enemy that his
small-arms fire becomes negligible.(58)

 The strength of Patton's methods, which,
incidentally, also called for firing .30-caliber
machine guns and automatic rifles from the hip, was
that maximum firepower was projected forward with
substantial shock effect. Resembling in many ways the
mass wave tactics of the Great War, however, they had
the weakness of being extremely vulnerable to counter-
action and machine-gun fire along fixed lines. Troops
so trained were, furthermore, less able to take
advantage of terrain factors and organic teamwork than
those trained in fire and maneuver tactics. Nonetheless,
given the rudimentary trained state of many American
inductees and their leaders, plus massive fire support
from aircraft, tanks, and artillery, such tactics
appear to have worked reasonably successfully. In
general, then, American infantry companies tended to
sweep along in accordance with factors beyond their
control, their tactics often dwarfed and twisted by the
presence of more powerful weapons. Although it usually
depended on cooperating artillery support more than
other national infantries, the American infantry did
become surprisingly battle effective.(59)

 As the American army was primarily organized for
offensive action, rifle-company defensive tactics were
not given equal emphasis before combat actually began.
There therefore developed a tendency to grow in
armament and to expand protective diggings at the
expense of concealment. It may have been for such
reasons that Patton forbade digging in until the final

objective was taken, as once troops dug in, they could not be "restarted."(60) Basically, a company in the front line was disposed in a resistance echelon and a support echelon, usually meaning two platoons forward and one in depth. Within a defensive area, squads were never split up; often, because foxholes were sometimes sited in two lines, they contained part of a platoon's depth in themselves. A company defensive area normally covered a frontage of from 400 to 700 yards and was sited in depth anywhere from 150 to 450 yards. Great emphasis was placed on mutual supporting fire, particularly that of machine guns, around which framework--along with antitank weapons--the defense was built. A rifle company, though responsible for its own immediate local security, was not ordinarily charged with responsibility for antitank defense, other than for local protection. Significantly, though the advantages and disadvantages of reverse-slope positions were known, the "most favourable position" was regarded as that "usually found on the forward slope of commanding ground, at or below the military crest."(61)

Some of the teething problems the American army first experienced in Tunisia were mentioned in a report by General McNair, commander of the Army Ground Forces:

> My study of operations in the North African theatre particularly, by both observation on the ground and from reports and dispatches, convinces me thoroughly that the combat forces there are too much concerned with their own security and too little concerned with striking the enemy. The infantry is displaying a marked reluctance to advance against fire, but they are masters of the slit trench--a device which is used habitually both in defense and attack. Regimental and higher commanders are not seen sufficiently in the forward areas, and battalions show the lack of first-hand supervision. Commanders are in their command posts. I found that infantry battalions in the assault have their command posts organized in forward and rear echelons, the former the stronger. . . . I maintain that our organization must be an offensive one, not cringingly defensive. We cannot provide thousands of purely defensive weapons with personnel to man them without detracting from our offensive power. Nothing can be more unsound than to provide a headquarters guard organically for a high command post. If a commander feels so much concern for his own safety, let him withdraw a battalion from the frontline for his own protection, but do not provide him with such a unit organically.(62)

With the invasion and conquest of Sicily, July-August 1943, the American ground forces attained a combat maturity and prestige equal to that of the British. The topography of that island was so favorable to the defense, however, that tortuous outflanking maneuvers were normally required to deal with German-Italian holding actions. In the steep-sided valleys and trackless hills, there was no way to mass and bring to bear the superior tank and artillery strength of the Allies. The struggle for Sicily became, in fact, a host of small-unit infantry engagements in which, ultimately, the Germans and Italians distinguished themselves more than either the British or Americans. With somewhat less than four divisions, little air strength and practically no naval support, the Germans opposed for five and a half weeks an Allied force of more than 12 divisions, enjoying absolute superiority in the air and on the sea.(63) Fortunately for the Allies, their initial landing went largely uncontested as they achieved almost total tactical surprise. Had they been forced to fight, it might have been difficult, for most of their infantrymen waded ashore grossly overburdened by their staffs, carrying "more on their bodies than could have been expended in the most desperate firefight."(64)

The fight for possession of the long boot of Italy was to be an even more wearisome slugging match than the battle for Sicily, which had been won primarily by the "resolution, endurance and fine fighting qualities" of the Allied infantry.(65) The "soft under-belly" of Europe turned out to be anything but soft geographically, and the struggle became a matter of "one more river, one more mountain." It was also a struggle in which a French force under General A. P. Juin, using only a few vehicles and pack animals, actually moved far faster in identical terrain than vehicle-clogged British divisions. Contrary to common expectation, modern equipment had not necessarily made an army more mobile. Road and weather conditions made it patently obvious once again that there is an everlasting difference between physical and tactical mobility. In Italy, during the winter months, armor could not maneuver off roads, the unceasing rains turning the countryside of "sunny Italy" into a quagmire. Most offensive action, therefore, had to be effected by infantry, sappers, and artillery, not always in sufficient numerical superiority. The battlefields of Italy soon began to bear a striking resemblance to those of the Great War, the conditions of the Somme, Verdun, and Passchendaele being in large part recreated in the battles of Anzio and Cassino.(66)

Savage city fighting, though on a lesser scale than that of the Eastern Front, also marked the Italian campaign. At Ortona, the First Canadian Division became locked in a costly "street-fighting" battle with the German First Parachute Division. Choosing to defend only the north part of the town but leaving the southern half a nightmare of trapped and mined houses, the Germans made its defense into a miniature Stalingrad of interconnected and heavily mined strongpoints. Here the <u>Faustpatrone</u>, or <u>Panzerfaust</u>,(67) an expendable infantry antitank weapon, made its appearance. Here, also, the Canadians developed an improved method of "mouse-holing," a technique taught in battle schools from 1942 on by which manner breaches were blown in house dividing walls with "beehive" demolition charges so that troops could clear rows of houses without once having to appear in the dangerous streets. The enemy had first hit upon this technique for defensive purposes, however, and captured buildings consequently had to be occupied in strength to prevent the Germans from recapturing them by infiltrating back through their own concealed "mouseholes." Such practices naturally resulted in extremely close quarter combat, a condition amply attested to by the fact that the frontage of Second Canadian Infantry Brigade was not more than 250 yards. Significantly, the Canadians did not break down their standard tactical organizations at "Little Stalingrad," as Ortona came to be called. It took eight days and 2,605 casualties to take the town, population 10,000.(68)

The amphibious landing at Anzio in January 1944 was intended to provide a tactical alternative to the tedious and costly advance of the Allied armies. Unfortunately, the Anzio bridgehead with its 70,000 men and 18,000 vehicles (one for every four soldiers) became hostage to German arms. Like a beached whale, this "army of chauffeurs" had to be rescued, which operation, in turn, involved taking the town of Cassino. Four major battles and over 8,000 casualties were eventually required to win this Somme-Passchendaele-type victory.(69) Again, it was primarily an infantry struggle waged in acres of rubble, from reverse slopes, caves, and concrete bunkers, with the mortar the most effective weapon. The "sangar," or breastwork, sprang up in place of the trench since the ground was so rocky it was difficult to dig. Tanks were very much restricted to a supporting role in the rugged terrain surrounding Monte Cassino.(70)

In the professional judgment of Fuller, the battle for Cassino was based more on "perspiration" than "inspiration." Noting that General R. Y. Malinovsky's Russian soldiers had been able to storm strong enemy positions by night with "cold steel and hand grenades" and without artillery support, he queried the wisdom and tactical value of subjecting Cassino to massive artillery and air bombardments "unsurpassed in the history of warfare."(71) As at Passchendaele, the effect of such "weight of metal" was to create an almost insurmountable obstacle to the movement of attacking Allied forces. It was a return to the unimaginative tactics of the Great War in which "artillery conquered and infantry occupied." Not surprisingly, the enterprising German infantry defending Cassino survived, in some instances by fitting up to six of their number at a time into two-man shelters called "crabs."(72) Unhappily for the Allies, on their reappearance after the bombing and shelling, the German infantrymen took full advantage of the tactical truism that "it is only when buildings are demolished that they are converted from mousetraps into bastions of defence."(73) The defense of Cassino by the infantry and paratroopers of the Wehrmacht was truly an epic tactical achievement.(74) The position itself was eventually taken by an outflanking maneuver.

For the Anglo-Saxon powers, the lesson was slowly driven home that, even in modern combat, infantrymen were often more relevant than either tanks or airplanes. To be sure, the proposal to ship the Fifth Canadian Armored Division to Italy drew from General Alexander the remonstrance that "we already have as much armour in the Mediterranean as we can usefully employ in Italy. I should have preferred another Canadian Infantry Division." It was also during the campaign in Italy that the Allies first began to experience critical shortages of infantry. By the end of 1943, every one of the rifle companies in the First Canadian Division had suffered 50 percent casualties.(75) Though the Canadians, volunteers all, resorted to a remustering policy to get more infantrymen, by 1944, the average company strength of some line battalions had fallen to 45. The shortage of trained infantrymen (few reinforcements arrived in units with a knowledge of elementary platoon and section tactics and some without having fired the Bren) eventually became so serious that it precipitated a political crisis within Canada itself.(76)

The British, faced with a similarly acute shortage of infantry, were forced to break down two divisions, though this measure solved only part of the problem.(77) The short of it was, they had been caught out; North African theater "rates of wastage" used by the War Office were simply inapplicable to either the Italian or North-West European campaigns. Whereas in North Africa during intense combat the infantry suffered 48 percent of total casualties (compared to 15 percent for the armored corps and 14 percent for the artillery), in Italy and Normandy, the infantry arm incurred 76 percent of all casualties, the armored and artillery accounting for but 7 and 8 percent, respectively.(78) The Americans calculated that during intense combat one of their infantry divisions suffered about 100 percent losses in its infantry regiments every three months. Though automatically replaced under the American system, the division normally suffered disastrously from the standpoint of efficiency with such a higher turnover in infantrymen. Significantly, by the first weeks of 1944, the American army shortage of infantry replacements had reached crisis proportions.(79)

The replacement requirement for General Patton's Third Army in that year reached 9,000, the average rifle company being at but 55 percent of its established strength. Patton's solution was characteristically radical. As no replacements were in sight at the time, Patton withdrew 5 percent of his corps and headquarter troops to train as infantry. This supposedly produced "loud wails" from many of his section heads who declared that they "could not run their offices if any cut were effected." Patton later wrote, with obvious delight, that, "As a matter of fact, even the ten per cent cut which we subsequently made had no adverse effect." He maintained that had such action been taken by General Dwight D. Eisenhower with troops of the Communications Zone, "we would have had enough soldiers to end the war."(80) As it was, however, less positive actions were taken. In an attempt to boost the sagging morale and prestige of the infantry, the rifle-squad leader was upgraded from sergeant to staff sergeant and the assistant squad leader from corporal to sergeant.(81)

Ironically, while infantry shortages plagued the Allied armies, their logistic trains remained virtual cornucopias of men and material. In this regard, it is extremely interesting to compare the "divisional slices" of Allied, German, and Russian armies. A rough and ready method of measuring the efficiency of army organization, the "divisional slice," is determined by dividing the total number of men in an army by the number of divisions it fields. Simply put, it is the total number of personnel required to man, supply, and keep a division in action. For the Canadian Army, the figure was a weighty 93,150; for the British Army, 84,300; and for the more affluent United States Army, a lower 71,100. A certain interpolation is, of course, required before a meaningful comparison can be made since Canadian divisions with strengths of 18,376 (infantry) and 14,819 (armored) were larger than their American equivalents of 14,037 (infantry) and 10,670 (armored). Still, whatever the measurement applied, no nation needed so many men to keep one man in action as did Canada. Taking the fighting arms all together, in the Canadian Army they made up but 34.2 percent of the whole, while in the American and British armies, they constituted 43.5 and 65.3 percent, respectively.(82)

By contrast, the German army divisional slice was roughly 23,000, based on an average divisional strength of 12,000. This meant that a Germany of 85 million inhabitants could mobilize successively 325 divisions, while an America of 140 million could barely maintain 89 divisions. The Soviets, with divisional slices of 22,000 and a population exceeding 170 million, were able to field better than 500 divisions with average strengths of about 10,300 each.(83) According to Colonel Nicholas Ignatieff, a Russian-born Canadian serving with British intelligence, it took seven noncombatants to keep one man fighting with the British Army, whereas in the Red Army the ratio was little more than two to one. He was particularly struck by the spectacle of one rain-drenched Soviet soldier who arrived on the banks of the Elbe River near Torgau, Germany:

> He was wearing a filthy bandage on his head and he was riding on the back of an ox, which was pulling a small ammunition truck, which had run out of gas. He maintained, with the sense of romance common to all soldiers of experience, that he had been riding the ox and pulling the truck all the way from Stalingrad, and if he hadn't been, the Allies might still be in fairly serious trouble. I asked him what he intended to do with the ox now that the war was nearly over. "Eat it," he said, and there can't be the slightest doubt that he did.(84)

The tendency toward oversupply in the Anglo-American armies was literally carried into action. The litter that reportedly accompanied the landing on the beaches of Normandy was caused less by human slothfulness than by slothful thinking. It was also such thinking, or lack thereof, which cost so many infantry lives on the bloody day of assault. Most infantry in the leading waves were, in fact, criminally overloaded. The American soldier carried more than 80 pounds, and any careful examination of photographs of British and Canadian troops waddling ashore on that day will reveal that they, too, were weighted down with roughly the same load. The American case will be considered more specifically, however, as it subsequently became the subject of a remarkably comprehensive study by the able army historian, S. L. A. Marshall.

One irrefutable fact established by this study was

that most wounds received by soldiers in the water on D-day ultimately proved to be mortal, as the weight of their equipment dragged them under and they drowned. Wounds received on the beach close to the water line also led to death in most cases since many soldiers--weighted down by water and kit--lacked even the strength to keep ahead of the tide, which moved inland at the pace of a man in a slow walk. Out of a total of 105 men lost on D-day by Company E, Sixteenth Infantry, which landed on Omaha beach, only one man was killed from the top of the beach inland; the rest perished in the water. It is indeed sadly ironic that many of these soldiers carried on their backs several cartons of cigarettes (the killing kindness of a concerned welfare officer no doubt) and sufficient rations for three days. Subsequent surveys showed that in the excitement of action most men did not even eat during the first day of fighting.(85)

The D-day assault definitely showed that a direct relationship exists between an infantry soldier's tactical performance and the load he carries on his back. Though obviously hampering to movement, the real curse of the overload is that it kills fire at the fire base, wasting soldiers who might otherwise be good fighting men. A vicious cycle is invariably set in motion since tired men take fright more easily and frightened men tire more quickly. The load carried by a soldier on a hard road during training is quite likely to be too heavy for him under fire, for at that point he is not the same man. The soldiers of Company E had but 250 yards of beach to cross when they went to ground, yet it took them one hour to negotiate this distance; even the most seasoned veterans among them lacked the strength to rush forward under the weight of their packs. On the beaches of Normandy, the fighting infantryman had absolutely no requirement whatsoever for an "axe in case he . . . [had] to break down a door."(86) He should have been as lightly equipped and fleet as an athlete.

Unfortunately, while many officers throughout history have claimed this principle to be axiomatic to infantry operations, they have nonetheless tended to apply it more to the loading of mules than to the loading of soldiers. The fact is--and it was demonstrated years earlier during German marching trials conducted by the Institute William Frederick--that no amount of training will ever condition an infantryman to carry excessive weight.(87) Nor is it a sensible staff solution to suggest that the innate common sense of the soldier will invariably cause him to discard unnecessary items after the shooting starts. On the beaches of Normandy, this was impossible, and the combat line consequently foundered under the weight of bangalore torpedoes that were never exploded, gas equipment that was never used, and ladders that would have been useful had there been a cliff.(88) Commanders at all levels may well have permitted their men to be killed by trying to protect them against all contingencies.(89)

After successfully breaching the Atlantic Wall of Fortress Europe in one eventful day, the armor-heavy Allied armies received their next shock in the bocage country of Normandy. Composed of small fields delineated by ditches and thick mounds of earth high as a man's head and crowned by trees and thickets whose roots had been undisturbed for years, the bocage was interconnected by numerous narrow roads and dotted with tiny hamlets.(90) It was anything but ideal for armored warfare; in crossing roads and climbing banks, tanks invariably heaved up, thereby dangerously exposing their soft undersides to enemy antitank fire. The deadly, short-range infantry Panzerfaust proved an extremely grave threat to any tank caught in such a posture. The "battle of the hedgerows" that subsequently developed thus became primarily an infantry contest, with armor in a supporting role. Strategy was therefore once again forced to wait upon the dictates of tactics.

In the struggle in the bocage, Allied infantry that had fought elsewhere found that it had to modify its tactics. For example, the Fiftieth (Northumbrian) Division discovered that normal battle drills for the attack were unsuitable for dealing with German defenses in the hedgerows. Sand was accordingly brought up from the beaches, and through improvising sand models of the countryside, the division was able to devise a new drill for attacking through bocage. Taught to all platoon commanders, it was credited with saving many lives.(91) The American reduction of Cherbourg offered a further example of the close-combat skills required by infantry in eliminating pillboxes and strongpoints:

> Dive bombers and artillery drove the defenders in the outer entrenchments to seek the shelter of the concrete. Then the infantry, covered by a light bombardment, advanced rapidly until they were 300 to 400 yards from their objective. From there, machine guns and anti-tank guns directed intense fire into the embrasures while demolition squads worked round to the rear of the pill-box. They then dashed in and blew down the steel door with "beehives" or "bazookas", thrust in pole charges and phosphorous grenades and left the explosives and the choking smoke to do the rest. It was a slow process, but it was sure and comparatively inexpensive.(92)

At a higher level, Operation "Goodwood/Atlantic" was launched by the British on July 18, with Lieutenant General Richard O'Conners's mighty Eighth Corps, of three entirely armored divisions, striking east and south of Caen on a narrow front. The object of "Goodwood" was to "engage the German armor in battle and 'write it down' to such an extent that it [was] . . . of no further value to the Germans as a basis of the battle."(93) Although it did ultimately tie up seven of the nine panzer divisions available to the Germans, thereby making Patton's future breakout possible, the operation actually "wrote down" more British armor than German. In 72 hours, Eighth Corps incurred 300 tank casualties. Strongpoint defense in depth had again proven to be more than a match for armor. In this instance, Rommel had arranged the defense in five zones based on fortified villages and well dug in gun positions. The third and by far most critical zone was essentially a "cushion" of 12 small villages, each garrisoned by an infantry company and three or four antitank guns. The fifth and last zone was also organized around defended villages. Despite an aerial "carpet" bombardment of unprecedented ferocity (2,000 bombers in two hours), the British could not clear the enemy infantry and antitank guns from such strongholds, and the German 88s consequently began to exact their toll.(94)

Unquestionably, Operation "Goodwood" failed for lack of sufficient British infantry, without which fortified strongpoints could not be mopped up quickly enough or German infiltrating counteractions prevented. The one infantry battalion on establishment of a British armored brigade at this time was found to be inadequate for carrying out the infantry tasks required. Furthermore, the motorized infantry brigades during the "Goodwood" action had been kept too far back to do much good. In this regard, it might be fairly charged that the British were not as flexible as the Americans in forming mixed battle formations.(95) This was certainly Rommel's opinion, as he thought that the Americans "showed themselves to be very advanced in the tactical handling of their forces" and that they "profited much more than the British from their experiences in North Africa."(96)

To be sure, the successful employment of antitank guns and mines, notably by the Germans in the 1941-1943 African campaigns (but also by the British and the Russians), had made it patently clear to the Americans that tanks would frequently have to be preceded by foot soldiers to destroy antitank defenses. Thus, by March 1942, all American armored divisions, except the Second and Third, were reorganized to comprise three tank battalions, three artillery battalions, and three armored infantry battalions of 1,001 all ranks each. All battalions were administratively self-contained and capable of attachment and detachment. This reorganization effectively doubled infantry strength in proportion to tanks, which were reduced in number from 390 to 263 in an armored division.(97) When the Germans launched their Ardennes offensive out of the fog on December 16, 1944, the American Seventh Armoured Division that so effectively slowed them down at St. Vith was organized in this manner.(98)

Notwithstanding these efforts and the pugnacious defense of the 101st Airborne at Bastogne, which hopefully put to rest the battlefield myth that the soldier perpetually risks running out of ammunition,(99) General Patton had some caustic words for the American infantry in general. Noting that the Germans were "colder and hungrier" than U.S. infantry but that "they . . . [fought] better," he remarked bitterly on "the stupidity of . . . [American] green troops." Recalling the old saw that the "poorer the infantry, the more artillery it needs," he admitted that the "American infantry need[ed] all it . . . [could] get."(100) Such statements would have certainly been supported by Alexander McKee, who charged that "Patton's infantry was poor," citing as evidence the example of its being fought off by a bakery company of the Ninth SS Panzer Division.(101)

Generalizations about inherent national fighting abilities are, of course, impossible to substantiate as unit and formation performances varied wildly even within the same armies. However, from all opinions offered, it would appear that the German infantry was universally regarded with great respect by the Allies. It must be remembered that in the Battle of Normandy the German soldier often had to operate without assistance from aircraft, the Allies possessing near-absolute air superiority. That Rommel opposed the grouping of German motorized reserves back from the Atlantic defensive front clearly indicates the serious limitations this situation imposed. So convinced was he of the overwhelming nature of Allied air supremacy that he believed there would be no possibility of ever moving large formations of troops again, even by night.(102) Bumper-to-bumper vehicle convoys on open roads in daylight, a blatant characteristic of the Allied campaign in Italy, were completely out of the question for the Germans.(103)

For the German infantryman, Allied air supremacy meant that, in the offense, he had to rely on foul weather, surprise, and sheer tactical resourcefulness to prevail. In the defense, he had to make maximum use of camouflage and digging, concealing himself as never before (not even on the Eastern Front, where air superiority was usually more of an ephemeral, highly localized phenomenon). These straits contrasted sharply with the situation of Allied infantry, who normally went out of their way to make themselves conspicuous from the air because they feared a case of mistaken identity on the part of the RAF or USAAF more than they did an attack by the Luftwaffe. The extreme severity of accidental bombing by friendly forces, most painfully experienced by the Canadian Regiment de la Chaudière,(104) tended to confirm the wisdom of such Allied infantry methods.

    A further limitation of the German army in the West, according to General Doctor Hans Speidel, was that battle-tried soldiers and officers were scarce. As a whole, the army was undernourished and its supplies were inadequate. The winter of Stalingrad had inflicted some half million irreplaceable casualties, a disastrous blow that virtually broke the back of the German army.(105) This same army nevertheless fought on with a tactical skill that was never really matched by either the British or the American armies in the West. There is even ample statistical evidence to indicate that the Germans, man for man, outfought the British and Americans. German soldiers on the average inflicted three casualties on the Allies for every two they incurred, even as late as the Ardennes offensive. By more conservative measure, that of historical combat modeling analysis, 100 German soldiers were determined to be the combat equivalent of 120 Americans or British. According to Colonel T. N. Dupuy, whose statistics these are, there is good reason to believe that this superior combat effectiveness resulted from German "institutionalization" of military excellence. In his assessment, a "unique trait of the German Army was its systematic efforts to make first-rate soldiers as well as independent-minded scholars out of any man who gave evidence that he could combine these characteristics."(106) In short, the German army encouraged an intellectualism and independent decision-making ability that devolved to the lowliest soldier. It may even have adhered to the precept, later espoused by Marshall, of teaching its soldiers <u>how</u> to think rather than <u>what</u> to think.(107) In any case, it can be reasonably argued that the German section commander was taught to think like an officer.

    At the same time, the German section was often allowed more freedom of action in the execution of offensive missions than was normal in other armies. According to General Frido von Senger und Etterlin, "All ranks of the German Army were well trained in leadership, for with us this was a tradition."(108) The widely accepted view that German soldiers excelled over their Allied opponents when operating alone or in pairs,(109) not only tends to substantiate this remark but points as well to a superior German leadership and tactical ability at the lower level. Contrary to widely held opinion, the German soldier was far from being an automaton; less gregarious than either his British or American counterpart, he was highly capable of individual thought and action. Comments such as "the infantry soldier is not trained to fight in twos and threes, whereas the German does so frequently," and "the German is first-class at infiltration, because he will work as a single individual" often appeared in Allied reports.(110) The problem of how to overcome the herd instinct plagued Allied infantry for much of the war.

In an associated vein, the Germans appear to have retained an almost unshakeable faith in the importance of small-group actions. In the Ardennes offensive "storm battalions" were again used as spearheads, rapidly infiltrating without artillery support between forward American positions.(111) According to a Tenth SS Panzer document, the "best counter-measure evolved" against Allied offensive action was "an inconspicuous filtering . . . in numerous small detachments" into an enemy area weakly occupied by infantry.(112) Small group actions were also the building blocks on which active German counterattacks were based, as the following Allied after-action report indicates:

> Experience has shown that the Germans will almost invariably launch a counter-attack to break up an attack made by small infantry units. You can expect such a counter-attack, usually by 10 to 20 men, not more than 5 minutes after you get close to the German positions. They are usually well armed with light machine guns and machine pistols, and counter-attack by fire and movement. They keep up a heavy fire while small details, even individuals, alternately push forward. The Germans almost always attack your flank. They seldom close with the bayonet, but try to drive you out by fire.(113)

Interestingly, from the start of the war, incessant talking and shouting characterized most German small-unit offensive actions. Erroneously interpreted by Allied soldiers as a sign of poor discipline, it was later ascertained that such chatter was, in fact, a most effective means of dispelling individual loneliness and heightening group cohesion. The Germans, like Marshall, were apparently convinced that the infantryman is "sustained by his fellows primarily and by his weapons secondarily."(114) That Allied infantry were invariably reluctant "to make use of their own fire to help themselves forward"(115) was probably less of a technical problem than a human one. That it was a major infantry problem, of course, there can be no doubt. In his classic study of American infantry, Marshall established that no more than one quarter of even "well-trained and campaign-seasoned troops" would actually fire their weapons at the enemy of their own volition. It should not be surprising, then, that the most common cause of local defeat in minor tactics is usually attributed to "the shrinkage of fire."(116) By recognizing the importance of their small primary groups, however, the Germans may have gone a certain distance in overcoming this problem, whether inadvertently or not.

As the war progressed, the Germans tended to reduce the size of company organizations. As many saw it, the task of command was to prevent the complete attrition of numbers, always to assemble new units and hold others in readiness, even though they were smaller in size. After 1943, the battle strength of a German infantry company was fixed at a maximum of 80, compared with 180 as formerly. According to Senger und Etterlin, in Italy, this figure was further reduced to 70, with "very experienced divisional commanders" limiting it to 40. One reason for restricting company strengths was to accommodate junior officers who were apt to lose control in handling larger companies. It was also discovered, however, that bigger losses occurred in larger companies without much difference in effect. Because of this, the Germans did not care to have large masses of infantry at focal points,(117) but what they lacked in numbers, they made up for in tactical skill. The following remarks by a veteran Canadian company commander who served in North-West Europe bear testimony to this fact:

> The enemy's defensive tactics are brilliantly conceived, and carried out with great tenacity by some of the best soldiers in Europe. No rigid defence: under attack, they hold on as long as possible in their excellently concealed slit-trenches, then they withdraw to prepared positions a little farther back. Instantly previously arranged mortar and artillery fire is poured on the positions they've just vacated--even if a few of their own men are still there. The shelling is coordinated with infantry assaults to retake the ground they've lost. Superb tactics.(118)

Whatever edge the Germans possessed in tactical prowess can be attributed in large part to their superior small group performance, for it is a military fact that the people who really count in battle are the commanders and fighters at battalion level and below. German divisions, for example, were normally considered "used up" when their victualing strength dropped from 12,000 to 11,000 men; those that were "totally battle-weary" rarely fielded less than 10,000.(119) The difference was invariably accounted for in infantry ranks. It was on the rifle strengths of its battalions that the condition of a division had to be judged. To quote a Canadian, Major General Chris Vokes, "the offensive power of an infantry division is bound to become spent, not for lack of offensive spirit, but simply because the quality of offensive team play within . . . rifle companies . . . deteriorated."(120) Thus, while divisional staffs and headquarters tended to grow more efficient as they gained experience in war, the overall combat effectiveness of such formations conversely began to decrease as small-unit leaders and commanders became casualties.(121) By maintaining numerous smaller units in action, however, the Germans appear to have actually created an ongoing school of experience for these vital personnel.

In comparison with the British, who seemed to stress keeping a tactical balance so as not to be caught unprepared by the enemy, the German army placed its emphasis on dash and deliberate acceptance of risk.(122) This particular trait was demonstrated most strikingly during the Ardennes offensive, at which point the German army was close to being on its last legs. To what extent celebrated British tenacity in defense accounted for this attack being directed against the American sector is, of course, difficult to determine. To their credit, however, for most of the war, the Americans generally exhibited a greater flair for the offense than the British,(123) although some Germans included them in a general criticism levied against the "hesitant and careful" attitude of Allied troops, particularly the infantry, in the battle of Normandy. To be sure, a report by Twenty-First SS Panzer Grenadier Regiment of the Tenth Panzer Division was not that laudatory of Allied infantry performance:

> The morale of the enemy infantry is not very high. It depends largely on artillery and air support. In case of a well placed concentration of fire from our own artillery the infantry will often leave its position and retreat hastily. Whenever enemy is engaged with force, he usually retreats or surrenders.(124)

The foregoing statement was obviously not universally applicable to Allied infantry, many units of which performed in splendid fashion. The Australians and New Zealanders were, in Rommel's opinion, outstanding infantry.(125) So, too, were the <u>Goumiers</u> of General Juin's <u>Corps Expeditionnaire Français</u>. The Canadian performance at Ortona against Germany's finest was of the highest standard, as was the epic stand of the 101st American Airborne at Bastogne. The "outstanding independent parachute battalion action of the war,"(126) of course, was that of Lieutenant Colonel John Frost's Second Battalion, First Parachute Brigade, at Arnhem. The high standard of marksmanship within this unit demonstrated just how effective modern infantry action could be.(127) To paraphrase General James M. Gavin, it was a good example of the airborne truism that the "offensive action of an airborne force can best be measured in terms of its ability to defend."(128) This was, of course, the essence of all infantry: to be able to hold, and by holding to disorganize and dislocate the enemy, throwing him off stride and preparing the way for a decision. From most reports, however, the Germans appear on the average to have appreciated this the best.

Ironically, the powers who originally put their least faith in the infantry arm were to suffer the most traumatic experiences because of it. According to Senger und Etterlin, though the numbers of infantry had steadily declined relative to other arms, "yet paradoxically the infantry remained more firmly established as queen of the battlefield."(129) To Rommel, this was but one further indication that the "fundamental superiority which all previous wars had shown, of a well dug-in defence, equipped with modern [antitank] weapons still held good."(130) Had Montgomery had more infantry, he may not have chosen to use his tanks with such reckless abandon as he did in Operation "Goodwood."

The root cause of the Allied shortage in infantry may, of course, lie in a decided Anglo-Saxon preference for technological solutions and the somewhat related pursuit of a high standard of living both on and off the battlefield. It is indeed most significant that the American military had extreme trouble in maintaining a ground force of 89 divisions mainly because of a shortage of physically fit replacements for the infantry, the strength of which arm grew to 1.8 million by April 1945. Though this represented a lower ratio of infantry to other arms than that found in allied armies,(131) the infantry nonetheless remained the mainstay of the American army. This was true also of the British Commonwealth. The weakness of the Allied infantry in quantity was partially reflected in quality. Fortunately, most of the infantry divisions required to win the war were available on the Eastern Front.

NOTES

1. Russel F. Weigley, "To the Crossing of the Rhine: American Strategic Thought to World War II," Armed Forces and Society, 5 (1979):305-16.

2. Ellis, The War in France and Flanders, p. 326.

3. Liddell Hart, The Rommel Papers, p. 33.

4. Arthur Bryant, "The New Infantryman," Canadian Army Training Memorandum, no. 44 (1944), p. 32.

5. I have been using the term "British army" (with a small "a") as all-inclusive of Commonwealth armies, that is, the British Army, Canadian Army, and Australian Army, etc. As General Pope remarked of the Canadian Army, "All our manuals were British and so was all our tactical training." "Lieutenant General Maurice A. Pope, Soldiers and Politicians (Toronto: University Press, 1962), p. 53.

6. Alexander, Montgomery, and Alan Brooke all served with the B. E. F.

7. Arthur Bryant, "The New Infantryman, "Canadian Army Training Memorandum," pp. 31-2; and Farrar-Hockley, Infantry Tactics, p. 19.

8. Robert H. Ahrenfeldt, Psychiatry in the British Army in the Second World War (London: Routledge and Kegan Paul, 1958), pp. 198; and G. R. Stevens, Princess Patricia's Canadian Light Infantry, 1919-1957 (Montreal: Southam, 1958), pp. 54-5; and Farrar-Hockley, Infantry Tactics, pp. 19-20.

9. Farrar-Hockley, Infantry Tactics, pp. 20-4.

10. Liddell Hart, "The 'Ten Commandments' of the Combat Unit," p. 288.

11. Farrar-Hockley, Infantry Tactics, pp. 24-5. The debate between the "Reactionary Infantry School" and the "Advanced Tank School" went beyond the British Isles to the Dominions. Young Captain G. G. Simonds of the Royal Canadian Artillery, whose terms I have used, argued very convincingly in 1939 that "the object of training is not to prove that 'Infantry is Queen of Battle' but to teach cooperation between the many varied fighting troops which are the essential components of a modern army." Captain G. G. Simonds, "The Attack," Canadian Defence Quarterly 16 (1939):382, 386-7 and 389.

12. Liddell Hart, The Rommel Papers, p. 184. By way of comment, Liddell Hart originally proposed a system of drills whereby the soldier was taught to reason rather than just react.

13. War Office, Current Reports From Overseas, no. 65 (November 29, 1944), p. 13.

14. Major General Sir Howard Kippenberger, Infantry Brigadier (Oxford: University Press, 1949), p. 180.

15. Farrar-Hockley, Infantry Tactics, pp. 28 and 41.

16. Kippenberger, Infantry Brigadier, pp. 173-4 and 336; and Messenger, The Art of Blitzkrieg, p. 157.

17. Palit, War in the Deterrent Age, p. 97.

18. Liddell Hart, The Rommel Papers, p. 520.

19. Liddell Hart, Europe in Arms, p. 35.

20. Young, World War 1939-1945, p. 111.

21. Mellenthin, Panzer Battles, p. 142; Liddell Hart, The Rommel Papers, pp. 91 and 262; and Albert Kesselring, Memoirs (London: William Kimber, 1953), pp. 106-7. According to Liddell Hart, the Italian army barely conquered Ethiopia. Although it used poison gas and air power, its mechanized forces were judged to be little more than the trimming to masses of infantry. Operation orders were supposedly so long and detailed that after reckoning the time for writing and reading them, little time was left for their execution. Liddell Hart, Europe in Arms, p. 36.

22. Kesselring, Memoirs, p. 107.

23. Liddell Hart, The Rommel Papers, pp. 130 and 133. Rommel described positional warfare as "always a struggle for the destruction of men" in contrast to mobile warfare "where everything turns on the destruction of enemy material." Ibid., p. 133.

24. J. R. Lester, Tank Warfare (London: George Allen and Unwin, 1943), p. 109.

25. Liddell Hart, The Rommel Papers, p. 201

26. Young, World War 1939-1945, pp. 230-4; and Mellenthin, Panzer Battles, pp. 90-109, and 112-17.

27. Young, World War 1939-1945, pp. 235-6 and 256-62; Liddell Hart, The Rommel Papers, p. 254; Mellenthin, Panzer Battles, pp. 52 and 136-42; and Macksey, Tank Warfare, pp. 193-5.

28. Fuller, Armored Warfare, p. 71; and Liddell Hart, History of the Second World War. (London: Pan, 1978), p. 188. The heavier 88-millimeter, on the other hand, though a formidable "tank killer" in its own right, had a high silhouette that made it more vulnerable to counterfire than standard antitank guns. Liddell Hart, History of the Second World War, p. 282.

29. Ogorkiewicz, Armoured Forces, pp. 59-60.

30. Messenger, The Art of Blitzkrieg, pp. 157-8.

31. Farrar-Hockley, Infantry Tactics, p. 28. This was partially a consequence of combat in the desert where lack of cover and natural obstacles necessitated basing defensive positions on large well-coordinated mine fields.

32. Weller, Weapons and Tactics, p. 107.

33. "A Few Tips From the Front," Canadian Army Training Memorandum, no. 29 (1943), pp. 19 and 21; and Kenneth Macksey, Crucible of Power: The Fight for Tunisia, 1942-1943 (London: Hutchinson, 1969), pp. 159, 240, and 262. Junior commanders were in general urged to learn how to do flanking attacks and "not to throw troops away by pounding straight ahead against well-organized resistance." Canadian Army Training Memorandum, no. 28 (1943), p. 25.

34. "A Few Tips From the Front," Canadian Army Training Memorandum, pp. 19-20.

35. "A Few Tips From the Front," Canadian Army Training Memorandum, pp. 20-21; and Macksey, Crucible of Power, p. 185.

36. J. A. English, "Confederate Field Communications" (M. A. thesis, Duke University, 1964), pp. 11 and 22-4. It should perhaps be reiterated that Europeans were little affected by the experience or lessons of the Civil War, its being largely ignored by the military establishments of most major powers. Moltke's description of the conflict as "two mobs clashing in the wilderness" summed up the initial European attitude. The Americans, on the other hand, were affected by their Civil War to a much greater extent than foreigners can imagine, even to this day.

190   A PERSPECTIVE ON INFANTRY

37. Though the U.S. Marines rejected the practice by certain French units of attacking in waves. Captain John W. Thomason, Fix Bayonets (New York: Charles Scribner, 1926), pp. 9 and 172.

38. Virgil Ney, Organization and Equipment of the Infantry Rifle Squad from Valley Forge to ROAD (Fort Belvoir: U.S. Army Combat Operations Research Group Memorandum 194, January, 1965), pp. vii, 13, 15-16, 18, 27, 29-30, 32-8, and 75; and Major Richard G. Tindall et al., Infantry in Battle (Washington: The Infantry Journal, 1934), p. 269.

39. The Army Lineage Book, Volume II: Infantry (Washington: Office of Military History, Department of the Army, 1953), pp. 43-43; and Ney, Infantry Rifle Squad, pp. 37-42 and 47-8. The automatic rifle (Browning 1918A2) was considered inaccurate with several serious defects. Interestingly, the British in 1907 took the same tack, concentrating on aimed rifle fire, after Treasury refused to allot funds to provide machine guns. Carrington, Soldier From the Wars Returning, p. 25.

40. The Army Lineage Book, Volume II: Infantry, pp. 46-7; and Ney, Infantry Rifle Squad, p. 44.

41. Kent Roberts, Greenfield, Robert F. Palmer, and Dell I. Wiley, The Organization of Ground Combat Troops: The United States Army in World War II; The Army Ground Forces (Washington: Department of the Army, 1947), pp. 271 and 300-1; Ney, Infantry Rifle Squad, pp. 43, 46, and 48-9; and The Army Lineage Book, Volume II: Infantry, pp. 47 and 53.

42. Greenfield, The Organization of Ground Combat Troops, pp. 271 and 301-2.

43. Greenfield, The Organization of Ground Combat Troops, p. 323; and Virgil Ney, The Evolution of the Armored Infantry Rifle Squad (Fort Belvoir: U.S. Army Combat Operations Research Group Memorandum 198, March 1965), pp. 31-7 and 39. Patton stressed that armoured infantry "should not attack mounted." Patton, War As I Knew It, p. 354.

44. Dupuy, Military Heritage of America, p. 635; and Greenfield, The Organization of Ground Combat Troops, pp. 189-90.

45. Eli Ginzberg et al., The Ineffective Soldier; Lessons for Management and the Nation: The Lost Divisions (New York: Columbia University Press,

1959), pp. 8 and 18. A peak strength of 12,124,000 was reached in 1945. Morris Janowitz and Charles B. Moskos, "Five Years of All-Volunteer Force: 1973-1978", Armed Forces and Society 5 (Winter 1979):179.

46. The Army Lineage Book, p. 48.

47. Liddell Hart, The Rommel Papers, p. 521.

48. The British, as previously mentioned, suffered from a similar shortage in junior officers; hence, the compensatory rank of WO III platoon sergeant major.

49. Greenfield, The Organization of Ground Combat Troops, pp. 48 and 51; and Farrar-Hockley, Infantry Tactics, p. 59.

50. Blumenson, The Patton Papers, vol. 2, p. 572.

51. Farrar-Hockley, Infantry Tactics, pp. 59-60.

52. He was recognized as an authority in the publication Infantry in Battle, pp. 1, 318, and 329.

53. Colonel P. S. Bond, ed., Military Science and Tactics: Infantry Advanced Course; A Text and Reference of Advanced Infantry Training (Washington: Bond, 1944), pp. 22 and 30.

54. Bond, Military Science and Tactics, pp. 31-4, 38-41 and 65; Small Unit Tactics Infantry (Harrisburg: The Military Service Publishing Company, 1948), pp. 52, 72, 115, and 117; and Small Unit Actions (Washington: War Department Historical Division, 1946), p. 152. There were later added platoon "vee" and platoon "wedge" formations.

55. Weller, Weapons and Tactics, p. 119.

56. Major General J. C. Fry, Assault Battle Drill (Harrisburg: The Military Service Publishing Company, 1955), pp. vii-xii, 54, and 64-76. He wrote: "From the platoon viewpoint, most attacks could be generally classified as frontal attacks, for it is the battalion and company commanders who are more responsible for manoeuvring units." Ibid., pp. 64-5.

57. Blumenson, The Patton Papers, vol. 2, p. 225.

58. Patton, War As I Knew It, p. 293. For further evidence of his opposition to rushing tactics, see Blumenson, The Patton Papers, vol. 2, p. 454.

59. Weller, Weapons and Tactics, pp. 119-20 and 121. According to Shelford Bidwell, the Americans "had no tactics as such at all but bashed and battered their way along, using their splendid technology to provide fire-power in a manner reminiscent of the British in 1916 or 1917. Every scrap of ground taken had to be retained, regardless of casualties." Bidwell, Modern Warfare, p. 140.

60. Blumenson, The Patton Papers, vol. 2, p. 225; and Weller, Weapons and Tactics, p. 120.

61. Bond, Military Science and Tactics, pp. 50-1 and 55-7; and Small Unit Tactics Infantry, p. 83. Defensive areas were formerly designated as combat groups (platoon areas), strongpoints (company areas), and centers of resistance (battalion areas). Bond, Military Science and Tactics, p. 49.

62. Greenfield, The Organization of Ground Combat Troops, pp. 31 and 316. Interestingly, in the spring of 1942, the chief of infantry and other heads of combat arms were merged in with the newly created "Army Ground Forces," commanded by McNair, an experienced practical combat arms officer who had pushed "triangularization." The chiefs of the combat arms were eliminated because they were thought to "foster schisms within the Army." The Army Lineage Book, p. 52.

63. Lieutenant Colonel G. W. L. Nicholson, The Canadians in Italy 1943-1945 (Ottawa: Queen's Printer, 1956), p. 175.

64. Stevens, Princess Patricia's Canadian Light Infantry, p. 72. The Patricia's were one of two Canadian assault battalions. Nicholson, The Canadians in Italy, pp. 68 and 75.

65. Nicholson, The Canadians in Italy, p. 175.

66. W. G. F. Jackson, The Battle for Italy (London: B. T. Batsford, 1967), pp. 122-3, 138, 139, 141, 167, 238, 263, and 321.

67. The Germans brought out the Panzerfaust and Panzerschrek between 1942 and 1943, the latter being a bazooka of 88 millimeters with a range of 100-150 yards. The bazooka had proven its worth in Tunisia.

68. "Ortona," Canadian Army Training Memorandum, no. 42 (September 1944), pp. 31-5; Current Reports From Overseas, no. 33 (April 15, 1944), pp. 1-11; Jackson,

The Battle for Italy, pp. 152-3; and Nicholson, The Canadians in Italy, pp. 325-9, 333 and 681.

69. Fred Majdalany, The Battle of Cassino (Boston: Houghton, Mifflin, 1957), p. 88; Jackson, The Battle for Italy, pp. 191, 198, 201, 212, 215, and 236; and Brigadier Peter Young, ed., Decisive Battles of the Second World War: An Anthology (London: Arthur Barker, 1967), pp. 264-71 and 274-82.

70. Majdalany, The Battle of Cassino, pp. 38-9 and 236; and The Monastery (London: John Lane, 1945), pp. 12-13; Liddell Hart, History of the Second World War, pp. 553-9; and Senger und Etterlin, Neither Fear Nor Hope, pp. 224-32.

71. J. F. C. Fuller, Thunderbolts (London: Skeffington and Son, 1946), pp. 54-5. On March 15, 1944, bombers dropped 1,100 tons of explosives on Cassino while 300 fighter bombers attacked targets in the vicinity. When the air attack ceased, 610 artillery pieces threw 1,200 tons of shells into the town. Operations in Sicily and Italy (West Point: U.S.M.A. Department of Military Art and Engineering, 1950), p. 75.

72. These were steel cylindrical cells, each seven feet deep and six feet in diameter, housing a two-man machine-gun crew and their weapon. Only the top 30 inches, which was of armor 5 inches thick, extended above the ground. They were nicknamed "crabs" from their appearance when being towed on removable wheels to the place of installation. Nicholson, The Canadians in Italy, p. 396.

73. Liddell Hart, The Other Side of the Hill, p. 374.

74. The spectacular performance of the German First Parachute Division in the third battle of Cassino sometimes obscures the fact that the task of defending the town in the first two battles fell to the soldiers of the Fifteenth and Nintieth Panzergrenadier Divisions, the Fifth Mountain, and the Seventy-first Infantry. Majdalany, The Battle of Cassino, p. 244.

75. Nicholson, The Canadians in Italy, pp. 338 and 344.

76. Ralph Allen, Ordeal by Fire: Canada, 1910-1945 (Toronto: Popular Library, 1961) pp. 442-5; and Colonel C. P. Stacey, The Victory Campaign: The Operations in North West Europe, 1944-1945 (Ottawa:

Queen's Printer, 1960), p. 385. According to Farley Mowat, men from every branch of the service were "drafted into the infantry to fill the gaps." He described the Service Corps, Anti-Tank Artillery and Ordnance as "the flesh of an army that was being forced to practise self-cannibalism in order to keep the fighting units in existence." Farley Mowat, The Regiment (Toronto: McClelland and Stewart, 1977), pp. 291-2.

77. Major L. F. Ellis, Victory in the West (London: H. M. Stationery Office, 1968) vol. 2, pp. 141-2.

78. Stacey, The Victory Campaign, p. 284.

79. Greenfield, The Organization of Ground Combat Troops, pp. 191, 193-4, 244-6, and 250.

80. Patton, War As I Knew It, pp. 163, 165, and 169; and Blumenson, The Patton Papers, vol. 2, pp. 583 and 588. Patton complained that people did "not realize that 92 per cent of all casualties occur in the infantry rifle companies, and that when an infantry division has lost 4,000 men, it has practically no riflemen left." Blumenson, The Patton Papers, vol. 2, p. 586. Patton's figures are borne out (or based on) the study by Stouffer et al. Examination of casualty rates of four infantry divisions in Italy showed that the infantry troops suffered 92 percent of battle casualties, although they constituted but 67 percent of the authorized strength of an infantry division. Stouffer et al, The American Soldier, p. 102. According to S. L. A. Marshall, many of the soldiers who were converted into riflemen "acted as if they had been betrayed by their country." Marshall, The Soldier's Load, p. 97.

81. Ney, Infantry Rifle Squad, pp. 50-1. Marshall also argued that the "propaganda that sought the practical elimination of foot forces as a major factor in mobile war was thoroughly injurious. . . . It reacted as a depressant upon the self esteem of . . . infantry . . . reducing . . . general combat efficiency. This fact is established by Army polls which show that . . . infantrymen . . . continued to hold a low opinion of the importance of their own branch." Marshall, Men Against Fire, p. 17.

82. Major General E. L. M. Burns, Manpower in the Canadian Army, 1939-1945 (Toronto: Clarke, Irwin, 1956), pp. 14, 18-19, 21 and 23; and Allen, Ordeal by Fire, pp. 441-2. The high British figure reflects the

very large proportion of antiaircraft artillery in British formations. This subcorps was reduced by conversion of its individual members to infantry all through 1944. Burns, Manpower in the Canadian Army, p. 21. In Vietnam 600,000 troops were committed to support 70,000 combatants. Beaumont, Military Elites, p. 13.

83. Miksche, Atomic Weapons and Armies, p. 163; and Senger und Etterlin. Neither Fear Nor Hope, pp. 196 and 223. In Sir John Slessor's opinion, one of the reasons for the Germans' astonishing ability to maintain their divisions in central Italy in apparently impossible circumstances, with every road and railway almost continually cut in several places behind them, was that the German division was able to subsist on a fraction of the daily tonnage required by an Allied division. Slessor, The Great Deterrent, p. 114.

84. Allen, Ordeal by Fire, pp. 439-40.

85. Marshall, The Soldier's Load, pp. 11, 22-3, 25, 35, 40; and Battle at Best (New York: William Morrow, 1963), pp. 53-4. On the two division front of Omaha beach only six rifle companies were judged to have been effective.

86. Marshall, The Soldier's Load, pp. 40, 47, and 52-3. General Scharnhorst is reputed to have remarked, "The infantryman should carry an axe in case he may have to break down a door." Ibid., p. 8.

87. Major N. V. Lothian, The Load Carried by the Soldier (London: John Bale, Sons and Danielsson, circa 1920), p. 56.

88. Marshall, The Soldier's Load, pp. 20, 23, 32, 47, 49 and 52. When the 153rd Infantry Regiment went ashore at Kiska in the Aleutians against a supposedly Japanese-held base, "each soldier carried the following: underwear, shirt, lined trousers, Alaskan field jacket, steel helmet and liner, raincoat, poncho, extra shoes, rifle belt, six grenades, 240 rounds of ammo, rifle, pack board, sleeping bag, two shelter halves with pole and pins, twelve cans of "C" rations, heat tablets, cook stove, two cans of Sterno, long knife, intrenching tool, bayonet, flashlight, maps, pocket-knife, change of clothing, wire cutters, waterproof matchbox, identification panel, ruck sack, four chocolate bars, three signal panels, compass, and a 'book of battle songs.'" One officer reported, "Had the enemy been there with only two machine guns, we would have been repelled. . . ." As in Sicily, enemy resistance was expected. Fortunately, none was encountered. Ibid., pp. 8 and 34.

89. The losses on D-day were not light, roughly 9,000, but they were less than anticipated, which may have given rise to the lie that they "were not heavy." Stacey, The Victory Campaign, pp. 119-20.

90. Dupuy, Military Heritage of America, pp. 535-6; Macksey, Tank Warfare, p. 229; and Charles B. MacDonald, The Mighty Endeavour: American Armed Forces in the European Theatre in World War II (New York: Oxford University Press, 1969), pp. 282-3 and 293.

91. Brigidier A. J. D. Turner, Valentine's Sand Table Exercises (Aldershot: Gale and Polden, 1955), p. 2.

92. Chester Wilmot, The Struggle for Europe (London: Collins, 1974), p. 375.

93. Ibid, pp. 405-6. "Atlantic" was the Second Canadian Corps operation to secure the "Goodwood" right flank. The Eighth Corps consisted of Guards Armoured, Seventh Armoured, and Eleventh Armoured Divisions.

94. Liddell Hart, History of the Second World War, pp. 578-9; Wilmot, The Struggle for Europe, pp. 405-11; and Stacey, The Victory Campaign, pp. 169-70.

95. Stacey, The Victory Campaign, pp. 169-70 and 176-80; Macksey, Tank Warfare, pp. 231-2; and Alexander McKee, Caen: Anvil of Victory (London: Souvenir, 1964), pp. 248-9, 251-2, 267, 274-7, 280-2, and 290.

96. Liddell Hart, The Rommel Papers, p. 523; and Ladislaw Farago, Patton: Ordeal and Triumph (New York; Dell, 1975), p. 494. It has been suggested that Montgomery's overall shortage of infantry forced him into attempting to "write down" German armor by using his own, rather recklessly.

97. Greenfield, The Organization of Ground Combat Troops, pp. 321-3 and 327-31. An armored infantry regiment had a driver complement of 544. The Second and Third Armored retained the old organization of six tank battalions, three infantry, and three artillery. Ibid., pp. 323 and 329.

98. Macksey, Tank Warfare, pp. 239-40.

99. The stand at Bastogne, in fact, made for more effective sharing and use of ammunition. Defeat because of an ammunition shortage is among the things least likely to happen on the battlefield, the reason basic first-line infantry ammunition loads should

probably be reduced. According to Marshall, there was only one instance of ammunition shortage within the 101st Airborne on the Normandy landing, namely, the three-day stand of 84 men under Captain C. G. Shettle at Le Port Bridge. Marshall, The Soldier's Load, pp. 18-19. Patton also spoke of the "ridiculous and widespread fear among all our troops that they will never run out of ammo." In his experience, it never happened. Blumenson, The Patton Papers, vol. 2, p. 459.

100. Blumenson, The Patton Papers, vol. 2, 521 and 615. Quoted originally to Patton by French General Koechlin-Schwartz of Langres Staff College as, "The poorer the infantry, the more artillery it needs; the American infantry needs all it can get." Ibid., p. 521.

101. McKee, Race For the Rhine Bridges, p. 123; and Young, Decisive Battles, pp. 430-1.

102. Guderian, Panzer Leader, p. 330; and General Dr. Hans Speidel, Invasion 1944: Rommel and the Normandy Campaign (Chicago: Henry Regnery, 1950), pp. 52-55. The stubborn defense of the German army in Italy during the last 18 months of the war was carried out under total Allied air supremacy. Slessor, The Great Deterrent, p. 113.

103. McKee, Race for the Rhine Bridges, p. 83.

104. Stacey, The Victory Campaign, pp. 223, 243-4, and 274; McKee, Caen: Anvil of Victory, pp. 338-9. The Chaudière's were attacked on July 17 by the Luftwaffe, on August 8 by the USAAF on August 14 by the RAF, and on 15 August by the USAAF again. The bombing by the RAF was judged by far the worst. McKee, Caen: Anvil of Victory, p. 335.

105. Speidel, Invasion 1944, pp. 170-1. As France was used as a training ground for formations going to the Ostfront, however, some crack units were available. Yet, overall, there had been a decline in quality. Between October 1942 and October 1943, 36 infantry and 17 armored and motorized divisions, including such good formations as the Seventh Panzer and S.S. "Adolf Hitler," "Das Reich," and "Totenkopf" left the West for other theaters. Stacey, The Victory Campaign, pp. 51 and 129.

106. Dupuy, A Genius for War, pp. 3-5, 292-4, and 305-6.

107. Marshall, Men Against Fire, p. 116.

108. Senger und Etterlin, Neither Fear Nor Hope, p. 219; and Weller, Weapons and Tactics, p. 143. Significantly, by maintaining a low ratio of officers to men throughout the war, the German army ensured a high standard and number of NCOs. They did not drain their NCO corps to supplement their officer corps. Gabriel and Savage, Crisis in Command, pp. 34-5.

109. Liddell Hart, The Other Side of the Hill, p. 425.

110. Current Reports From Overseas, no. 66 (1944), p. 10; and Current Reports from Overseas, no. 36 (1944), p. 11.

111. Liddell Hart, The Other Side of the Hill, pp. 452 and 459.

112. Current Reports From Overseas, no. 61 (1944), p. 14.

113. "The Germans--How They Fight," Canadian Army Training Memorandum, no. 39 (June 1944), p. 39; and Current Reports From Overseas, no. 34 (1944), pp. 11-12.

114. Marshall, Men Against Fire, pp. 42-3, 127, and 136. Marshall elaborated: "That there was a direct connection between these methods and the phenomenal vigor with which our enemies organized and pressed their local counter-attacks seems scarcely to have occurred to our side." Ibid., p. 127.

115. Current Reports From Overseas, no. 61 (1944), p. 12.

116. Marshall, Men Against Fire, pp. 19, 39, 50 and 54-8. This figure may even be as low as 15 percent. Defeat because of ammunition shortage in these circumstances is certainly not likely to occur.

117. Liddell Hart, History of the Second World War, p. 253; and Senger und Etterlin, Neither Fear Nor Hope, pp. 228 and 232.

118. Ben Dunkelman, Dual Allegiance (Toronto: Macmillan, 1976), p. 133. Dunkelman went on to command a brigade in the Israeli army in the 1948 war. Here he applied the German defense tactics he learned in Europe. Ibid., p. 198.

119. Senger und Etterlin, Neither Fear Nor Hope, pp. 196 and 223.

120. Nicholson, The Canadians in Italy, pp. 338-9.

121. McKee, The Race for the Rhine Bridges, p. 465.

122. Ibid., p. 83.

123. Patton had a high opinion of American ability to maneuver (though he was worried that American troops' "ability to fight is not so good"). He was convinced that the American "method of attacking all the time [was] . . . better than the British system, of stop, build up, and start." In Africa, he intensely disliked the "condescending attitude of Alexander and Monty [Montgomery] and other British officers towards American troops, whom they regarded as second rate at worst, inexperienced at best." Blumenson, The Patton Papers, vol. 2, pp. 294, 307, and 633.

124. Stacey, The Victory Campaign, pp. 119 and 274. They made no distinction between British and American infantry. This division had little if any contact with Canadians at the time.

125. Liddell Hart, The Rommel Papers, pp. 130 and 133.

126. Major General James M. Gavin, Airborne Warfare (Washington: Infantry Journal Press, 1947), p. 120.

127. Cornelius Ryan, A Bridge Too Far (London: Book Club Associates, 1975), pp. 234-5.

128. Gavin, Airborne Warfare, pp. 33-4. According to Alexander McKee, airborne units generally tended to be less battle experienced than those in the line: "one moment they are living in safety and peace . . . the next they are in the middle of an actual enemy . . . But their ordeal is brief and in no way compares to that undergone by the ordinary infantryman, whose chances of survival in the long term are exceedingly small." McKee, Caen: Anvil of Victory, p. 68. Generally speaking, the British airborne program was a waste, the airborne divisions spending too much time out of action. The First Division, for example, was in reserve from June through September 1944. Good potential leaders were thus diverted to minimal roles. Beaumont, Military Elites, p. 93.

129. Senger und Etterlin, Neither Fear Nor Hope, p. 228.

200 A PERSPECTIVE ON INFANTRY

130. Liddell Hart, The Rommel Papers, pp. 453 and 478.

131. Canadian Army Training Memorandum, no. 49 (April 1945), p. 34; Greenfield, The Organization of Ground Combat Troops, pp. 243-50; and Farrar-Hockley, Infantry Tactics, p. 59.

7. Cinderella of the Army

Infantry East of Suez, West of Pearl

The fall of Malaya in February 1942 was a dismal chapter in British military history. The surrender by Lieutenant-General A. E. Percival of 85,000 British, Indian, and Australian troops at Singapore to a smaller force of Japanese contrasted sharply with the gallant American defense of the Philippines.(1) Yet both were catastrophic defeats for Western arms: one, the worst "at the hands of yellow men since the days of Genghis Khan,"(2) the other the most calamitous ever suffered by America. In retrospect, they were but the final announcement to the world at large that Asiatic peoples had come of modern military age and were henceforward capable of teaching lessons to European armies. Ironically, the British, who originally regarded the Japanese as second-rate troops at best, now went to the other extreme of extolling their invincibility and superhuman qualities. This was eventually to pose a major problem for the commander of the British army in Burma, General W. J. Slim, for although some units never subscribed to the view of Japanese superiority at all, the idea was not to be completely eradicated until late in 1944.(3) The tactical performances that prompted this assessment obviously deserve examination.

The ensuing war fought against the Japanese in both the Pacific and Burma assumed a multifaceted character. In addition to major land campaigns, massive naval, amphibious, and air operations were required to drive the Japanese from the areas they had conquered. In many ways, it was the most modern war, encompassing as it did a broad technological spectrum. Yet, at the same time, there was a primitiveness about it; the fighting qualities of the Japanese soldier and the jungle terrain of the Pacific islands and Burma made the war on land primarily an infantryman's struggle. The battleground of Kohima-Imphal in many ways came to resemble that of Passchendaele, where the British army in taking 45 square miles lost 370,000 men, or 8,222 per square mile. At Iwo Jima, 28 years after Passchendaele, the cost was a comparable 5,500 casualties per square mile, albeit to both sides;(4) astoundingly, out of a Japanese garrison of 25,000 men only 216 were taken prisoner. It was already widely known, however, that at the beginning of 1944--by which time many thousands of Allied prisoners had been taken--the number of Japanese captured unwounded, in all theaters of war, did not total 100. On the Burma front, the number was estimated at six.(5)

When Japan surrendered to the Allies, there were only 27 American divisions in the Pacific. As late as the Battle of the Philippine Sea (June 19-20, 1944), American ground forces had been no more numerous than the British Commonwealth forces arrayed against the Japanese. It is additionally important to recall that the Japanese had also been fighting the Chinese since 1937, in terrain, incidentally, far different from that of the jungles of the Arakan and Guadalcanal. Furthermore, the Japanese never once lowered their guard against their old enemy, Russia. In this regard, the final disposition of Japanese forces is significant: 2,115,000 in Japan, Sakhalin, and the Kuriles; 1,310,000 in Manchuria, North China, and Korea; 953,000 in South and East China, Malaya, Burma, and Indo-China; and 772,000 in Formosa, the Philippines, the Mandated Islands, and Indonesia.(6) This, if nothing else, should indicate quite clearly that the Japanese army was anything but a jungle-oriented force. On the contrary, it was originally as much concerned with conducting cold-weather operations in Manchuria as it was with assaulting beaches in the South Seas.

In essence, the Japanese army remained a general purpose force very much like the armies of Great Britain and the United States. Traditionally prepared to wage offensive war against the Russians, it had gradually been drawn deeper into a protracted struggle with China since the 1937 incident at the Marco Polo Bridge. The open hostilities that briefly flared with the Soviets during the 1938 Chungkufeng and 1939 Nomohan incidents led, consequently, not only to an increased respect for the Red Army but also to the eventual conclusion of a neutrality pact with the Soviet Union on April 13, 1941. The Japanese army was thus freed to pursue a forcible settlement of the China problem, though it never relaxed in its vigilance toward Russia. In any event, the Japanese never expected anything less than that they would have to fight outnumbered; logic dictated, therefore, that quick decisions should constantly be sought. Logic did not necessarily dictate, however, that the Americans and British had to be tackled in order to solve the China problem.(7)

Within the Japanese army itself, the German model had always been emulated. The smashing success of blitzkrieg naturally made an even deeper impression as to the wisdom of this choice, though a mechanized headquarters was not established until April 1941.(8) The experience gained at Nomonhan (or Khalkhin Gol) had prompted the notion that ten fully equipped tank divisions should be raised. For whatever reason, however, none were activated before Pearl Harbor.(9) The Japanese army by 1941 did possess, nonetheless, 51 divisions (each 18,000 strong) and four tank regiments. The armament and equipment of most of these divisions reflected to a great degree Japanese army involvement in China. Years of mobile operations against poorly armed but clever and stubborn Chinese forces had led the Japanese to depend increasingly on light mortars, light tanks, bicycles, and local transport and supplies in areas with poor ground communication.(10) At the minor tactical level, this proved to be initially advantageous. In the battle for Singapore, the lightly equipped, individually mobile Japanese soldier literally ran circles around his British counterpart, who was overloaded with all the paraphernalia of European warfare, including steel helmets and gas masks.(11) Interestingly, the Japanese divisions earmarked for Malaya had previously fought only in the cold climate of Manchuria.(12)

As previously indicated, the Japanese army basically neglected tank development, though it retained a general amphibious capability. The infantry remained most definitely the main striking element within the land forces. A predominance of light-caliber and easily man-handled weapons characterized this particular arm. Its general dependence on light machine-gun and mortar firepower accounted, in large part, for a comparative weakness in overall artillery organization. Accordingly, there was a lack of finesse in the coordination of all arms. Japanese infantry attacks were often successful, nonetheless, as they were usually carried out quickly and with energy. In both the Philippines and Malaya, Japanese movements were much faster than expected, their speed attributed in the main to the Japanese soldier's light equipment, fitness, and a simplicity of rations.(13)

The Japanese infantryman was relatively self-contained except for ammunition. Each man carried one day's emergency rations and five days' supply of rice. Since individuals were responsible for their own cooking, normally done on a section basis, there was no waiting for hot meals to be brought up. Stoves were not provided, and soldiers were encouraged to live off the country.(14) The Japanese soldier in North China reputedly "fought and froze and made . . . terrific marches on a ration which . . . consisted of a half-pound of rice and some blackish potatoes."(15) Further evidence of the spartanism of the Japanese infantryman was found in his physical training, which was extremely rigorous. Special emphasis was placed on marching, distance being progressively increased during training until after eight months the soldier was capable of sustaining 25 miles a day in full battle order. Bayonet training was also stressed, and the honor of being the best bayonet fighter in a company was eagerly sought after. Xenophobic to a startling degree, the Japanese soldier was fanatically patriotic.(16)

The individual Japanese infantryman was a master at infiltration and camouflage. Many even carried pieces of matting of the same color as the terrain background; running behind these in a crouching run, they were practically invisible through the sights of a weapon. Thoroughly professional at digging, they normally constructed individual foxholes in the shape of an inverted boot, the "toe" facing the enemy and containing the fire step, the deep "heel" in the rear, providing security from enemy artillery and mortar bombardment. The roof of the boot, if not underground itself, was usually heavily protected by logs and earth.(17) In the area of individual weapon handling, the Japanese soldier was renowned for his excellence in the bold and efficient employment of the mortar, which appeared to be his favorite weapon. He also used machine guns with imagination and effect, sometimes even positioning them in trees to gain longer fields of fire. In other instances, they were dug in on open ground but protected by rifle fire, in which case assaults on them became virtual death traps for attacking infantry. Surprisingly, however, the Japanese infantryman was discovered to be a "notoriously poor marksman."(18) Even Japanese snipers, expert at camouflage, were only trained up to ranges of 300 yards.(19)

The Japanese infantry platoon consisted of 42 all ranks, organized in four sections of ten men each. It was commanded by a second lieutenant supported by a noncommissioned communications officer or <u>Renrakukashi,</u> responsible for maintaining communications by visual means or through runners between sections. The first three sections were organized as rifle sections, each comprising seven riflemen armed with 6.5-millimeter or 7.7-millimeter rifles, and a 6.5-millimeter Nambu light

TABLE 7-1. JAPANESE RIFLE SECTION

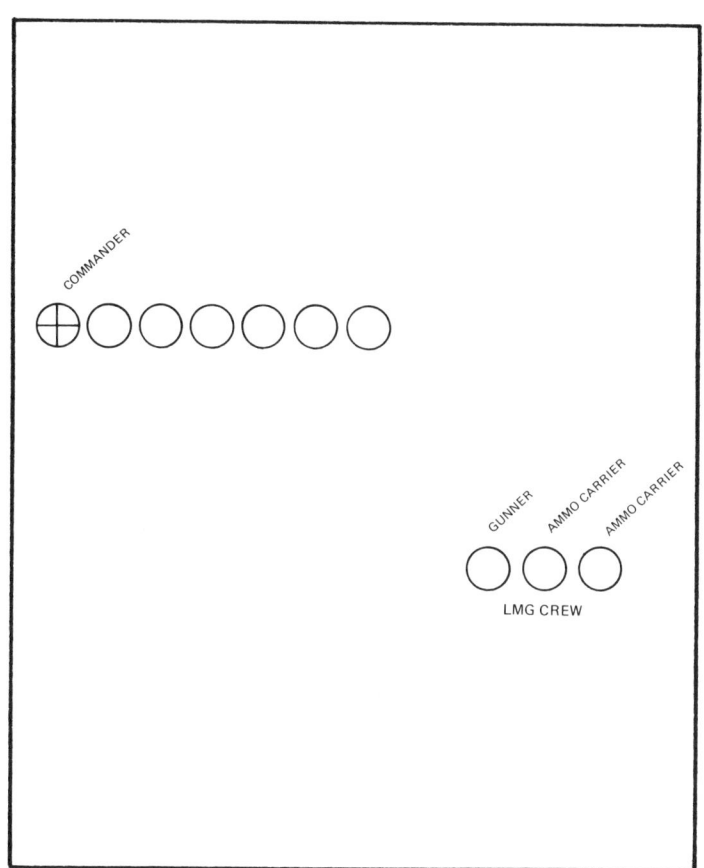

machine gun, served by a gunner armed with a pistol and two riflemen acting as ammunition carriers. There were no organized subsections, and the light machine-gun group was not intended to be divided from the rest of the section for action. No submachine gun was included in the section armament, but hand grenades were carried by all. (The Japanese did have an 8-millimeter fully automatic submachine gun however.) The fourth section was called the Tekidanto "knee mortar" (grenade discharger) section, and it included three 50-millimeter grenade dischargers,(20) each carried and operated by one man assisted by two riflemen ammunition carriers. Again, this section was not intended to be split up but was to be used as a reserve of firepower. The platoon commander gave all orders for the deployment and engagement of sections.(21)

A Japanese rifle company commanded by a captain comprised three rifle platoons. Three such companies plus a machine-gun company made up an infantry battalion, commanded by a major. The machine-gun company consisted of three platoons, each of two sections of ten men and one M-92 (1932) 7.7-millimeter heavy machine gun. Three battalions, a 75-millimeter infantry gun battery, a 47-millimeter antitank gun battery, and a headquarters comprised a regiment of just over 2,000 men, commanded by a colonel; a lieutenant colonel served as second in command. Japanese divisions were organized along both square and triangular lines, the latter type consisting of about 12,000 all ranks. Interestingly, with practically no Chinese tank forces to worry about, Japanese antitank guns remained essentially dual purpose.(22)

Tactically, when a Japanese rifle platoon contacted an enemy, the first two sections immediately engaged in frontal holding actions, while the third rifle section attacked either the left or right flank. Double envelopment was rarely carried out for danger of hitting friendly troops by crossfire from both flanks. The Tekidanto section usually remained in the center rear, providing additional fire support as required. At company level, a leading platoon would normally be preceded by six scouts, ranging about 350 yards ahead. On contact with an enemy, the two following platoons usually attempted a double-envelopment maneuver.(23) This was normally accomplished through infiltration, in which technique the Japanese infantry was extremely well versed. In action against the British in Burma, the Japanese "fought the jungle, whereas the British fought the roads."(24) According to General Slim, the Japanese ability to move through the jungle more freely than the British, whose mechanical transport system made them essentially road bound, gave the Japanese "every advantage--advantages which they had earned and deserved." The only remedy, as far as he was concerned, was to learn how to move on a light scale, "to shake loose from the tin-can of mechanical transport tied to our tail."(25)

How much Japanese tactics were dictated by British methods of combat is difficult to say. It is known, however, that the Japanese held that "quick, decisive

TABLE 7-2. JAPANESE INFANTRY REGIMENTAL ORGANIZATION (WORLD WAR II)

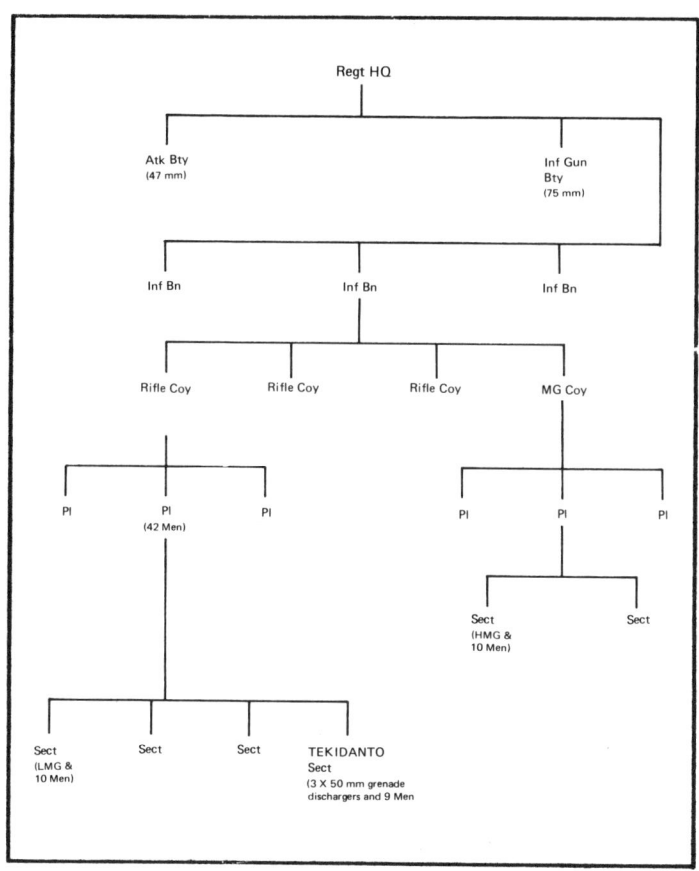

battle should be sought by flanking and encirclement" because the British army, though possessing "some mechanical mobility in general," did "not have much manoeuverability." It was also noted that British firepower was preponderantly aimed to the front in defense, an additional reason for stressing flanking and enveloping maneuver. Night fighting, always considered a Japanese specialty since the Russo-Japanese War, was given even stronger emphasis, as the British were not considered particularly good at it. Because the British were also regarded as methodically exhaustive in the offense and cautious in effecting encirclements of their own, the Japanese advocated the adoption of surprise countermeasures such as spoiling attacks and "encircling the enemy's encircling force."(26)

The flanking and encircling tactics of the Japanese, which were probably derived originally from the German model, were termed "road-block" tactics by General Slim and credited with dominating all British operations in Malaya and Burma during the period of defeat.

> Tactically we had been completely outclassed. The Japanese could--and did--do many things that we could not. The chief of these and the tactical method on which all their successes were based was the "hook". Their standard action was, while holding us in front, to send a mobile force, mainly infantry, on a wide turning movement round our flank through the jungle to come in on our line of communications. Here, on a single road, up which all our supplies, ammunition, and reinforcements must come, they would establish a "road-block" sometimes with a battalion, sometimes with a regiment. We had few if any reserves in depth--all our troops were in the front line--and we had, therefore, when this happened, to turn about forces from the forward positions to clear the road-block. At this point the enemy increased his pressure on our weakened front until it crumbled. Time and again the Japanese used these tactics, more often than not successfully, until our troops and commanders began to acquire a road-block mentality which often developed into an inferiority complex.(27)

In Malaya and Java, Japanese formations often made sweeps as wide as three miles and as deep a four to six miles. These were normally effected at a slow but steady pace. Advance guards, in the meantime, pressed forward in small groups, usually of platoon and no greater than company strength, attempting to get forward between enemy positions in order to get behind any organized defenses. Each group was normally given a definite mission, such as seizing a particular point or attacking a located headquarters or flank. When counterattacked, these groups held their fire, letting the attack go through them and ultimately engaging it from the rear and flanks. On numerous occasions, Japanese companies hid in swamps and rivers up to their necks in water waiting for such situations to develop.(28)

Although the Japanese army was steeped in the offensive, it rapidly gained a reputation for being equally formidable in the defense. Strong believers in active defense, the Japanese were often known to crawl close to enemy lines when bombarded by hostile artillery and mortars, not just to enhance their own security--as such action did--but to bring light mortar fire to bear on the enemy lines as well.(29) The machine gun, of course, remained the principal weapon of Japanese defensive operations, and heavy models were often sited well forward to cover main lines of approach. Reverse-slope positions were commonly used, and extensive digging always proved the rule. As the war progressed, the Japanese turned more and more to defensive systems in depth. All positions and trenches throughout the war were made as mutually supporting as possible.(30)

> It was something of a wonder to British officers [at Kohima] to learn how few men there were in some positions. The secret was that the Japanese did not fight to their front if they could fight to a flank. This meant that they had to rely on neighbouring bunkers for the protection, that is "to cover them", while they covered their neighbours. This system involved a good deal of training and discipline, and a consistently high standard in the siting of posts; but it did make the maximum use of firepower. British and Indian troops were psychologically incapable of such tactics, each man preferring to fight to his front and remaining responsible for his own protection. Also, of course, though recognizing the need for head cover, the British hated being entombed in bunkers and liked the free use of their weapons, denied by Japanese-type bunkers. But these bunkers did allow the Japanese to bring down mortar fire on their own positions, when under attack, and time and again drove the British and Indians from them before they could dig in. And in Burma the Jap mortarmen were the counterparts of the German machine-gunners in the First World War.(31)

According to Slim, the strength of the Japanese army lay not in its higher leadership but in the spirit of the individual Japanese soldier, who "fought and marched till he died." In the end, he was shattered and beaten through the combination of superior air and firepower and basic infantry tactics. Under Slim's

guiding hand, the myth of Japanese superiority was
dispelled. The British infantryman learned "to move on
his own feet and to look after himself." Mastering the
absolutely vital art of patrolling and realizing that
mobility and survival were synonymous, he quickly
gained in confidence. Slim built further on this by
launching a series of minor offensive operations in
which only attainable objectives were assigned to
troops participating. Like the Germans in their "snail"
offensive in Russia, he avoided confidence-sapping
defeat by overinsuring for victory, often sending
brigades to attack companies, battalions to assault
platoons. Emphasis throughout was placed on skilled and
determined leadership at the junior level, firm
discipline, and physically tough, self-reliant troops
capable of effecting long marches and withstanding
hardship generally. Reduced scales of transport and
equipment were accordingly introduced, to the point
where an Indian division required but 120 tons of daily
supplies in lieu of the 400 per day considered normal
for sustaining a standard division in the field.(32)
Slim noted with some relish that "as we removed
vehicles from units and formations which joined us on
European establishments, they found to their surprise
that they could move farther and faster without
them."(33)

In actual operations, greater tactical freedom was
given to subordinate commanders. Companies, even
platoons, under junior leaders became the basic units
of the jungle. Out of sight of one another, often out
of touch, their wireless blanketed by hills, they
marched and fought on their own. Instead of retreating
when outflanked, troops now tended to withdraw into
"keeps" or strongholds, dubbed hachi-no-su or "beehives"
by the Japanese; maintained by airborne supply these
functioned somewhat like an anvil on which Japanese
intrusions were hammered by mobile reserves. The
Japanese army never fully recovered from its defeat in
the Imphal-Kohima battle, which, significantly, was
inflicted by a regular land force. Slim was personally
very much opposed to the formation of special forces,
which he considered "wasteful." For a commander who was
often critically short of his most important commodity,
infantrymen,(34) this observation is not particularly
surprising:

> The result of these . . . special units was
> undoubtedly to lower the quality of the rest of
> the Army, especially of the infantry, not only by
> skimming the cream off it, but by encouraging the
> idea that certain of the normal operations of war
> were so difficult that only specially equipped
> corps d'elite could be expected to undertake them.

> Armies do not win wars by means of a few bodies of super-soldiers but by the average quality of their standard units. . . . The level of initiative, individual training, and weapon skill required in, say, a commando, is admirable; what is not admirable is that it should be confined to a few small units. Any well-trained infantry battalion should be able to do what a commando can do; in the Fourteenth Army they could and did.(35)

The "forgotten war" in Burma was gradually given less and less attention by the Japanese as they began to focus on the greater menace of the inexorable American juggernaut rolling across the Pacific. Strategically a two-pronged effort, the brilliant, comparatively low-casualty Southwest Pacific campaign of General Douglas MacArthur contrasted sharply with the battering-ram style naval advance across the Central Pacific Theater.(36) The tactics of the Americans in the Central Pacific could not be termed indirect, subtle, or "tricky." Unlike the Japanese, who effected landings through surprise, the Americans, with their preponderant naval and air resources, effected them through enormous firepower followed by frontal "storm landings." On the average, they enjoyed a numerical superiority of 2 or 5 to 1: at Attu and the Gilberts, it was 5 to 1; at Saipan 2.5 to 1; and at Guam 2 to 1.(37) MacArthur, not unnaturally, was highly critical of Central Pacific methods, believing that thousands of American soldiers and marines were sacrificed needlessly. He argued that at Okinawa, once sufficient area had been conquered for American purposes, instead of driving the Japanese off the island, American commanders should have "had the troops go into a . . . [defense] and just let the Japs come to them and kill them from a defensive position, which would have been much easier to do and would have cost less men."(38) His argument appears reasonable in light of the fact that the Japanese had begun by this time to forfeit their advantages in tactical skill--for example, superior night-fighting capability--by a growing conformity to an unrealistic military tradition, namely, the sacrificial "banzai" attack.(39) This tactic, and the "stand and die" policy adopted by many Japanese commanders, undoubtedly accounted in large part for the tremendous difference in battle-casualty figures between American and Japanese forces. It was a hard way to learn the central fact of modern war that once a battle is lost, casualties tend always to mount for the loser as combat continues.

Although the American victory in the Pacific was predicated on the overwhelming power of naval and air forces, the war on land was waged essentially by infantrymen. It is interesting to note, therefore, that in both the Southwest and Central Pacific theaters, two American infantries were committed to battle. While both infantries were much materially superior to the Japanese infantry in machine-gun and small arms firepower, they were mutually different in their basic organization. The army infantry that fought in the Pacific was for the sake of expediency organized along the same lines as the infantry serving in Europe. Marine infantry, on the other hand, had evolved and continued to evolve in another fashion, for which reason it will be examined in the greater depth in this chapter.

Experience in Nicaragua in the early part of the century had persuaded the Marine Corps of the importance of the automatic rifle as a base of fire. Involvement in quelling street riots in Shanghai in 1938 had already led to the adoption of the four-man fighting team as the basis for a Marine riot company. By World War II, the Marine infantry platoon comprised a seven-man headquarters, an eight-man BAR squad, and three nine-man rifle squads. Each squad consisted of a squad leader, a BAR man, six riflemen, and a rifle grenadier armed with a grenade launcher. The riflemen were armed with the M 1903 .30-caliber Springfield. The M-1 was not issued to the Marine Corps until after Guadalcanal. Any Marines who used the M-1 during that particular battle picked them up from members of the 164th Infantry.(41)

The platoon organization described above was not found to be completely suitable for Marine operations, however, as the squad was the lowest fire-control level and could not be broken down into a smaller tactical unit. Traditional Marine experience and interest accordingly came into play at this juncture, and the "most dramatic revolution" in Marine Corps organization subsequently resulted. Three additional tables of organization were eventually introduced during World War II, during which process squad organization, to the eternal credit of the Marine Corps hierarchy, received the most detailed scrutiny by Marine general officers. First, the BAR squad disappeared from the platoon and the rifle squad was increased in size to 12 men; a squad leader, an assistant squad leader, six M-1 riflemen, two assistant BAR-man armed with M-1s, and two BAR-men. So organized, the squad could be broken down into two six-man units, each containing an automatic rifle and five semiautomatic rifles. Experimentation continued, however, particularly among paramarine units and Marine raider battalions, especially Major Evans F. Carlson's Second Raider Battalion. Carlson had spent six months with the Communist Chinese army in North China in 1937 and had

CINDERELLA OF THE ARMY   213

been highly impressed with their organization in three-man fire groups. The squads he initially organized in his battalion resembled the Chinese pattern inasmuch as they consisted of a corporal commander and nine men, organized into three fire groups of three men each. Emphasis in training was on

TABLE 7-3. U.S. MARINE "FIRE TEAM" SQUAD ORGANIZATION (WORLD WAR II)

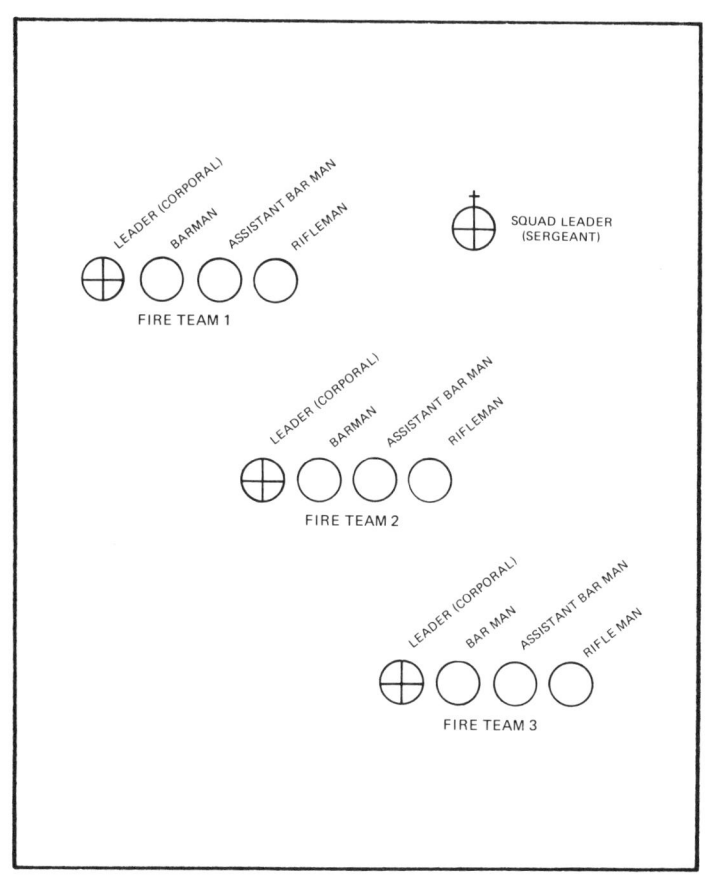

speed of movement on foot, endurance, self-sufficiency, and maximum firepower compatible with such mobility. Total squad armament included five Thompson submachine guns, four M-1 rifles, and one BAR. Each group, led by a scout armed with an M-1, was supported by two automatic weapons; all three groups were mutually supporting. Carlson argued that a squad so organized could cover a front of from 100 to 300 yards, as opposed to the former 50-yard frontage covered by the orthodox infantry squad armed with Springfields and one BAR. In March 1943, the squads of the First Raider Battalion were reorganized into three groups of three men each, with a corporal squad leader. Each fire team was equipped with one BAR, one carbine, and one M-1; the senior man was appointed fire-team leader. The value of such organization was proved in actual combat in New Guinea and elsewhere.(42)

At Camp Pendleton in July 1943, further experiments were conducted, based on lessons learned in action. At first, a 2-BAR, 13-man squad was tried. It was later recommended that the squad be organized into four three-man fire groups. This organization was rejected, however, in favor of a three four-man division, which was thought to be more in line with Marine triangular organization, better able to absorb battle casualties, and easier to control. The organization that finally evolved in March 1944 called for a squad of 13 organized into three four-man fire teams built around a BAR. The squad thus comprised a squad leader armed with a carbine, three fire-team leaders, and three riflemen (armed with M-1s and M-7 grenade launchers), three assistant BAR men armed with carbines and M-8 grenade launchers, and three BAR men.(43) The squad leader was a sergeant, and the fire-team leaders, corporals.

The rifle squad so formed seemed to give the commander the requisite control and additional firepower found to be so necessary in both jungle and island fighting. Independent action by small units was essential in the bush-choked terrain of Bougainville and New Britain. According to General Gerald C. Thomas, Marine Director of Plans and Policies at the time, the switch to fire-team organization was made "because a leader could not control seven men in combat, so we certainly would not expect him to control twelve." Under the fire-team concept, the squad leader was responsible for the training, control, and general conduct of his squad. He was to coordinate the employment of his fire teams in a manner that would accomplish the mission assigned by his platoon commander. He was also responsible for the fire control, fire discipline, and maneuver of his fire teams as units. The fire-team leaders were similarly responsible for their fire teams. By the delegation of command authority to squad and fire-team leaders, it was believed that military leadership would be more widely disseminated and that the rifle squad would become more aggressive and efficient.(44)

With Marine emphasis on firepower and more firepower, particulary in automatic weapons at lower levels, the rifle company grew from 183 in 1942 to a strength of 242 in 1945. This growth rate was, in part, caused by the three-platoon battalion machine-gun company being parceled out to rifle companies on a permanent basis. A rifle company by 1945 thus consisted of three platoons (each of three 13-man squads and a 6-man headquarters), a 56-man machine-gun platoon of 14 guns (air and water cooled), and a company headquarters that included a 20-man 60-millimeter mortar section of three mortars. Three such companies, a 58-man, 81-millimeter mortar platoon of 12 tubes, a 55-man assault platoon, and a headquarters group made up a battalion. The marine assault platoon was added to forestall the necessity of denuding rifle companies of men to reduce Japanese strongpoints. The assault platoon consisted of three sections of two seven-man squads, each comprising a squad leader, flamethrower team of two, bazooka gunner and assistant, and two demolition men. Three battalions, two 37-millimeter gun platoons, a 105-millimeter (formerly 75-millimeter) self-propelled howitzer platoon, services company, and headquarters made up a marine regiment of 3,412, all ranks.(45)

During World War II, the U.S. Marines unquestionably brought amphibious warfare to a peak of tactical perfection, becoming in the process the foremost experts in the world in this form of warfare. General Holland M. Smith, Commanding General, Fleet Marine Force Pacific, during the war, has even been referred to as the "father of amphibious warfare." Obviously a highly specialized style of warfare, amphibious operations call for a tremendous amount of detailed planning and coordination of all arms. It also called for overwhelming superiority of force. In the Central Pacific campaign, seaborne assaults against strongly fortified positions represented in grand fashion "truly amphibious warfare a l'outrance."(46) Yet, as the experiences of Tarawa and Peleliu taught, after the landings were effected, time-consuming and costly assaults on fortified positions and pillboxes became the hallmark of operations. It is perhaps worth noting at this point that MacArthur refused to describe his campaign as "island-hopping," referring instead to his policy of "hit 'em where they ain't--let 'em die on the vine" as exactly the opposite.(47)

At small-unit level, "storm"-style landings called

TABLE 7-4. U.S. MARINE INFANTRY REGIMENTAL ORGANIZATION (WORLD WAR II)

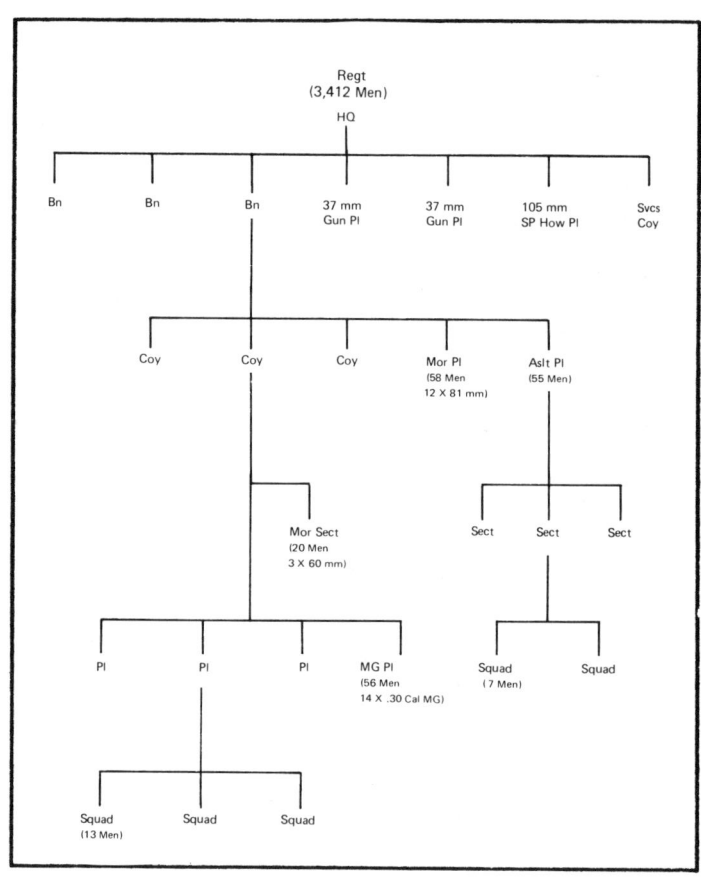

for maximum fire support from ships, artillery, and planes and, in addition, the closest coordination of tanks and demolition teams. Once leaving the protection of special landing craft, however, foot soldiers--if they were to survive--were also required to project maximum small arms and machine-gun fire forward. The Marines, with their preponderant number of automatic weapons, were much better able to do this than a standard army unit attacking from the seaward in waves.(48) As most of these actions tended to take on a frontal character, however, they may have colored Marine tactical thinking generally. The legendary Marine general, Lewis B. Puller, went so far as to advise that there was "mighty little room for fancy tactics below division level." In his view, it was simply a matter of:

> The enemy are on the hill. You go get 'em. In the end you'll save more. There are times when you'll have to flank, but don't forget that the shortest distance between two points is a straight line.(49)

In the attack on Peleliu island, where the Japanese for the first time defended in depth as well as on the beach, Puller's First Marines lost in total casualties 56 percent of their strength, the highest regimental losses to that time in the history of the corps.(50) The Japanese resistance was extremely heavy in this instance, however, particularly on Umurbrogal ("Bloody") Ridge, so the high casualties cannot necessarily be attributed to a penchant for frontal attacks. No strangers to jungle warfare, the Marines tended to regard jungle tactics as simply the common-sense application of standard tactical principles and methods in tropical terrain and vegetation. Whether in Manchuria or Guadalcanal, infantry combat appears to have been much the same. According to Marshall, the Marines at least outstripped the army in getting down to carrying only basic combat essentials. In general, then, the Marines fought well in the Pacific, exacting more enemy casualties than they normally incurred themselves. This was also to prove the case in Korea against other Oriental enemies, who also had beaten the Japanese.(51)

If World War II demonstrated that basic infantry skills were still critical to success on all fighting fronts, it also confirmed that the Japanese were not the only Asiatics who could be made into good soldiers. Generals MacArthur and "Vinegar Joe" Stilwell had thought as much for some time, of course, and the Indian Army had long since ably demonstrated the fact.(52) However, this in no way lessened the acute shock felt in Western circles by the rout of modern United Nations (UN) forces in Korea by a Chinese Communist army composed primarily of infantry. It was the most cataclysmic defeat ever inflicted on an American field army. In the hour of its anguish, the Eighty Army was a "wholly modern force technologically," the Chinese army but a "peasant body composed in the main of illiterates."(53)

218 A PERSPECTIVE ON INFANTRY

The infantry arm was--and still is for that matter(54)-- the pride of the People's Liberation Army (PLA), an organization that had been fighting steadily since 1927 against both the Chinese Nationalists and the Japanese until its victory over the former in 1949. By the time of the Korean conflict, the PLA was an indoctrinated force endowed with a rich experience and equipped with a dynamic doctrine of guerrilla and mobile warfare. It also possessed a tradition of victory. Its soldiers, though not all ideologically pure, were generally confident and convinced of the righteousness of their cause. Only the best and most politically reliable units had been selected to fight in the war to "Resist Aggression and Aid Korea." The leadership of the PLA was of a high quality, practically all of the officers having fought with the Eighth Route Army against the Japanese. Most battalion, company, and platoon commanders, as well as NCOs, were veterans of the civil war. All were imbued with the tactical doctrine that the Communists had applied with such success against both enemies. Based on mobility, deception, distraction, and surprise, this doctrine called for concentration of superior force at a vital point, a "short attack," and speedy disengagement. Thorough reconnaissance an essential preliminary for all attacks, accounted for the almost uncanny Chinese ability not only to strike along boundaries between enemy units but to flow along neglected avenues of approach deep into the rear of enemy positions.(55)

The Chinese entry into Korea was accomplished in outstanding military fashion. A misplaced Western trust in air surveillance ensured their achievement of total surprise when they eventually descended from the hills on road-bound UN forces in the autumn of 1950. Marching only by night and resting by day to avoid detection, the PLA managed to concentrate 300,000 troops in North Korean hills: 19 divisions (180,000 men) in the west and center opposing the UN Eighth Army and 12 divisions (120,000 men) in the east confronting the U.S. Tenth Corps. Most movement was accomplished on foot, and one division at least averaged 18 miles a day for 18 days. The march normally began after dark around 1900 hours and ended at 0300 the following morning. During daylight, only scouting parties moved forward to select bivouac areas for the next day's rest. Strict camouflage and march discipline were rigidly enforced.(56) All in all, it was a magnificent military feat.

The Chinese soldiers who crossed the Yalu were armed with every conceivable type of weapon: American M-1 rifles and carbines, old Japanese rifles, and new Soviet submachine ("burp") guns. Individuals carried 80 rounds of rifle ammunition and four or five "potato masher"-type grenades. In addition to his basic combat load, the PLA soldier carried a few extra clips for automatic rifles and "burp' guns, loaded belts for machine guns, and one or two mortar shells or TNT for satchel charges. (The Chinese had no antitank weapons, so each platoon carried enough TNT to make 8 to 10 five-pound satchel charges). The Chinese soldier was reasonably attired for operations in temperate winter conditions inasmuch as he was issued with a heavily quilted cotton uniform for wear over his summer dress. Usually mustard brown in color--although some varieties were dark blue--this uniform was white on the inside and often reversed for fighting in the snow.(57) The Chinese soldier wore no helmet, only a heavy cotton cap with big ear flaps. His shoes were usually rubber or canvas sneakers fitted over layers of cotton socks, although many of the first men across the Yalu had been issued fur boots.(58) For sustenance, every man carried emergency rice, tea, and salt for five days: an "iron ration," to be supplemented by requisitioning from the natives the Korean staple of millet seed, rice, and dried peas ground into a powder. Whenever possible, the soldier cooked these rations; otherwise, he ate them cold. Such spartanism on the part of the individual infantry soldier meant, of course, that the PLA had practically no administrative tail.(59)

There are indications that the Chinese army expected a quick and relatively easy victory in Korea. Steeped in the "man-over-weapons" military philosophy of Mao Tse-tung, Chinese leaders launched their major offensives with confidence that the superior doctrine, tactics, and morale of their best armies could defeat the better-equipped foe.(60) While recognizing U.S. superiority in naval, air, and artillery firepower and overall coordination of arms, the Chinese appreciated that the Achilles' heel of the system might lie in the opposing infantry. One Chinese estimate of the American foot soldier was not that favorable:

> Their infantrymen are weak, afraid to die, and haven't the courage to attack or defend. They depend on their planes, tanks, and artillery. At the same time, they are afraid of our fire power. They will cringe when if on the advance, they hear firing. They are afraid to advance farther. . . . They specialize in day fighting. They are not familiar with night fighting or hand to hand combat . . . If defeated, they have no orderly formation. Without the use of their mortars, they become completely lost . . . they become dazed and completely demoralized. . . . They are afraid when the rear is cut off. When transportation comes to a standstill, the infantry loses the will to fight.(61)

After analyzing the strengths and weaknesses of the UN forces, the Chinese refined the essential tactical principles of Mao Tse-tung(62) to accommodate the reality of the Korean situation. The avoidance of highways and flat terrain became central to their operations, which always had as their object the interposition of force on an enemy's line of retreat.(63) Like Chuikov's Russians at Stalingrad, the Chinese chose to adopt "hugging" tactics of getting in as close as possible to the enemy. Night attacks became so much the rule for them that any exception came as a surprise. By this approach, the Chinese army nullified to a substantial degree UN close air-support capabilities and its preponderance in heavy weapons and combined arms. At the same time, Chinese primitive means of transport largely negated UN aerial interdiction capabilities, though UN formations, heavily dependent on roads for resupply, remained very much affected by Chinese tactics of wide envelopment and infiltration.(64)

In general, the inhospitable terrain of Korea further served to minimize the disparity between forces. Whereas UN troops found the countryside a handicap to their operations, the Chinese turned the rugged hills and desolate valleys to good advantage. The Chinese infantryman was an excellent camofleur, and many UN troops were often taken under fire at almost pointblank range by skillfully concealed machine guns and automatic weapons. Equally serious, the broken mountainous landscape, particularly on the east coast, severely restricted UN tactical wireless communications, while guerrilla remnants of the North Korean army and Communist Chinese infiltrators cut rear-area wire lines almost as fast as linesmen could lay them. An inability to communicate in moments of crisis spelled disaster for many an isolated UN platoon and understrength company. The Chinese, obviously, were not so critically affected, as their radio nets only extended down to regimental level and telephones only to battalions or occasionally to companies. Below battalion, PLA communication normally depended on runners or such signaling devices as bugles, whistles, flares, and flashlights. Even then, by a seemingly strange twist, the resultant cacophony that invariably accompanied a Chinese attack had the advantage of being psychologicaly extremely unnerving to an enemy being attacked at night.(65)

Organizationally, the PLA was generally structured in accordance with the triangular concept. An army comprised three 10,000-man divisions, each composed of three infantry regiments, a pack-artillery battalion, and engineer, communications, transport, and medical companies. An infantry regiment had a strength of roughly 3,000 men and comprised three battalions, each of three companies of three platoons. In Korea, the Chinese also employed a number of independent brigades, regiments, and detachments. Like the Russians during the battles for Moscow and Stalingrad, the Chinese were extremely flexible in their organizations, utilizing the ad hoc or "task force" method of shifting subordinate units freely to compose forces or groups deemed appropriate to the mission assigned.(66)

The most noteworthy organizational feature of the PLA remained, however, the "three-by-three" structure of its 12- to 16-man squads. No doubt linked to the politicization of its army leadership, each squad in the Chinese army consisted of three small teams of three, four, or five men plus a squad leader. The squad leader was normally a Communist Party member and the team leaders either members or aspirants to membership; they were also the most combat experienced. This squad organization, largely ignored by many Western observers, was effective in both its tactical and institutional aspects. Tactically, a squad so organized provided great flexibility at the lower level. It permitted a squad leader to exploit the terrain and to take advantage of the enemy situation by classic methods of fire and movement. It also provided a mechanism for assured control and constant surveillance. As members of the squad ate, lived, slept, studied, marched, trained, and fought together, they developed a distinct cohesion, cemented by commitment to a common ideology and shared aspirations and experiences. Any deviations from the political or motivational norm could be recognized early by team and squad leaders and corrected before they assumed problem proportions. It is conceivable, therefore, that this basic organization, with its stress on supervision, reflected a desire to solve a long-standing problem of warfare that modern weapons have merely exacerbated, namely, the problem of getting everyone in combat units to fight. General James Van Fleet, one of the few observers struck by this organization, noted that though most soldiers in the PLA were not Communists at all, they fought well. This he attributed to the "three-by-three" organization.(67) Interestingly, Ardant du Picq foresaw a century earlier that lower-level supervision would be required on an ever-increasing scale:

> In modern armies where losses are as great for the victor as for the vanquished, the soldier must more often be replaced. In ancient battle the victor had no losses. To-day the soldier is often unknown to his comrades. He is lost in the smoke, the dispersion, the confusion of battle. He seems

to fight alone. Unity is no longer insured by mutual surveillance. A man falls, and disappears. Who knows whether it was a bullet or the fear of advancing further that struck him! The ancient combatant was never struck by an invisible weapon and could not fall in this way. The more difficult surveillance, the more necessary becomes the individuality of companies, section, squads. Not the least of their boasts should be their ability to stand a roll call at all times.(68)

Contrary to many press reports, the Chinese in Korea did not employ "human sea tactics," frequent reference to which inspired the derisive Marine comment: "How many hordes are there in a Chinese platoon?"(69) In reality, the Chinese Communists seldom attacked in units larger than a regiment, and even these were usually reduced to a seemingly endless succession of platoon infiltrations. According to the U.S. Marine Corps official history, it was not mass but deception and surprise that made the PLA a formidable opponent in the field. Individually, Chinese soldiers liked to get "in close," crawling toward the enemy under cover of darkness and then, to the blaring of bugles and shrilling of whistles, jumping up to hurl grenades and charge. Lightly equipped and clad, the Chinese infantryman was capable of great battlefield mobility. A master of stalking and fieldcraft technique, it was not uncommon for him to "rise out of the very earth" in the vicinity of UN positions, often around midnight, and launch--to the very great shock of all--a short, vicious surprise attack. Such offensive tactics prompted one American officer to describe a Chinese attack as a virtual "assembly on the objective." Masterful use of terrain, deception, infiltration during darkness, and close combat in which superior Chinese numbers were applied at vital points in the deep flanks and rear (like the Japanese they employed the "road block") was to spell disaster for many UN units and formations.(70)

According to an official U.S. Army report, the Chinese made excellent use of terrain during their attacks, but unlike their North Korean allies, they were not prepared to defend it to the death. Withdrawal was as important to their tactics as the advance. Additionally, PLA troops were always maneuvered (even in patrolling) regardless of the size of the unit, and attacks were habitually launched from more than one direction. Everywhere the method was the same: first, probing by 8- to 15-man groups to feel out the UN defensive position, define its outline, create confusion, and draw fire, particularly from automatic weapons; then, while "short attacks" of fierce intensity struck fronts and flanks, the Chinese in small groups seeped simultaneously into the rear where roadblocks turned withdrawals into bloody debacles. The attacks followed a pattern developed earlier by Lin Piao, known as the "one point-two sides method," basically a V-shaped maneuver incorporating a frontal fix at the base with simultaneous double envelopment executed suddenly and precisely. The object of these enveloping tactics was to surround, isolate, and destroy piecemeal separated enemy elements. The attacks were characterized by their intensity, a quality the Chinese described san-meng kung-tso, or the "three fierce actions" of fierce fires, fierce assaults, and fierce pursuits. When taken under effective enemy fire, the Chinese infantryman went to ground, crawling from one position to another, maintaining the attack by continually seeking open or exposed flanks. As a rule, the Chinese soldier was "well and courageously led at the small unit level."(71)

One week after the Chinese armies lashed out at UN forces, the Eighth Army was forced back 50 miles in its center, precipitating the longest retreat in American military history. The effect of surprise and the incredibly high standard of infantry skills among its soldiers enabled the PLA to decisively wrench the initiative away from the UN. Furtive, light of foot, and highly elusive, the Chinese soldier again raised the bogey of the "superman." The lessons of Pacific fighting against a technically inferior but determined Asiatic enemy had evidently, and most unfortunately, been largely forgotten. Generals Ridgway and Puller would afterward remark that the Eighth was a "fleeing army,"(72) that "the Communists had seriously defeated U.N. forces." Puller later recounted his discovery and impressions of an American army artillery battalion that had been surprised by Chinese soldiers:

> They had fought hardly at all; the Reds had worked so fast that few shots were fired from our weapons. . . . It was a disgrace to American arms. . . . this was not the only such incident. If I saw one shot-up American battalion in Korea I saw fifty, and I mean fifty.(73)

Puller, regimental commander of the First Marines in Korea, was convinced that "fancy weapons systems" could not in themselves guarantee success in war. What was required, in his view, was hard training and the development of a certain attitude, "the fundamental spirit that alone can produce great armed services." The philosophy he espoused was, in fact, very much akin to that of the "man-over-weapons" idea subscribed to by the Communist Chinese. During training at Camp

Pendleton in preparation for Korea, Puller spent "most of his waking hours on the range," constantly striving to ensure that firepower would be provided and that it would be accurate. A strong believer in the value of day and night forced march training, he ranted: "I want 'em to be able to march twenty miles, the last five at double time, and then be ready to fight."(74)

When the First Marine Division, as part of X Corps, was surrounded in the area of the Chosin Reservoir by General Sung Shin-lun's Chinese Ninth Field Army, which had as its object the annihilation of the Marine body, the correctness of Puller's training philosophy was demonstrated. Under appalling conditions and a seemingly hopeless situation, the First Marine Division--its dead lashed to vehicle hoods, running boards, and gun barrels--advanced "in another direction" for 13 days and 35 miles through strong Chinese resistance out of the "Frozen Chosin" and into legend. Behind, on naked hills and snow-filled valleys, it left the battered and demoralized remnants of seven Chinese divisions. The First Marines suffered 7,500 casualties, half of whom were frostbite cases, but they had saved the Tenth Corps.(75)

It is worth noting that Marine squad organization at this time was roughly identical to that of the Chinese enemy. The most common Marine attack formation was also V-shaped, two rifle platoons in line with a third in close support. The weapons company had been restored to a Marine battalion before Korea, but the Marine rifle battalions that fought there had only two rifle companies. Although the Marines had never added a weapons squad to a rifle platoon, as the U.S. Army did after World War II, their rifle company machine-gun firepower in Korea remained six to three times more powerful than that of the army. (Rifle companies in both organizations possessed three 60-millimeter mortars.) However, support weapons firing ammunition above rifle caliber generally tended to be either present in smaller numbers or at a higher level in Marine organization. Marine philosophy dictated that heavier weapons be controlled at battalion level. Major attention in rifle companies was primarily given to rifle and machine-gun fire. In the Marines, first emphasis was always placed on fire, second on maneuver.(76)

Interestingly, American army infantry organization had undergone a radical change before the Korean conflict. World War II battle experience had shown rather conclusively that it was extremely difficult for one man to effectively control a group as large as 11. Consequently, in 1947, the rifle squad was reduced by three men (the ammunition bearer and two scouts). The rank of the squad leader was not accordingly reduced,

TABLE 7-5. U.S. ARMY INFANTRY PLATOON ORGANIZATION
(KOREAN WAR)

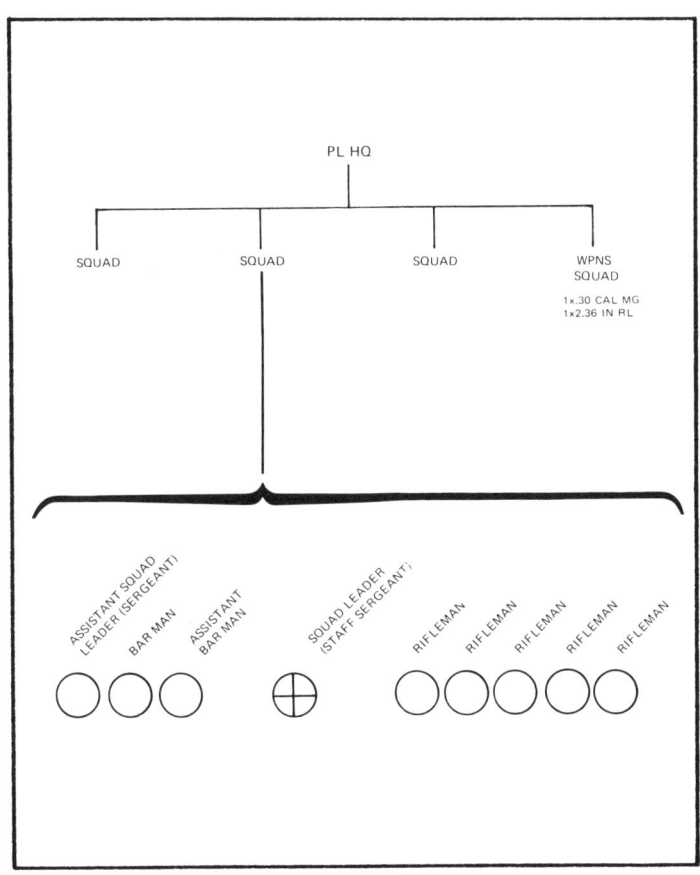

however, and the nine-man squad thus comprised one staff sergeant (squad leader), one sergeant (assistant squad leader), one corporal (BAR automatic rifleman), one private (assistant automatic rifleman), and five privates (riflemen). This squad structure, built around the concept of an automatic weapons team supported by riflemen, bore a striking resemblance to British Commonwealth organizations except for rank differences. At about the same time, a weapons squad was added to the rifle platoon. Commanded by a staff sergeant, it consisted of a corporal (machine gunner), private (assistant machine gunner), two privates (ammunition bearers), one .30-caliber light machine gun, and one 2.36-inch rocket launcher. In the infantry company weapons platoon, two light machine guns were replaced by a section of first one and then three 57-millimeter recoilless rifles.(77)

In general, Korea offered a bitter school of experience for the American army. After reaching a peak strength of 8,290,000 in May 1945, it had dwindled to but 592,000 by the summer of 1950. (By comparison, at the time of Pearl Harbor, there were 1.6 million men in the army). Even worse, the 1950 army of 592,000 was top-heavy with technicians and service people, for the "myth of the push-button war" had "downgraded the foot soldier."(78) The American conscript soldier who eventually bore the brunt of fighting in Korea was therefore relatively untried and inexperienced. The effects of the infamous Doolittle Report had also eroded leadership and the dyadic buddy system replaced the primary group. The latter phenomenon was brought on by the American system of personnel replacement, which additionally saw to it that officers often served shorter combat tour lengths than their men.(79) In the wake of the Chinese onslaught, "bug out" fever not surprisingly characterized many army units.(80) Ultimately, General Matthew B. Ridgway's policy of instilling "toughness, skill at arms, and a determination to get in and win" raised the morale and fighting performance of the American infantrymen.(81)

By the end of the first year of hostilities in Korea superior firepower had taken its toll; the Chinese army had been administered a sound drubbing, and its advance was severely blunted. We now know that at one point the morale of the PLA was so badly shattered that it was close to collapse. There were limits, obviously, to overreliance on a "man-over-weapons" doctrine. Though extremely mobile at divisional and lower tactical levels, the PLA soon discovered that it could not sustain the forward momentum of large-scale offensives due to its inadequate system of supply. Nor could high commands exploit local Chinese successes. Despite magnificent infantry efforts, many attacks had to be cut short for lack of artillery and mortar ammunition.(82) Finally, much against its will, the PLA was forced to fight a war of position, a form of warfare totally unsuited to its mobile doctrine and fluid method of combat. Yet, even in this situation, the PLA performed creditably, organizing a superior system of strongpoint defense in depth. Unlike UN forces, which tended to adopt forward slope positions, the Chinese army was forced by the magnitude of UN firepower to develop more cleverly concealed positions on reverse slopes. Ever conscious of the value of overhead protection, the Chinese soldier built his trenches and fortifications "much closer to the specifications set forth in the U.S. Army field manual than most Eighth Army soldiers did."(83) The PLA was, in fact, the first army ever to fight a war under nuclear conditions.(84)

In the last two years of war, the Korean conflict bogged down into a deadly stalemate, a two-sided contest played out by infantry patrols in a no man's land reminiscent of the Great War. In an atmosphere of negotiation, costly attacks on opposing lines were mounted with ever-decreasing frequency. The effectiveness of the PLA nonetheless improved during this period, particularly in artillery technique. Also, unlike the Americans who rotated their troops regularly, the Chinese retained their veteran units in the line thereby gaining certain advantages. They remained to the end of the war a formidable enemy. According to Marshall, on the Korean "training ground," the Chinese army "became as skilled as any in the world in the techniques of hitting, evading, and surviving."(85)

The Korean campaign was clearly an infantryman's war; supported by the gunner, he dominated the battlefield.(86) On the Chinese side, the foot soldier prevailed against tremendous odds, proving that a modern air force could not necessarily isolate a battlefield and that a high degree of combat effectiveness could still be attained by an army despite inferior artillery support. In the end, the Chinese had to be stopped by infantrymen. Fortunately, there were sufficient infantry units of good quality; had there not been, the UN most assuredly would have lost the field. Ironically, success in the first major war of the Atomic Age hinged not on high technology but on the performance of the old-fashioned soldier on foot, the ancient and unglamorous "Cinderella" of the army.

NOTES

1. Young, Brigadier Peter World War 1939-1945, pp. 212-13 and 221-5. Percival entered the army in 1914 and was once described as a "brilliant staff officer." Norman Dixon, On the Psychology of Military Incompetence (London: Jonathan Cape, 1976), p. 144.

2. "An Era of Empire Ends at Singapore," Life February 23, 1942, p. 17.

3. Field Marshal Sir William Slim, Defeat into Victory (London: Cassell, 1956), pp. 187-9; Major E. B. Stanley Clarke and Major A. T. Tillott, From Kent to Kohima: The History of the 4th Battalion The Queen's Own Royal West Kent Regiment (T.A.), 1939-1947 (Aldershot: Gale and Polden), p. 39; and Young, World War 1939-1945, p. 222.

4. Ropp, War in the Modern World, p. 250.

5. Young, World War 1939-1945, p. 477; and John Masters, The Road Past Mandalay (New York: Bantam, 1979), p. 169.

6. Ropp, War in the Modern World, pp. 381-2. A million men finally recovered Burma in "the largest single action fought against the Japanese." Ibid., p. 384.

7. Saburo Hayashi and Alvin D. Coox, Kogun: The Japanese Army in the Pacific War (Quantico: The Marine Corps Association, 1959), pp. 2-3, 9-12, 14-15, 19-20, 26, and 158. The Pacific War was viewed by the Japanese High Command as an extension of the (Marco Polo Bridge) China Incident. Ibid., p. 47.

8. Some observers found the lack of tanks "a strange absence in an army which according to legend . . . is German-trained and indoctrinated." Lieutenant-Colonel Paul W. Thompson, et al, How the Jap Army Fights (New York: Penguin, 1943), p. 26.

9. Hayashi, Kogun, pp. 5 and 25.

10. Ropp, War in the Modern World, pp. 360-1; and Philip Warner and Michael Youens, Japanese Army of World War II (Reading: Osprey, 1973), pp. 14-16 and 25. The Japanese did introduce tanks into Malaya to the surprise of the British. The heaviest tank weighed 30 tons. Ibid. The British insisted tanks could not operate in the jungle. Dixon, On the Psychology of Military Incompetence, p. 138.

11. Young, World War 1939-1945, pp. 211-12.

12. Ivan Simson, Singapore: too little, too late (London: Leo Cooper, 1970), p. 151. They lacked tropical and jungle experience. The Japanese soldier was given a book entitled Read this alone--and the War Can be Won, which set forth what conditions he could

expect in tropical jungle and how he could combat them. Forty thousand copies of this book were produced. Ibid., p. 152.

13. "Japanese Tactics and Jungle Warfare," Canadian Army Training Memorandum, no. 35 (1944), pp. 6-7; and Simson, Singapore, pp. 150-1. In 1938, the individual Japanese infantryman ordinarily carried 65 pounds (knapsack, weapons and ammunition, tools, supplies, and clothing). In action against the Russians, however, the helmeted soldier carried but 60 rounds of ammunition, a haversack containing two grenades, canteen, and gas mask. He wrapped all of his equipment to reduce noise, and sometimes, instead of boots, he wore web-toed rubber-soled ground socks to muffle footfalls. Alvin D. Coox, The Anatomy of a Small War: The Soviet-Japanese Struggle for Chungkufeng/Khasan, 1938 (London: Greenwood, 1977), pp. 142-3.

14. "Organization and Tactics of the Japanese Platoon and Section on the Malayan Front," Canadian Army Training Memorandum, no. 14 (1942), p. 23; and Ropp, War in the Modern World, p. 360. The Australian soldier's daily ration provided 4,300 calories, the British 3,700, the Indian 2,700, and the Japanese 2,200. (The Japanese gave their prisoners 2,050 calories.) Simson, Singapore, p. 45.

15. Thompson, How the Jap Army Fights, p. 15.

16. "Notes on the Japanese Soldier," Canadian Army Training Memorandum, no. 16 (1942), pp. 6-7. There were three grades of Japanese private: senior, first class, and second class. Warner, Japanese Army in World War II, p. 4.

17. "Know Your Enemy," Canadian Army Training Memorandum, no. 33 (1943), p. 32; "Prepared Defenses" Canadian Army Training Memorandum, no. 51 (1945), pp. 19-21; and Masters, The Road Past Mandalay, p. 341.

18. "Japanese Mortars," Canadian Army Training Memorandum, no. 50 (1945), pp. 38-39; "Japanese Soldier," Canadian Army Training Memorandum, no. 51 (1945), pp. 24-26; and "Jungle Fighting," Canadian Army Training Memorandum, no. 34 (1944), pp. 27-29.

19. Warner, Japanese Army of World War II, p. 38.

20. Described by Allied troops as a "knee mortar."

21. "Organization and Tactics of the Japanese

Platoon and Section on the Malayan Front,"Canadian Army Training Memorandum, no. 14 (1942), p. 21; Coox, The Anatomy of a Small War, pp. 136-7 and 154-5; and Warner, Japanese Army in World War II, pp. 18-23. The standard-issue rifles for the Japanese army between 1941 and 1945 were 6.5 and 7.7 millimeters. The 7.7-millimeter Arisaka rifle was considered a superb weapon. So were the 50- and 81-millimeter mortars, the 7.7-millimeter heavy machine gun, and the 6.5-millimeter Nambu light machine gun. A 7.7-millimeter light machine gun was introduced during the last two years of the war. William Manchester, American Caesar: Douglas MacArthur, 1880-1964 (Boston: Little, Brown and Company, 1978), p. 331; and Warner, Japanese Army in World War II, pp. 22-23, and 40. In China, the Japanese platoon often divided into six eight-man groups, three of which were armed with a Nambu light machine gun. Thompson, How the Jap Army Fights, p. 36.

22. Warner, Japanese Army in World War II, pp. 3-4 and 19; Coox, The Anatomy of a Small War, pp. 137, 143, and 154; and Thompson, How the Jap Army Fights, pp. 25-26 and 40. The 7.7 HMG fired 600 rounds per minute and had a maximum range of 4,000 meters.

23. "Organization and Training of the Japanese Platoon and Section on the Malayan Front," Canadian Army Training Memorandum, no. 14 (1942), pp. 21-22. To avoid casualties to their own troops during company double envelopment maneuvers, the Japanese fired wooden bullets that burned up after a hundred meters or so but that could at short range kill as "dead as the finest American lead." Burke Davis, Marine! The Life of Lt. Gen. Lewis B. (Chesty) Puller (Toronto: Bantam, 1964), p. 161.

24. "Information on Japanese Tactics in Malaya," Canadian Army Training Memorandum, no. 16 (1942), pp. 8-9.

25. Slim, Defeat into Victory, pp. 32-33.

26. "How the Japanese See Us," Canadian Army Training Memorandum, no. 30 (1943), pp. 10-11. Slim claimed that the British could not initially effect strong encirlements "by reason of our complete dependence on motor transport and the unhardiness of our troops in the jungle." Slim, Defeat into Victory, p. 119.

27. Slim, Defeat into Victory, pp. 33 and 119.

28. "Japanese Tactics and Jungle Warfare,"

Canadian Army Training Memorandum, no. 35 (1944), p. 7; and Slim, Defeat into Victory, p. 120.

29. "U.S. Enlisted Men Discuss the Jap," Canadian Army Training Memorandum, no. 44 (1944), p. 6

30. "Japanese Tactics and Jungle Warfare," Canadian Army Training Memorandum, no. 35 (1944), p. 8; "Breakneck Ridge", Canadian Army Training Memorandum, no. 50 (1945), pp. 24-25; and "Japs Deepen Defences," Canadian Army Training Memorandum, no. 52 (1945), pp. 13-14.

31. Arthur Swinson, Kohima (London: Cassell, 1966), p. 187.

32. An Allied division in Europe consumed 650 tons per day. Martin van Creveld, Supplying War: Logistics From Wallenstein to Patton (Cambridge: University Press, 1977). pp. 214-15.

33. Slim, Defeat into Victory, pp. 187-91, 540 and 549. The fewer the vehicles on tracks and roads, the quicker they can travel. Monstrous traffic jams often slowed down movement in Italy. This is why Juin's French Expeditionary Corps, with less motor transport, could move faster than most British divisions.

34. Slim, Defeat into Victory, pp. 33, 293, 368, 376, 542, and 549; Hayashi, Kogun, p. 100; and Liddell Hart, History of the Second World War, pp. 539-40.

35. Slim, Defeat into Victory, p. 547. He excepted parachute forces. Everybody in Fifteenth Corps, including Slim himself, went through weapons qualifying courses on the rifle, pistol, Bren gun, bayonet, and grenade. Route marches also formed a vital part of training. Ibid., pp. 138-40.

36. The Australians were the first of the Allies to beat the Japanese. Except for his brutal frontal approach with Australian and green American troops--losses 5,698 and 2,848, respectively--in New Guinea-Papua, MacArthur directed forces that "flowed into . . . Japanese weaker points . . . as water seeks the weakest entry to sink a ship." Ropp, War in the Modern World, p. 373; and Samuel Milner, United States Army in World War II; Victory in Papua (Washington: Office of the Chief of Military History, 1957), pp. 370-7. The losses incurred in his campaign were unbelievably small. For example, Anzio cost 72,306 American casualties, Normandy 28,366; however, between MacArthur's arrival in Australia and his return to

Philippine waters over two years later, his troops suffered but 27,684 casualties, including those incurred at "bloody Buna." Manchester, American Caesar, p. 339. When MacArthur invaded the Philippines with 280,000 men, he faced a Japanese force of 275,000; in the main battle for Leyte, the Japanese lost 70,000, the Americans 15,584. The battle was still going on when the Japanese surrendered, and it ultimately yielded 192,000 Japanese dead with a mere 9,700 wounded possibly the highest ratio of killed in any major campaign since the sixteenth century. Macksey, The Guinness History of Land Warfare, p. 207.

37. Hayashi, Kogun, p. 110-11.

38. Manchester, American Caesar, pp. 431-2.

39. Liddell Hart, History of the Second World War, pp. 536 and 544.

40. Thompson, How the Jap Army Fights, p. 28; and Hayashi, Kogun, p. 60. One of the most explosive controversies of the War occurred when Lieutenant-General Holland Smith, then Marine Corps General commanding the Fifth Amphibious Corps, relieved Army Major-General Ralph C. Smith from command of the Twenty-seventh Infantry Division, which was at the time part of Fifth Corps. The army resented this peremptory action, which may have been brought about by different approaches to combat. DuPuy, Military Heritage of America, p. 599.

41. Benis M. Frank and Henry I. Shaw, Jr., Victory and Occupation: History of U.S. Marine Corps Operations in World War II (Washington: Historical Branch, G-3 Division, Headquarters, U.S. Marine Corps, 1968), pp. 695-7. The Marine Corps in January 1945 comprised six activated divisions and a strength of 421,605, the bulk of which served in the Central Pacific. Ibid., pp. 23-24.

42. Frank, Victory and Occupation, pp. 696, 698 and 700; and Alexander L. George, The Chinese Communist Army in Action: The Korean War and its Aftermath (New York: Columbia University Press, 1967), p. 52. Carlson was a personal friend of President Roosevelt, who, probably because of Carlson's influence, appeared interested in special "raider" units. These were disbanded by the Corps in 1944 as there "was insufficient justification . . . to maintain special units organized solely to conduct hit-and-run raids." Frank, Victory and Occupation, pp. 708 and 711.

CINDERELLA OF THE ARMY   233

43. Frank, Victory and Occupation, pp. 698-9, 720 and 849-50.

44. Frank, Victory and Occupation, p. 701.

45. Frank, Victory and Occupation, pp. 702 and 849-50.

46. Frank, Victory and Occupation, pp. 23-24, 657, 666 and 720-3. An indication of American efficiency in landing masses of troops was the putting ashore of 8,000 Marines at Saipan in 20 minutes. Liddell Hart, History of the Second World War, p. 647.

47. Manchester, American Caesar, p. 336.

48. Weller, Weapons and Tactics, pp. 134-5; and Frank, Victory and Occupation, pp. 85-86, 111-13 and 718-24. The Marines stressed tank infantry cooperation and training in small-unit tactics against fortified positions. In attacking such a position, "pin-up" teams of bazooka-men, BAR-men, and riflemen would direct intense fire at the target while demolition teams moved in for the kill with bangalore torpedoes and various other explosives such as pole, satchel, or shaped charges. It was ultimately concluded that all Marines, regardless of their specialties, had to be taught something about the use of demolitions. In other cases, tanks and flamethrowers were used. The Canadian Army-developed "Ronson" long-range flamethrower appeared in the Pacific in 1944. Frank, Victory and Occupation, pp. 704, 718, and 720.

49. Davis, Marine!, p. 99.

50. Davis, Marine!, p. 206. The First Battalion lost 71 percent, the Second 56 percent, and the Third 55 percent. Seventy-four percent of the officers were casualties, indicating that the Marines did not lack for personal example. Ibid. pp. 206 and 241. The Twenty-ninth Marines lost 2,821 out of 3,512 in 82 days at Okinawa, the highest price ever paid by a U.S. Marine regiment in a single battle. Manchester, American Caesar, dedication.

51. Frank, Victory and Occupation, p. 719; and Marshall, The Soldier's Load, p. 33. Total Marine casualties were 18,420 killed and 67,207 wounded. Frank, Victory and Occupation, p. 730. One habit the Marines got into during World War II was that of going to the immediate aid of a fallen comrade, regardless of the danger, because of the extremely cruel treatment often meted out to wounded by the Japanese, who in

participating in their own bushido death orgies, were only too happy to finish off wounded enemy soldiers (only 2,274 Marines were taken prisoner during World War II, four of them by Germans. Frank, Victory and Occupation, p. 731). The habit was to become traditional, and in Vietnam it sometimes proved disastrous. The Viet Cong and North Vietnamese often sited snipers to purposely engage Marine rescuers going to the aid of a stricken comrade. According to one disapproving company commander, it also helped to bog down attacking troops at "precisely the worst moment." Captain Francis J. West, Small Unit Action in Vietnam, Summer, 1966 (Washington: Historical Branch, G-3 Division, Headquarters U.S. Marine Corps, 1967), pp. 7 and 111.

52. Ropp, War in the Modern World, p. 385; and Barbara W. Tuchman, Stillwell and the American Experience in China, 1911-45 (New York: Bantam, 1979), pp. 4, 206, 218 and 532. Stillwell's consistent absence from headquarters gave rise to gibes about his being "the best three-star company commander in the U.S. Army" fighting the "platoon war in Burma." Ibid., p. 533. History has been kind, however, in its assessment of "Vinegar Joe" as a field commander.

53. S. L. A. Marshall, The River and the Gauntlet: Defeat of the Eighth Army by the Chinese Communist Forces, November, 1950, in the Battle of the Chongchon River, Korea (New York: Morrow, 1953), p. 1.

54. Harvey W. Nelsen, The Chinese Military System: An Organizational Study of the Chinese People's Liberation army (Boulder: Westview, 1977), p. 115.

55. George, The Chinese Communist Army in Action, pp. 5 and 7; and Samuel B. Griffith, II, The Chinese People's Liberation Army (New York: McGraw-Hill, 1967), pp. 56 and 130-1. Griffith was the commanding officer of the First Marine Raider Battalion when Carlson's squad organization was "copied" from the Communist Chinese. Ibid., p. 336.

56. Robert Leckie, Conflict: The History of the Korean War, 1950-53 (New York: G. P. Putnam's Sons, 1962), p. 195; and Roy E. Appleman, United States Army in the Korean War: South to the Naktong, North to the Yalu (Washington: Office of the Chief of Military History, 1961), p. 60.

57. General Puller made particular note of the "square of dirty white cloth and . . . straw mat" carried by many Chinese soldiers, remarking: "They're a

hell of a lot smarter than we are in the field. . . . They cover themselves with that cloth when there's snow, and a plane comes over. They can hide a whole damned division from us, right along this road. They use the straw on open ground. It's too sensible an idea for American forces--and too cheap, so we'll never have the advantage of it." Davis, Marine!, p. 276.

58. Appleman, South to the Naktong, North to the Yalu, pp. 688 and 719.

59. Marshall, The River and the Gauntlet, pp. 19, 85 and 174; Griffith, The Chinese People's Liberation Army, pp. 131 and 138-9; Leckie, Conflict, pp. 185-6; and Appleman, South to the Naktong, North to the Yalu, p. 688. Brigadier C. N. Barclay reported that a "rather terrifying number of administrators were required for every Commonwealth fighting soldier." Brigadier C. N. Barclay, The First Commonwealth Division: The Story of British Commonwealth Land Forces in Korea, 1950-1953 (Aldershot: Gale and Polden, 1954), p. 196. The Chinese soldier required but 8-10 pounds of supply per day, whereas the U.S. soldier required 60 pounds. A reinforced Marine division of 22,000 consumed more than 700 tons of supplies per day, while a smaller U.S. Army division required almost 600; a Chinese front-line division consumed but 40-50 tons. Griffith, The Chinese People's Liberation Army, p. 157. According to Neville Brown, the Chinese divisional slice was 12,500. Brown, Strategic Mobility, p. 212.

60. George, The Chinese Communist Army in Action, pp. vii and 7.

61. Appleman, South to the Naktong, North to the Yalu, p. 720. General Matthew Ridgway recorded that "firepower in the hands of their infantry is more extensively used than our own." General Matthew B. Ridgway, The Korean War (New York: Doubleday, 1967), p. 82.

62. Mao's main principles were concentration, surprise, and deception. His tactical thinking was originally concerned with basically defensive postures for the guidance of a weak force in a hostile environment. Stuart Schram, Mao Tse-tung (New York: Simon and Schuster, 1966) pp. 142-5.

63. Appleman, South to the Naktong, North to the Yalu, p. 720.

64. Barclay, The First Commonwealth Division, p. 33; Griffith, The Chinese People's Liberation Army, p.

142; and George, The Chinese Communist Army in Action, pp. 3-4. A western division in the front line needs, in order to ensure an adequate flow of supplies, at least one two-way axial earth road forward of its rear boundary. A corps needs two double-lane improved roads perpendicular to the front and a series of lateral routes. Brown, Strategic Mobility, p. 215.

65. Griffith, The Chinese People's Liberation Army, pp. 132 and 142-3. Official reports did not reveal the full psychological effect of intense "short" attacks on UN forces, particularly South Korean troops, many of whom were literally terror stricken at the very mention of the word "Chinese." Ibid., p. 133. Once the bugle and whistle signals became known to UN troops, however, they served to "telegraph" Chinese intentions. Leckie, Conflict, p. 186; and Lieutenant-Colonel Herbert Fairlie Wood, Strange Battleground: The Operations in Korea and their Effects on the Defence Policy of Canada (Ottawa: Queen's Printer, 1966), p. 79.

66. Griffith, The Chinese People's Liberation Army, pp. 72 and 131; and Walter G. Hermes, United States Army in the Korean War: Truce Tent and Fighting Front (Washington: Office of the Chief of Military History, 1966), p. 79. The current Chinese infantry company has a strength of 9 officers and 160 men. It consists of a headquarters, a 60-millimeter mortar squad, a machine-gun platoon, and three 38-man platoons of three 12-man squads. One-third of all present-day PLA training is conducted at night, and "hugging" tactics are still practiced. Nelsen, The Chinese Military system, pp. 123 and 226.

67. Griffith, The Chinese People's Liberation Army, pp. 72-73; and George, The Chinese Communist Army in Action, pp. 51-53. The North Vietnamese under Vo Nguyen Giap also organized their squads into groups of three.

68. Colonel Charles J. J. J. Ardant du Picq, Battle Studies, trans. Colonel John N. Greely and Major Robert C. Cotton (Harrisburg: The Military Service Publishing Company, 1947), p. 116. He also argued that "four brave men who do not know each other will not dare attack a lion. Four less brave, but knowing each other well, sure of their reliability and consequently of mutual aid, will attack resolutely. There is the science of the organization of armies in a nutshell." Ibid., p. 110.

69. Lynn Montross and Captain Nicholas A. Canzona, U.S. Marine Operations in Korea, 1950-1953:

The Chosin Reservoir Campaign (Washington: Historical Branch, G-3 Division, Headquarters U.S. Marine Corps, 1957), p. 93.

70. Lynn Montross, Major Hubard D. Kuokka, and Major Norman W. Hicks, U.S. Marine Corps Operations in Korea, 1950-1953: The East-Central Front (Washington: Historical Branch, G-3 Division, Headquarters U.S. Marine Corps, 1962), p. 35; George, The Chinese Communist Army in Action, pp. 3-4; Leckie, Conflict, pp. 186 and 215; and Griffith, The Chinese People's Liberation Army, p. 134. On a smaller scale with platoons and companies, however, the Chinese did occasionally employ "human wave" tactics on positions considered critical to further penetration. In these circumstances, they appeared prepared to sacrifice lives, though this could also have stemmed from a tactical rigidity that left battalions and below little choice but to keep attacking until ammunition ran out. George, The Chinese Communist Army in Action, p. 5; and Hermes, Truce Tent and Fighting Front, p. 5.

71. Appleman, South to the Naktong, North to the Yalu, p. 720; Hermes, Truce Tent and Fighting Front, pp. 79 and 511; Griffith, The Chinese People's Liberation Army, pp. 143-4; Davis, Marine!, p. 287; and Montross, The Chosin Reservoir Campaign, pp. 92-93.

72. Leckie, Conflict, p. 215; Griffith, The Chinese People's Liberation Army, pp. 139, 144 and 146-8, and 151-2; and Marshall, Pork Chop Hill: The American Fighting Man in Action, Korea, Spring, 1953 (New York: William Morrow, 1956), p. 20.

73. Davis, Marine!, p. 309 and 316. The U.S. First Cavalry with tanks dug in was overrun by Chinese horse cavalry (they initially used two such regiments in the invasion); the Americans broke and "the Chinese sabred them, hundreds of them, from horseback." Ibid., p. 313. See also Edgar O'Ballance, Korea: 1950-1953 (London: Faber and Faber, 1969), p. 74.

74. Davis, Marine!, pp. 97. 217, 224, 314, and 319. Puller described PLA troops as "well-trained and led." Ibid., p. 287.

75. Montross, The East-Central Front, p. 38; Leckie, Conflict, pp. 208-11 and 218-25; and Griffith, The Chinese People's Liberation Army, p. 146.

76. Weller, Weapons and Tactics, pp. 128, 131 and 133-5. In 1957, the Marine weapons company was again incorporated into the headquarters company and an extra

238 A PERSPECTIVE ON INFANTRY

rifle company added. In 1964, a fourteenth man, armed with a grenade launcher (M79), was added to the Marine squad. Commanded by a sergeant squad leader, the squad continued to consist of three four-man fire teams, each of a corporal team leader, a lance corporal (automatic rifleman), and two privates, first class (rifleman). The Marine rifle platoon thus comprised 47 all ranks, three 14-man squads, and a 5-man headquarters. A rifle company consisted of three platoons, a nine-man headquarters, and a support weapons platoon of 53 men, six rocket launchers and six MMGs. Ibid., p. 133; and Ney, The Infantry Rifle Squad, pp. 79-82.

77. Marshall, The River and the Gauntlet, p. 152; and Ney, Infantry Rifle Squad, pp. 52-56 and 98-101. The reorganization of the army grade structure eliminated the rank of staff sergeant. The squad leader was consequently given the rank of sergeant, first-class (one level below master sergeant). In April 1953, the riflemen were designated as three privates first-class (one armed with a grenade launcher and two ammunition bearers) and two corporals (one armed with a sniper rifle). In 1956, the U.S. Army adopted the fire-team concept. The rifle squad was reorganized to comprise 11 all ranks, divided into two five-man fire teams and commanded by a first-class sergeant (E-7). Each fire team was composed of a sergeant (E-5) team leader, a corporal (E-4) automatic rifleman, a corporal (E-4) grenadier, and two PFC (E-3) riflemen. Ney, Infantry Rifle Squad, pp. 55, 57-62 and 101.

78. Leckie, Conflict, p. 59.

79. Richardson, Fighting Spirit, pp. 166-7. Buddy relationships were the basic element of infantry social organization in the Korean War. Roger W. Little, "Buddy Performance and Combat Performance," The New Military, pp. 195, 197, 204, 218, and 221. Two "new" buddy groups would often not get on with two "old" buddy groups in a squad. For a squad or platoon to cohere it must remain a constant group. See also Gabriel and Savage, Crisis in Command, p. 41.

80. This was foreign to American small-unit infantry tactical doctrine of the time, which stressed that a platoon "never withdraws except upon the verified order of a higher commander." In defense, companies were normally disposed with two "front line" platoons and one "support" platoon in rear. Squads were generally sited abreast of one another, depth being provided by staggering men within the respective squads. Tactics and Technique of Infantry: Infantry Tactics, Squad, Through Battalion And The Military Team

(Harrisburg: The Military Service Publishing Company, 1950), pp. 47, 67, 53, 123-5 and 128. The Americans also emphasized employment of substantial numbers of soldiers in combat outpost and patrol base positions to engage an enemy further out. Hermes, Truce Tent and Fighting Front, p. 75.

81. Ridgway, The Korean War, pp. 87 and 97; and O'Ballance, Korea, p. 150. Twenty-two UN contingents fought in Korea. However, of about 35,000 UN troops, exclusive of Americans, the Commonwealth accounted for 20,000 and Turkey for 5,000. The weight of the burden of fighting fell on the American and South Korean armies. Ibid., p. 149.

82. George, The Chinese Communist Army in Action, pp. viii-ix, 4, 9-10, 171, and 173-89; and Barclay, The First Commonwealth Division, p. 33.

83. Hermes, United States Army in the Korean War: Truce Tent and Fighting Front, pp. 180, 370 and 510-11. Marshall was quite critical of UN entrenchments, noting, for example, that many bunkers were too heavily built, a surplus of sandbags often weakening roof timbers, which consequently tended to break down under artillery fire. Sleeping bunkers on the forward slope in direct sight of the enemy were common but unwise. The main object of most entrenchments appeared to be to gain the fullest possible protection against enemy artillery rather than to afford the garrison reasonable cover without limiting the employment of its own weapons. Many forward trenches wre sited so poorly that they could not cover the immediate downslope. Marshall, Pork Chop Hill, pp. 21-26. The Americans called their main positions built along the lower slopes of hills the "Main Line of Resistance," whereas Commonwealth troops used a system of hilltops developed into "Forward Defended Localities" with all-round defense and interlocking fire plans. The Americans argued that the latter system resulted in inaccurate "plunging fire"; the Commonwealth held that this was a small price to pay for making the enemy climb the steep open slopes. Wood, Strange Battleground, pp. 215-16.

84. O'Ballance, Korea, pp. 127-8. The Chinese appear to have burrowed deeper and more thoroughly because they lived under the threat of nuclear weapons possibly being used against them.

85. O'Ballance, Korea, pp. 117, 127, and 146; and Marshall, Pork Chop Hill, pp. 20-21.

86. Barclay, First Commonwealth Division, p. 195.

8. Beyond Deterrence

The Contemporary Requirement for Infantry

A major lesson of World War II was that infantry on both sides, while incurring roughly three-quarters of total battle casualties and inflicting but 30 percent,(1) remained the basic fighting arm in the combat zone. This became especially evident during the second half of the war as the defense--Clausewitz's "intrinsically stronger . . . form of combat"(2)-- gained an increasing ascendancy over the offense.(3) Events in Korea tended to confirm this experience, as that particular conflict was ultimately decided or stalemated, depending on one's viewpoint, primarily by the actions of small units of soldiers on foot. The war in Indo-China and its bitter sequel in Vietnam were also won and lost mainly by the varying successes and failures of battling infantrymen. In fact, any serious analysis of the approximately 30 major military engagements that have taken place since World War II(4) must take cognizance of the central fact that the infantry arm has played a dominant, often vital role in the outcome of each. The man with the rifle in his hand and the idea in his head was proved to be one of the most effective military instruments in the last three decades.

The Arab-Israeli wars are of particular interest, as they reinforce several of the basic lessons of World War II, most of which had to be relearned by the victorious Israeli forces. Once again, the featureless terrain of the desert threw up certain tactical mirages; made to appear even more convincingly clear in the light of modern technology, they led to a resurgence of the "all-tank" idea. The role of the infantry, the standards of which have been judged "one of the most serious problems facing the Israeli Defense Force

[IDF] since the War of Liberation,"(5) was subsequently questioned, much as it had been in British and German armored circles in the years before World War II. For this reason alone, not to mention the East-West military confrontation it portends, the Israeli experience is probably worth fuller treatment than can be offered here.

In the 1948-49 War, the Israelis, inferior to the Arabs in artillery and armor, were quick to gain and maintain air superiority. This was singularly fortunate, as the 1948 Israeli army remained essentially an irregular infantry force lacking in tanks and artillery (although it became increasingly motorized and mechanized as the war went on, acquiring armored trucks, jeeps, and M-3 half-track carriers). Only one tank battalion was employed in action, but it performed in such lackluster fashion that right up to the Sinai Campaign of 1956 Israeli armor was used only in support of the infantry. The "mobile infantry-tank exploitation" concept propounded by Generals Yigael Yadin and Moshe Dayan, which held that battlefield mobility was best achieved in terms of vehicular speed (the half track was faster than the Sherman tank), thus became the accepted school of thought. Neither the combined arms theories of General Haim Laskov, who saw tanks operating independently in support of the main force, nor those of Lieutenant Colonel Uri Ben Ari, whose close study of Wehrmacht operations led him to advocate Guderian's methods, were adopted. Dayan's "typical formation" of an infantry battalion plus a company of tanks and artillery support confirmed the relegation of the tank to an infantry support role.(6)

With the experience of 1956 under their belts, however, and the Guderian-style dash of Colonel Ben Ari's Seventh Armored Brigade to the Suez Canal freshly in their minds, Israeli military leaders began to believe that a better measurement of battlefield mobility might be the ability to advance in the presence of enemy fire. This change in thought resulted in a very decisive victory for the "tank school"; Dayan and his followers of the "mobile infantry school" were thoroughly converted. When General Israel Tal took over as commander of the armored corps in 1964, he was convinced that the heavy tank was the true "queen of the battlefield." He flatly rejected the validity of internationally accepted armor-infantry doctrine for Israeli conditions, arguing that massed formations of tanks moving in open desert needed no mechanized infantry to protect them: Sinai was not Europe.

In Tal's view, the lack of natural cover on the desert meant that tanks had little to fear from most antitank weapons. His answer to the <u>Pakfront</u> was to train Israeli tank gunners to shoot accurately from long range, thereby enticing the enemy to reveal his position by shooting too soon. Tal refused to sanction the acquisition of armored personnel carriers(7) (the armored corps was responsible for mechanized infantry) and devoted the resources of the corps to the tank battalions of the Israeli armored brigades. In his opinion, infantry could be provided with inferior manpower and inferior equipment, as its only task would be to mop up and guard the tanks. Here was the manifestation of the "all-tank" idea par excellence.(8)

The astounding success of Israeli arms in the Six Day War of June 1967 appeared on the surface to verify the theories of Tal. His long-range sniping tactics worked with devastating effect as Israeli tank gunners picked off Egyptian antitank guns in Pakfronts at ranges of 1,000 meters or more. Except for the Golan infantry on the Syrian front and the paratroop brigade that fought in Jerusalem, no infantry brigade was given an independent operational role. The mechanized infantry battalions organic to Israeli armored brigades were used mainly for mopping-up operations. With high mobility maintained through a "conveyor belt" logistics system, complete air superiority, and a hefty ingredient of surprise, the Israelis achieved perhaps the most classic blitzkrieg ever. Yet the war was not without its hard foot slogging. General Ariel Sharon's division of one infantry brigade, one armored brigade, two parachute battalions, and a large artillery component was forced to resort to a rigidly centralized set-piece infantry attack by night to overcome Egyptian defenses at Abu Agheila. It took 24 hours of tough fighting before the battle was won. In general, however, Egyptian infantry gave way in 1967 as the French had given way in 1940., Had the Egyptians held on more doggedly to key positions, like the Russians at Kursk, the outcome may have been different. After the war, critics of Tal's "all-tank" theories argued that it was courting disaster to bypass and leave unreduced pockets of resistance behind, as troops more resilient than the Egyptians might have closed the breach on Israeli tanks and cut them off.(9)

A further charge levied against Tal was that he had bled the mechanized infantry of its better manpower to the point where even he doubted its fighting quality. His action on the eve of the 1967 June War in giving the half tracks of a mechanized infantry brigade to a parachute brigade was cited as proof of this. The paratroopers of the latter formation had undergone no armored training but were considered "better fighters." In any case, the very age of the M-3 carrier--standard equipment until 1967--and a critical shortage of mechanized infantry antitank weapons, presented a picture of some neglect.(10) The Israeli habit of

TABLE 8-1. SHARON'S TACTICS FOR SEIZING A STRONGHOLD

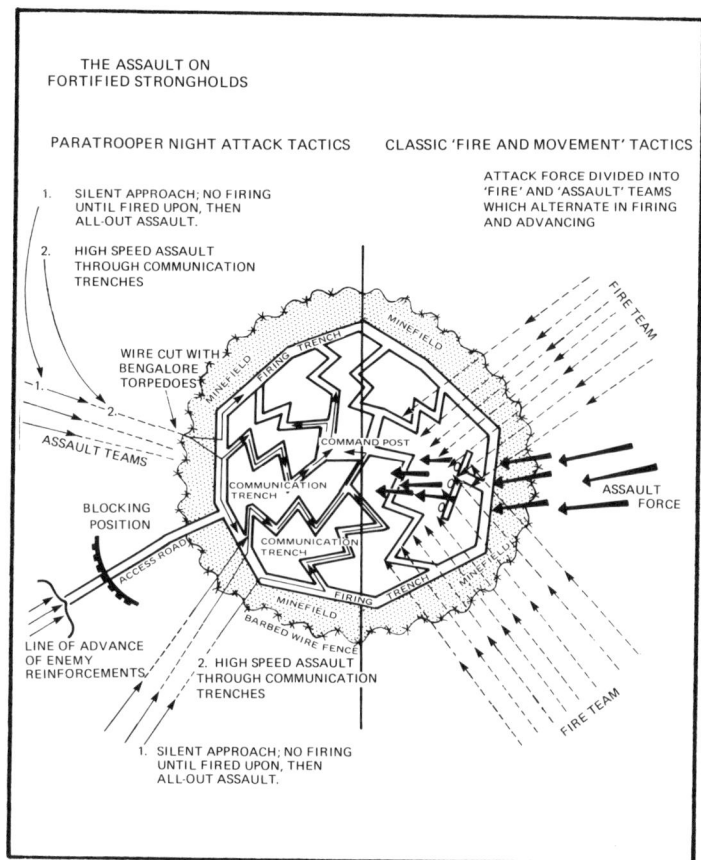

Source: Fig. 3, "The Assault on Fortified Strongholds, Two Tactical Systems," from Edward Luttwak and Dan Horowitz, The Israeli Army (Allen Lane, 1975), p. 115. Copyright © Edward Luttwak and Dan Horowitz, 1975. Reproduced by permission of Penguin Books Ltd.

charging in half tracks, especially with headlights blazing at night,(11) further reflected a certain degree of tactical naïveté.

It should perhaps be explained at this point that the armored corps, as well as wielding authority over mechanized infantry, constituted the most important of four Israeli "functional" commands. As such, it was primarily responsible for doctrine and tactics. Not surprisingly, by 1973, most Israeli infantry brigades had been converted to armor, leaving but three parachute and a few first-line infantry brigades available for action. The standard Israeli combat formation became the armored brigade of two tank battalions and one mechanized infantry battalion. The tank battalions were stripped of some of their organic infantry and mortars (among the most effective of weapons for suppressing antitank fire). The armored brigades were, in fact, almost pure tank formations. The mechanized infantry--what there was of it--was manned, trained, and equipped to fight mainly in a subordinate role.(12)

Like mechanized infantry, the Israeli parachute corps tended also to be separate from the standard infantry (though since 1967 there has been but one Chief Infantry and Paratroop Officer for both paratroopers and standard infantry). The manning of this volunteer airborne force, originally established as an independent commando unit to execute reprisal action for Arab terrorist activities, furthermore detracted from the overall effectiveness of Israeli infantry. The 1956 paratroop expansion, intended by Dayan to exert a reinvigorating influence on the mass of infantry, instead exacerbated a serious skimming effect whereby the standing force Golani Brigade (training brigade of the conscript infantry) began to receive a lower quality of soldier. Fortunately, Yitzhak Rabin, who took over as Chief of Staff in 1963, diluted the paratroops by systematically upgrading the Golani to the extent that both forces were by 1967 of roughly equal standard. The performance of Israeli conscript infantry in the Golan later attested to the wisdom of Rabin's action.(13)

In October 1973, the brilliantly executed crossing of the Suez Canal by Egyptian forces caught the IDF completely off guard. The Israelis were also technologically surprised by the Egyptian deployment of an effective antitank screen composed of Soviet long-range PUR-61 "Snapper" and PUR-64 "Sagger" antitank guided missiles (ATGM) and RPG-7 rocket launchers. With the virtual destruction of the Israeli quick-reaction 190th Armored Brigade, which counterattacked unsupported by infantry and lightly supported by artillery, the "all-tank" idea was once again shown to have serious limitation. It is interesting to note, moreover, that in the battle on the Canal many Israeli tank casualties were inflicted by massed RPG-7 fire from entrenched infantry. That the RPG-7 is but a super-<u>Panzerfaust</u> indicates, to some extent, that Egyptian infantry had learned to face tanks at close range.(14) The Israelis later admitted that many of their tank casualties were caused by single Arab soldiers lying behind cover and waiting until a tank came sufficiently close for a certain hit with the RPG-7.(15) The intent here is not to belittle the obviously lethal and possibly revolutionary effect of the ATGM but to point out that this was not the first vain attempt in history to throw unsupported armor against well-prepared defensive positions.

A further adverse effect of overreliance on the "all-tank" concept of operations was that the Israelis virtually abandoned their traditional specialty of night fighting, a valuable legacy left them by Orde Wingate.(16) In the Yom Kippur War, the Israelis hardly fought at night, with the result that armored forces were often called on to solve tactical problems that would have been better left, for reasons of economy of force, to other means. A classic example was the unsuccessful and casualty intensive attempt of the Israeli Seventh Brigade to take Tel Shams in Syria by frontal armored assault; the same position was taken the next night by a parachute battalion with four wounded.(17) In short, the "all-tank" school displayed a decided preference for daylight operations and fair-weather war. In addition, because armored officers under Tal were measured on their technical competence and gunnery skills, they tended to show a lack of tactical inventiveness and mental adaptability. This inability to rise above the "machine solution" contrasted sharply with the superior tactical performance of parachute officers and units, particularly as demonstrated on the Golan Heights. Sharon's highly criticized performance in crossing the Canal by night also appears more brilliant than reckless in this light. Innovative and a master of the set-piece attack, he was extremely well versed and practiced in night operations.*

---

*Luttwak, The Israeli Army, pp. 139, 335, 365-72 and 376. In 1956, Sharon's men took three police fortresses (at Rahwa, Ghirandal, and Husan) by night. The tactics he devised for seizing a stronghold consisting of concentric lines are worth noting. Rejecting standard fire and movement action, Sharon had his men approach the trench system in darkness, walking slowly and in absolute silence until fired upon. When discovered, they ran forward as fast as possible, firing on the move, while barbed-wire fences were breached by bangalore torpedoes. On reaching the trench line, the men split into small assault groups, and without pausing to clear the fire trenches, burst into the communication trenches, running and shooting all the

Clearly, what the IDF needed most in the October War was a good infantry, composed of high-grade skilled foot soldiers with good-quality armored carriers suitable for desert conditions. The Israeli infantry lacked mobility, and, with few exceptions, its weapons were no match for their Soviet equivalents; its antitank capabilities had also been drastically reduced because the erroneous assumption made in the IDF was that the best answer to a tank was another tank. The idea that armored forces could "operate freely without close infantry support" proved, in the opinion of General Chaim Herzog, to be "one of the dangerous concepts that had entered Israeli military thinking since the Six Day War." In 1973, there was not enough infantry, and what there was overall did not match the quality of the tank forces.(18) The harsh truth was that mesmerization with firepower and armor had induced, if not a myopic view of the worth of infantry in general, at least a benign neglect of valued infantry skills. The IDF was nonetheless able to persevere and ultimately prevail, its training methods and depth of tradition, even in a neglected infantry, eventually counting for more than mere weaponry and equipment. The Israelis are undoubtedly worth examining further in this light.

The IDF, like the Red Army and the PLA, was born of revolutionary combat and "baptized" in blood. The tradition of the Zahal(19) is rooted in the Wingate-inspired "Special Night Squads"(20) and guerrilla field companies of the underground Haganah, the full-time striking force of which the Palmach was formed in May 1941 to cooperate with the Allies during and after their invasion of Syria and Lebanon. Unlike the Jewish Brigade, which was an integral part of the British Army at that time, the Palmach was a completely independent military organization. Essentially a force of guerrilla infantry--its "ultimate unit" the soldier and his weapon--it comprised by the end of World War II four well-trained battalions. Always relatively weak numerically, the Palmach stressed group cohesion and combat leadership at all levels but most particularly

---

way to the center and out again. Sweeping one trench line after another, the assault teams kept moving until all defenders were captured or killed. The essence of Sharon's tactics was the shock effect of relentless movement and surprise to confuse the enemy and break down his resistance. The method was, of course, highly vulnerable to enemy counterattack; for this reason, Sharon prescribed isolating all strongholds prior to attack and blocking them all-round. Ibid., pp. 113-15.

at the section level. Significantly, it was <u>Palmach</u> commanders who rejected the concept of battle drill (propounded by British-trained Jewish officers) as "an over-schematic and artificial way of acquiring habits in the field."(21)

Though Prime Minister David Ben Gurion dissolved the <u>Palmach</u> command in October 1948 for political reasons and a personal desire to see a British-style military force established for Israel, the <u>Palmach</u> and its adherents retained a substantial amount of real influence within the IDF. The egalitarian "one-stream" system of induction (there is no separate officer entry scheme in the IDF)(22) and pronounced stress on "internal" discipline as opposed to compulsion must both be attributed to the influence of <u>Palmach</u>-inspired thinking. An important consequence of this emphasis is that a more intellectual rather than authoritarian approach (the former reflecting the greater Jewish tradition down through the years) has been fostered in the IDF. Specifically, it has meant that NCO courses for section commanders have tended to be more mentally challenging than those conducted in other modern armies.(23)

IDF special emphasis on elementary section and platoon training was reflected in its organization during the Sinai Campaign; it was, in effect, an army built around the eight-man infantry section commanded by a corporal. Within this basic organization, there were two light machine guns and four riflemen, one of whom was a specialist antitank grenadier. One other soldier and the section commander were armed with submachine guns. Section tactics for advancing in daylight called for the section leader, grenadier, and one machine gunner to act as a three-man "point." The remainder were deployed some paces to the rear, two men on one flank, three on the other. During darkness, the formation retained the same posture in front, but the five other men trailed behind in column. If forward scouting was required, it was done by the section leader; platoon and company commanders abided by the same rule of thumb at their respective tactical levels. In Israeli practice, light machine guns were similarly pushed as far forward as possible in the attack; fire bases set up to cover the forward movement of the rest of the section did not normally operate at more than 200 yards' range from the objective. Cutting this distance in half was considered ideal. When the section rushed an enemy position under cover of light machine-gun fire, one rifleman remained back to protect the gunners.(24)

In the opinion of Yigal Allon, ranked the outstanding field commander in the 1948-49 War, the great battles of that particular war, the Sinai

Campaign and the Six Day War, were "won in the NCO's courses of the Haganah and the Palmach." Allon rejected the argument that only in an army of partisans, fighting under special conditions and in independent small groups, was it necessary to train a section leader as a tactical commander capable of sizing up situations, making decisions, and putting them into effect on his own authority. He clearly considered the section leader's position to be the linchpin of operational effectiveness:

> The most brilliant plan devised by the most capable general depends for its tactical execution on the section-leaders. Poor section-leaders may ruin the best-laid plans; first-rate section-leaders will often save badly devised plans. This for one simple reason: the section-leader is the sole level of command that maintains constant and direct contact with the men who bear the brunt of the actual fighting. It follows, then, that the section-leader is to be trained as a tactical commander and as an educator of his men. [In the IDF] . . . section-leaders are trained to command independently in the field in every instance in which they are required to operate alone with their units. In "regular combat", moreover, when the section-leader acts within the framework of his platoon and under orders from his superior officer, he still requires a high standard of knowledge and an ability to sum up the situation. Modern fire-power and the development of tactical atomic weapons may compel armies to operate in small, dispersed formations both in attack and defence. . . . All levels of command must therefore be trained to think and act independently whenever circumstances demand that they should, and section-leaders are no exception to this rule. Besides, modern weapons which provide small groups of men with greater firepower and more flexibility of movement, call for a high standard of command at all levels. The section-leader is therefore to be trained technically as an officer, not as a corporal.(25)

Traditionally, Israeli army recruit training has been rigorous and demanding, "threefold tougher than in the United States Army," according to Marshall. Teaching the private soldier to think clearly, observe keenly, and report accurately has received great emphasis, though relatively little importance has been attached to parade square drill and routines familiar to other peacetime armies. Prior to 1956, sharpshooting was strongly stressed and physical training given a high priority, a standard test for NCO candidates being the completion of a 40-mile march in eight and one-half hours. As regards the latter, it is worth noting that the Israelis wasted no time in road marching, preferring to condition troops by cross-country movement.(26)

The Israeli soldier remains one of the most highly motivated fighters in the world. Heir to a national revolutionary tradition, he realizes full well that his country, friends, and family will not long survive if he is defeated. Though more egalitarian than militaristic, the IDF is probably the closest modern approximation of the old Prussian model of an "army with a state." The strong sense of Kameradschaft and cohesion within the IDF is, of course, immeasurably enhanced by the exceptionally high standard of leadership fostered at all levels of command. The deeply ingrained Israeli conviction that all men's lives are equally important has further served to ensure that commanders remain continually in the forefront, making determined efforts to keep casualties down while accomplishing their respective missions. Rooted in the experiences of past small-unit actions, which normally had to be conducted in the face of much superior numbers, this attitude has permeated the IDF. In short, the Israeli experience largely represents the triumph of quality over quantity.(27)

Since Israel as a nation could not tolerate high casualties, Yigael Yadin and other Israeli leaders were drawn to adopt Liddell Hart's "indirect approach" on the strategic plane, while embracing Rommel's counsels on the importance of small fighting units on the tactical level. Consequently, Israeli commanders were urged to avoid the expensive head-on attack, strategically and tactically, as there was usually a better way. In an associated vein, commanders were also strongly advised to rest exhausted troops as they were generally expected to be at the limits of their endurance. To Yigal Allon, battle was not just a matter of fire and movement but fire and movement and "consciousness," the mental preparation of soldiers for action. In the matter of motivating troops, the Israelis have traditionally seen small requirement for medals; hardly more than a score were awarded for bravery in the Sinai Campaign, and only 51 citations were given in the Six Day War. Significantly, almost half the total Israelis killed in both the Sinai Campaign and the Six Day War were officers.(28)

Notwithstanding the action-oriented training and dynamic leadership of the IDF, from which all armies could certainly learn something, there is a degree of danger in copying some Israeli methods, particularly those employed during the period of infantry decline. For example, in order to avoid becoming bogged down in house-to-house fighting in the 1967 battle for Jerusalem, the Israelis--supposedly acquainted with Chuikov's tactics--chose to advance down the streets mounted, shooting their way through as they went, leaving pockets of resistance behind.(29) In 1973, in Suez City, a similar situation developed, reputedly because the Israelis did not expect more than sporadic enemy resistance; the reluctance of their infantry to dismount prior to entry subsequently resulted in heavy casualties and the IDF's repulse from the city.(30) Without getting into the argument as to whether or not the Arabs were on these occasions an "inferior enemy," it is probably safe to say that such tactics would likely have backfired disastrously at both Ortona and Stalingrad where the streets were killing grounds swept by murderous fire. It is also questionable whether the mobile form of warfare made possible by desert conditions will ever yield appropriate lessons for the waging of war in less open lands. One of the major lessons of Eighth Army operations in North Africa and Italy in World War II was that techniques learned in the desert did not necessarily apply to other regions.

The technological impact of the Yom Kippur War with its pronounced East-West overtones nevertheless attracted the serious attention of the great powers. According to General William E. DePuy, the three major lessons of the war were the vastly increased lethality of modern weapons ("if it can be seen, it will be hit"), the reaffirmation of the worth of combined arms, and the importance of individual and collective training. In his opinion, the tank and the airplane had merely joined the ranks of the infantryman, long vulnerable to hostile fire and unable to move without its being suppressed--but having nonetheless learned to live with it. He concluded that improved tactics and techniques of movement, coupled with complementary suppressive fire on antitank weapons, were therefore the preferred means for ensuring success on the battlefield of the future.(31)

That infantry had been able to effectively take on tanks as never before, at long as well as close ranges, also tremendously impressed the Soviets, though they did not necessarily perceive the tank as the most threatened weapons system. Apparently, during 1974, the Red Army conducted a number of conferences and extensive field exercises to determine the actual impact of the new weapons on attacking forces. Apart from finding that the defense was greatly strengthened, they discovered somewhat to their horror that their main infantry carrier, the BMP-76PB, rather than the tank was the weakest link in Soviet combined arms formations.(32) A tactical debate was subsequently triggered, which, as far as is known, has remained largely unresolved to date. A modern version of the most authoritative Soviet reference, Taktika (which last appeared in 1966), has yet to be reissued and presumably will not appear until solutions to the BMP problem, among others, are worked out.(33)

By way of background, the Soviets have since World War II continued to emphasize the offensive, holding that in any conflict initiated by the West the "Communist socio-economic system" would have the best chance of survival if industrial Western Europe were seized quickly and intact. It is this availability of Europe that allows Russia to plan on the assumption that a general war need not imply mutual suicide. Given possession of a relatively undamaged Western Europe, the Soviets would be in a position to rebuild a "new socialism."(34) The United States has no comparable option. She cannot count on moving into Canada because it will be a Soviet decision whether or not Canada is to be spared. The Americans, on the other hand, might be reluctant to rain nuclear havoc on the lands and peoples of their erstwhile European allies. The absolutely critical aspect of this Soviet strategy, of course, is the Russian ability to "withhold" sufficient nuclear strike capacity to continue to threaten the Americans. Another important facet of this strategy, which is perhaps more to the point, is that the Soviets have for many years been thinking beyond deterrence; they do not reason, as do certain Western strategists, that "if the deterrent is used, it will have failed." In short, the Soviets do not separate deterrence from the general concept of defense; they are--and always have been--ready to fight a nuclear war if necessary, their ultimate object being the survival of the socialist system and not just the avoidance of war.(35)

In accordance with this strategic theory, the Red Army therefore began to accentuate high rates of offensive maneuver as "the backbone of its tactics."(36) Noting that weapons of mass destruction would be available to reduce effective defense to a minimum, Soviet planners thought to limit their own vulnerability through increased speed and maneuver. By avoiding initial concentration and making deep penetrations on multiple dispersed axes of advance and interposing forces between NATO defensive positions, it was hoped that concern for defending troop safety and rear-area stability would inhibit the Western allies from using tactical nuclear weapons. In order to improve the chances of accomplishing this, the Soviet General Staff began in 1967 to introduce an infantry vehicle expressly designed for rapid combined arms offensive operations under nuclear, or threat of nuclear, conditions. This vehicle was the BMP.(37) Manned by a crew of three, it carried an eight-man section. With its sloped armor, 73-millimeter smooth-bore gun, Sagger ATGM rail, coaxial 7.62 machine gun, and rifle firing ports, it was the first true infantry fighting vehicle (IFV).(38) Equipped with an air-filtration system, it was ideally suited for running roughshod over the nuclear battlefield.

On closer examination of the strategic scenario, it became obvious to the Russians that if for some reason a major war did start, it would be clearly in the interests of the Soviet Union to win it before NATO could reach a decision to use nuclear weapons and perhaps precipitate a global holocaust. Furthermore, with NATO committed to fighting a conventional battle for several days before going nuclear, a rapid Soviet advance might well hasten a European political collapse and perhaps stave off a nuclear war. A gradual shift of emphasis in Soviet military thinking consequently occurred, and the study of conventional operations under nuclear-threat conditions began to receive as much attention as the study of purely nuclear operations. It was soon discovered, however, that without initial nuclear bombardment enemy defenses would remain essentially unreduced and much more formidable than before. Attacking forces, therefore, instead of having merely to exploit success, would have to rely to a larger degree on the concerted action of combined arms. Thus, whereas the ratio of tank to motorized rifle divisions in the Red Army was 1 to 1.8 in the early 1960s, it had been reduced to 1 to 2.2 by 1974 (20 additional divisions having been raised--all motorized). This decreased ratio was a clear indication that the Soviets regarded the success of any conventional offensive as being highly dependent on the protection and mobility of motorized infantry. Up to this point, moreover, the Red Army assumed that the infantry, mounted in combat vehicles and with the support of massed indirect-fire artillery, could overrun NATO defenses and thus maintain an acceptable rate of advance. However, the recently perceived high vulnerability of the BMP to modern antitank fire has threatened to unhinge the entire structure of this Soviet land-force doctrine.(39)

Since the BMP was originally designed for exploiting nuclear strikes, there has naturally been a strong incentive for the Soviets to return to their earlier dependence on these weapons as a means of overcoming antitank defenses. There are indications, however, that many Soviet ground commanders are clearly unwilling to discount the antitank problem simply through reliance on nuclear weapons. There is apparently no argument about the effectiveness of modern antitank weaponry in Soviet circles; in fact, there seems to be a universal recognition that a tactical revolution is being brought about by the deployment of ATGM. Specifically, this means that tanks without infantry support attacking an unreduced defensive position sited in depth will be destroyed. The same fate will likewise befall infantry if it attacks mounted, as small-arms fire from BMPs is judged by the Soviets to be so inaccurate that it can be expected to suppress only the weakest of defenses. Thus, unlike in nuclear operations, Soviet infantry participating in a conventional attack on a prepared defensive position will have to dismount from their BMPs. It is this consideration, which essentially raises the age-old problem of the last 300 yards, that has sparked the most heated debate.(40)

Apart from considering the "one-variant" nuclear approach toward resolution of the attacker's problem, which virtually makes it go away, the Soviets have given thought to two other options, which sometimes merge into a single "multivariant" form. The first, strongly backed by armored officers, advocates exploiting the BMP's superior speed by carrying out daring surprise raids deep into the enemy rear, the aim of such a preemptive maneuver being to attack the defense with mounted infantry before it can deploy a dense antitank defense. In short, this school argues that surprise attacks with conventional weapons offer the same opportunities as nuclear strikes for ensuring a low enemy antitank "density" (that is, ratio of force to space), long held by certain Soviet writers to be the key variable influencing rate of advance. This option, which has such ominous undertones for NATO, is not expected to work against prepared positions, particularly those protected by minefields. Its aim remains to catch an enemy on the move and defeat him in an encounter battle before he can establish himself in defense.(41)

The second option, espoused by Soviet artillery officers, calls for employing more artillery in direct as well as indirect fire modes to suppress antitank weapons.(42) In the former mode, however, considered by Soviet artillery and armored experts to be the more effective method of suppressing enemy antitank fire, a substantial degree of tactical decentralization and mobility is required. The recent introduction of new self-propelled 122- and 152-millimeter howitzers thus presaged a major shift in Soviet artillery thought, which formerly stressed massive indirect fire barrages from towed artillery. While such artillery was retained as a divisional resource, the new self-propelled 122-millimeter howitzers were organically introduced (18 at a time) to most motorized rifle regiments.

Combined arms integration was lowered even further by a policy of frequently attaching self-propelled batteries to maneuver battalions.(43) Ideally suited to the preemptive maneuver option, which calls for regimental size actions, this system resembles somewhat the German World War II use of "shock artillery" or infantry guns.

An additional reason for the greater decentralization or combined arms within the Red Army appears to be the very real fear that vital communication links will inevitably be seriously disrupted during any European war. Colonel A. A. Sidorenko is not alone in perceiving radio command as a weakness of NATO nuclear artillery, and he has written that "the disruption of control of these means also comprises an important element of the combating of enemy means of nuclear attack."(44) Soviet emphasis on electronic warfare is correspondingly strong. The Fullerian concept of using partially disrupted radio nets to spread alarm and despondency, as the French artillery inadvertently did in 1940, has probably not been overlooked by the Soviets, either. The major point to note is that while NATO continues to develop and rely to an ever-increasing degree on more complex systems of command, control, and communications (C3), the Soviets appear intent on readying themselves for the primitive and chaotic form of struggle that they apparently feel will continue to characterize the battlefield.

Ironically, in their unrelenting search for a fluid and mobile style of warfare, the Soviets triggered a sort of imitative behavior in certain NATO circles. The advent of the BMP was matched by the introduction in 1969 of the German Marder mechanized infantry combat vehicle (MICV)(45) and the issue of American development contracts for an improved MICV (carrying 11 men) to be armed with the "Bushmaster" 25-millimeter high-performance automatic cannon and TOW(46) ATGM. In the restructuring efforts (Model 4) of the German army and the American army Division Restructuring Study, the tactical employment of the MICV figured prominently. Both proposals reduce the size of squads, platoons, companies, and battalions but increase the overall number of these units and subunits. In the German New Army Structure, the basis of all types of combat is formed by armored fighting troops, particularly panzer battalions, the main battle tank still being recognized as the best means of defense against a strong armored enemy.(44) The new armored brigade contains three tank battalions of 33 tanks each and one armored infantry battalion of 33 MICVs; the new armored infantry brigade has two tank battalions and three MICV battalions, one at cadre strength. (Both are supported by one field artillery battalion of 18 self-propelled howitzers.) Based on World War II experience, close cooperation between tanks and Panzergrenadiers (armored infantry) is strongly emphasized. So long as terrain considerations permit, one Panzergrenadier company will normally be attached to tank battalions and vice versa; interestingly, this does not preclude a mix of tanks and infantry down to platoon level if ground and visibility require it. Again, terrain considerations will dictate whether the infantry is to fight in the mounted or dismounted mode. In the attack or in a delaying action, the Panzergrenadiers will normally fight mounted, the fire fight being conducted with all turret weapons. The Germans have for some time advocated fighting from armored vehicles, holding that really accurate automatic fire is of little tactical consequence since enemy soldiers are seldom seen on the modern battlefield.(48)

In the defense, Panzergrenadiers will dismount from their MICVs with their mortars and light infantry weapons and prepare defensive positions. As far as possible, these will be on reverse slopes so that the enemy will not detect them too soon. Tanks and MICVs, on the other hand, will open fire early from forward or lateral positions of support to fully exploit the ranges of their heavier weapons. As the enemy keeps approaching, the armored vehicles will move back, constantly jockeying for position and continuing to engage, supporting the fires of the dismounted infantry. In this concept, reserves will be kept small (one or two platoons of Panzergrenadiers and panzers at battalion level) in order to bring maximum fire to bear on the enemy. The Germans estimate that a Panzergrenadier battalion reinforced by a panzer company, given a three-to-one defensive advantage, could quite reasonably expect to hold off a Soviet regiment of about 100 tanks.(49)

The American restructuring effort parallels that of the Germans, its object being to make the best use of high-technology weapons while at the same time decentralizing tactical authority. Specifically, it seeks to have companies specialize in single-weapon systems, whether they be tanks, infantry, or ATGM. Recommended are tank companies of 10 tanks and 50 men, ATGM companies of 12 vehicles and 50 men, and mechanized infantry companies of 13 IFV and 100 men. Mortars and weapons platoons would be withdrawn from rifle companies and concentrated at battalion. A mechanized infantry company would thus comprise three platoons of four MICVs each and one in company headquarters. Squads would be reduced to nine men. In this manner, there would be more experts and leaders per weapons system, more leaders per soldier,(50) and

TABLE 8-2. INFANTRY ANTITANK WEAPONS

MODERN NATO AND WARSAW PACT MAN-PORTABLE ANTI-TANK MISSILES

Weapon	Country	Manufacturer	Type	Weight	Range	Remarks
Dragon	USA	McDonnel Douglas	Wire guided missile (auto tracking)	Overall 13.9 kg Missile 11.2 kg	1000 m	In service in US Forces and other countries
Milan	France Germany	Aérospatiale MBB	Wire guided missile (auto tracking)	Overall 27 kg Missile 6 7 kg	75 to 2000 m	In service in NATO
Mamba	Germany	MBB	Wire guided missile (optical tracking)	Overall 44 kg Missile 11.2 kg	300 to 2000 m	In service
Sparviero	Italy	Breda	IR guided missile (no wire)	Overall 69 kg Missile 16.5 kg	75 to 3000 m	Under development
Swingfire	UK	British Aerospace	Wire guided missile (auto gathering, optical tracking)	Overall 61 kg (one loaded trolley)	4000 m	Mk II is a complete system and requires 3 trolleys. Mk III is a complete one missile system with one trolley only.
Sagger	USSR		Wire guided missile (auto tracking)	Overall 22 kg Missile 12 kg	4000 m	Normally vehicle mounted but there is a single man-portable unit which can be fired from the ground.

MODERN NATO AND WARSAW PACT MAN-PORTABLE ANTI-TANK UNGUIDED WEAPONS

Weapon	Country	Manufacturer	Type	Weight	Range	Remarks
ACL STRIM	France	Luchaire Sold by CFTH-Brandt	Rocket	Launcher 4.5 kg Round 2.2 kg	400 m	In service with French and other armies
Sarpak	France	CFTH-Brandt	Rocket	Launcher 1.9 kg Round 1.07 kg	200 m	In service with French and other armies
Armbrust	Germany	MBB	Projectile	Overall 6.3 kg	Man 1500 m Tank 300 m	Expendable launcher undergoing trials
Folgore	Italy	Breda	Rocket	Launcher 15 kg Round 8 kg	700 m	Shoulder launched version Can also be fired from a tripod when the round weighs slightly more and has a range of 1000 m. Under development
Carl Gustav M2-550	Sweden	FFV	HEAT Shell	Gun 15 kg Shell 2.5 kg	700 m	In service in NATO
Miniman	Sweden	FFV	HEAT Shell	Overall 2.9 kg	200 m	In service
LAW M72	USA	—	Rocket	Overall 2.5 kg	250 m	Expendable launcher. In service in NATO
RPG-7	USSR	—	Grenade	Launcher 5 kg Grenade 2.5 kg	300 m	In service in Warsaw Pact and other countries

Source: *Military Technology and Economics*, Vol. III, No. 7 (January-February 1979).

ergo, in theory at least, more performance per battle. The task of coordinating combined arms actions would, however, be shifted back from company to battalion in order to better utilize the available combat power on the battlefield,(51) particularly that of new long-range weapons.

The latest British Rhine Army reorganization, with

its fascination with fives, tends to confirm the growing disenchantment with the company (combat team) tactical deployment pattern of recent years. The new British armored division is organized into five battle groups(52) plus supporting arms, the battle groups put together as required by the tactical situation from the eight squadrons and 12 companies of the division's two armored regiments and three mechanized battalions. This new organization, it must be stressed, is underpinned by an unabashed confidence in the new Milan ATGM, without which the "organization would be a non-starter because there would be insufficient anti-tank potential" for infantry elements. A major consideration in the operational deployment of such a battle group, of course, will be to avoid breaking them up "into a series of independent and largely ineffective combat teams."(53) By adopting this organization, the British appear to agree with General DePuy, who reasons that the demise of the combat team has been brought about because "the application of our combat power depends too much on one man--the company commander--who is already overburdened far forward on a lethal battlefield. Company teams consisting of tanks, infantry, ATGM, and mortars supported by artillery, helicopters, and air simply do not dispose effectively enough the available combat power."(54) The recommended battle group system of tactical organization, on the other hand, strikes one as highly reminiscent of the older regimental configuration employed so successfully by the Germans in the mounting of blitzkrieg operations.

No matter what form of tactical grouping is utilized, however, the dilemma for the Western infantryman in the attack or counterattack remains essentially the same as that facing his Soviet counterpart. Both must decide where to dismount in relation to an objective that has to be secured by physical occupation. Against a reasonably well prepared defensive position mined and manned by an enemy armed with ATGM and protective machine guns, only the most reckless would choose to dismount from IFVs on the objective. Amidst the cacophony of battle and the strangeness of terrain, tiny sections(55) of infantry issuing from their respective IFVs in an understandably disorientated state would doubtless be cut to pieces very quickly by opposing dug-in infantry. For those foolish enough to dismount beyond the objective, a similar fate would likely await them in the gaping jaws of local antiarmor defenses in depth. In short, it is highly improbable that all resistance from dug-in infantry can be neutralized by supporting artillery fire, direct fire from tanks, or from advancing infantry combat vehicles themselves.(56) It goes almost without saying, of course, that attacking tanks would not last long on any hostile position without the protection of friendly infantry.

As near as can be determined, the Russians recently ruled that in any attack against a prepared position, the infantry should attempt to get as close as possible, normally dismounting between 400 and 300 meters from the enemy. It is further laid down that infantry should never dismount in front of tanks but always as near as possible behind them, not more than 200 meters distant so that they can protect the latter by small-arms fire. Companies are to attack in one echelon only, maintaining 50 meters between sections.(57) In this manner, BMPs following at about 300-400 meters in a second echelon will be able to provide intimate support by firing through the gaps provided. (Overhead fire is recommended only in undulating country, as it would otherwise adversely affect troop morale.) Interestingly enough, at least one Soviet high-ranking officer has become so disenchanted with the BMP that he has actually advocated a return to mounting infantry on tanks, as was Soviet practice in World War II.(58)

That the Soviets have not yet introduced a successor to the 1967 BMP within their normal 20-year procurement-replacement cycle is seen as further indication that the Red Army may be having some difficulty in deciding how best to employ infantry on the modern mobile battlefield. IFV design parameters are not just a Soviet problem, however; in Western circles, too, there is a growing divergence of military opinion over what is considered to be the ideal infantry vehicle. The French, with their recent AMX-10P (without side-firing ports), appear to have opted for an APC, whereas the Germans apparently remain convinced of the need for a MICV. Even on the matter of tracks versus wheels, lessons of the Ostfront notwithstanding, there is no longer any automatic agreement on the desirability of the former.(59) At the moment the only genuine consensus appears to be on the primary requirement for some type of armored protective vehicle to transport infantry rapidly from one defensive position to another. In holding to this consensus, military thinkers may well have reminded themselves that the infantryman riding in a modern vehicle has never truly represented a new weapons system, rather an ancient one advancing to battle by more efficient means. The essential difference today is that technological improvements in his weapons inventory have made the man fighting on foot a more dangerous adversary than ever before. The smallest target and most universally mobile of all weapons carriers, the foot infantryman with his "computer-brain" has proved a tougher species than Fuller ever imagined him.(60)

In Europe, however, the revolutionary impact of modern antitank weaponry is not the only factor bearing on the employment of infantry, or indeed on land force tactical operations generally. What has passed practically unnoticed to many military eyes since World War II is the gradually altered nature of the very terrain over which the next war may be fought. The phenomenal and relentless process of the urbanization of Europe, already with 374 cities of 100,000 or more, has radically changed the face of the potential battlefield. With roughly 250 people per square kilometer, the strategically centered Federal Republic of Germany (FRG) has one of the highest population densities in Europe; 70 percent of its citizens live in cities, of which 60 have populations exceeding 100,000 and four of over 1 million. Moreover, with Central European urban growth projected to continue at a rate of two to three times population rise, there will be, by 1995, a 50 percent increase in total urban area.(61)

As a result of such proliferation of European urban areas, many large cities and surrounding regions have already grown together to form conurbations. The largest conurbations are found in West Germany: Rhine-Ruhr, Rhine-Main, Hamburg, Stuttgart, Rhine-Neckar, Munich, Hanover, Nuremburg, Bremen, and Saarbrucken-Volkeingen, to label but a few. The Rhine-Ruhr is particularly interesting, as it is projected to converge with the Dutch Randstad sometime in the 1980s; together, they will form a contiguous urban barrier some 300 kilometers long from Bonn to the Hook of Holland. Strategically, this would block traditional avenues of approach such as those used by the Germans in the Great War and by Eisenhower, going the other way, in World War II.(62) It should not, therefore, require too much imagination to realize that traditional conceptions such as the "North German Plain . . . being a vast billiard table of agricultural land"(63) are similarly no longer valid. European conurbations both of the present and the future will doubtless present strategists and tacticians with totally new problems to face in the event that deterrence fails.

While noting that urban growth has significantly reduced the extent of open areas potentially suitable for mobile armored operations, it is also worth recalling that almost 30 percent of the FRG is, at present, woodland. Along with Austria and Luxembourg, it is one of the most richly forested areas in Europe. Moreover, government afforestation programs increase this sizable area by about 0.8 percent overall each year. By current measurement, forests and villages together would account for nearly 60 percent of the terrain in a typical defensive position occupied by a

NATO armored brigade on the East German border. As the spatial distribution of such villages and forested areas further restricts the breadth and width of open spaces, it has been suggested that advancing mobile forces would not be able tactically to bypass one village without almost immediately running into another. In short, open areas are rapidly shrinking in Europe. Urban concentrations, to the extent that they exert a distinct controlling effect over their surrounding areas, are fast becoming the dominant features of all terrain.(64)

Quite clearly, technological advances in antitank weapons systems and the greatly increased urbanization of the Central European countryside have served to largely enhance the role of infantry in land force operations. According to Brigadier A. J. Trythall, we are seeing perhaps for the first time an infantry tactics and technology capable of defeating the tank.(65) In the scenario of the armored battlefield, therefore, infantry will for the foreseeable future be required to overcome ATGM and other antiarmor weapons impeding the advance of tanks. Traditional infantry skills lost with the advent of the APC (but happily maintained in light infantry and mountain units) will consequently regain their former importance. Just as accurate rifle fire was necessary for picking off machine-gun crews in the Great War, so will selective fire and stalking likely be required to deal with ATGM detachments. The impact of urbanization can be expected to exert an even more powerful influence on the employment of infantry, of course, as NATO military thinkers come to realize that the best antitank defense is not another tank but rather one based on Fuller's "great town."

To the people over whose territory a future European war may be fought, however, the thought of city fighting appears to have little more appeal than the thought of having tactical nuclear bombs dropped at random over their country. From its inception, West German military strategy has been based on a forward defense aimed at keeping both friendly tactical nuclear strikes and the enemy's forces farthest away from the nation's largest urban industrial regions. Though the aim of this defense is to prevent the destruction of most of the country whose defense is at issue,(66) its potentially fatal flaw is that like the French defense of 1940, it tends to be linear in nature.(67) Given a short-war scenario, such a posture may make sense since hostilities could conceivably be terminated through negotiation before any appreciable damage was done to the West German countryside. Thinking beyond deterrence to the scenario of a longer war, however, forward defense appears to be but a thin veneer, easily cracked. To utilize the urbanization of West Germany for purposes of defense in depth, on the other hand, would turn a nation relatively lacking in geographical depth into an imposing array of urban "hedgehogs."

Interestingly, the "active defence" tactical doctrine championed by the Americans for effecting the strategy of forward defense essentially places everything up front. Exploiting terrain and maneuver as never before, defending forces are to engage the attacker at maximum range, "attriting" his combat power as he approaches the main battle area. As the mass armored formations of the enemy continue to concentrate despite heavy losses, friendly forces will fall back to successive alternate positions until such time as the enemy thrust is eventually blunted.(68) Apparently very little thought has been given to defending in depth, using village strongpoints and fortified cities to create a "web"-style defense, long known to be the most effective method of halting armored and massed attacks. As one German general put it, "We are prepared to fight in front of cities and between cities . . . [but if] we had to fight an enemy force in the Ruhr area, the war would be lost." Obviously, such a defense would be infantry intensive:

> Fighting in urban terrain takes infantry on foot above all. In highly built-up terrain, armour and vehicles face special handicaps through the protection which buildings and ruins offer to the opposing infantrist (!) My troops sit in vehicles, are trained to fight from vehicles, and their weapons are specially suited to fighting a mobile enemy in open country. I don't have the manpower, the training, the equipment for city fighting.(69)

Since military operations in built-up areas (MOBA) could well mean destroying cities in order to save them, the unenthusiastic attitude of West Germans and other Western Europeans toward urban warfare can readily be understood. At the same time, however, there is probably no more certain operational method of arresting an all-out armored attack than by fighting like the Russians did in their Great Patriotic War. If history tells us anything, it is this: the most successful defense is one in depth that incorporates a blend of static and mobile resistance.(70) In the Great War, von Lossberg's system of elastic defense, with its defended villages, strongpoints, and mobile counter-attack forces, proved time and again to be both extremely effective and relatively sparing of German lives. French failure to fortify cities and defend in depth in 1940 led to a disastrous and humiliating defeat for the "best army in Europe." The Russians, on the other hand, were able to defeat German tank forces by falling back on a weblike system of "islands of resistance" and fortified cities. Blitzkrieg, it can be said, was laid low outside the great fortresses of Moscow and Leningrad and pulverized in the meat grinder that was Stalingrad. There is no reason to believe that an elastic-style defense based on central European built-up areas and even supplemented with <u>Werwolf</u>-style guerrilla action(71) could not inflict a similar punishment on invading Soviet armies.

In any event, NATO allies could conceivably be forced into an infantry-intensive urban struggle through deliberate action on the part of the Red Army. Though the Soviets have long realized that urban areas are ideally suited for defense, it hardly seems realistic to assume that they would dogmatically risk bypassing such large and potentially symbolic urban concentrations as exist in NATO central Europe. Furthermore, the Soviets could seriously attempt to limit NATO's use of tactical nuclear weapons by purposefully adopting city "hugging" tactics.(72) That Soviet experience in urban warfare is extensive is indicated by the fact that during January and February 1945, the Red Army forced the Wehrmacht out of 300 towns and cities. While the Americans took ten days to take Aachen, the Red Army reduced Berlin in seven.(73) Today there is good reason to believe that no NATO army is better prepared for urban combat.(74) The following comment by General Chuikov amply illustrates the profound depth of current Soviet urban warfare doctrine:

> On the basis of experience gained in the fighting on the Volga and in the towns of the Ukraine, Byelorussia and Poland a manual was produced for the use of soldiers and officers, giving an exposition of the tactics of offensive fighting in small storm groups through city streets. . . . In such operations the decisive factor was not great masses of troops, but the soldier's skill and intelligent action on the part of officers commanding small groupings--platoons, sections. . . . Tanks are needed . . . not as an independent force, but for joint action with other ground arms, and in assault groups. . . . An advance through a city goes in jumps, from one captured building to the next. . . . Where manoeuvre to surround a city is an operational art, the storming of a city is a matter of tactics carried out by small units. It is the officers commanding platoons, companies, and battalions who organize the assault, who carry out the reorganization of their units into assault groups and detachments, the latter being made up according to the nature

of the objectives to be taken. The role of
officers commanding small units, and the
initiative in action of the sergeant and the
soldier, thus become of prime importance in a city
battle. It is they who have to solve the tactical
problems . . . and it is upon them that the
success of the whole battle entirely depends.(75)

Like the Germans in the Great War, who applied their Sturmtruppen counterattack tactics of elastic defense to the offense, the Russians in World War II adapted their defensive storm-group tactics to offensive urban warfare. "Storm groups" were formed to deal with enemy "strong points" while "storm detachments" were tailored to reduce larger "centres of resistance."(76) Again like the Germans, the Soviets learned by rude experience not to use armored formations in city streets where tanks tended to become helpless dinosaurs waiting to be finished off by Faustpatrone. As late as the battle for Berlin, certain Soviet formation commanders, who were "concerned for their prestige" and reluctant to decentralize tactical control, sent in masses of tanks "to the storm of Berlin in column."(77) Between April 25 and May 2, 1945, the Second Guards Tank Army fighting in this manner lost 64 percent of its tanks, half of them destroyed by man-portable antitank weapons in the hands of German infantrymen.(78) The lesson remains clear: in the short-range and chaotic environment of city fighting where streets and squares are no man's land, the "group" of fighting men on foot still reigns supreme. Since it takes far fewer troops to defend an urban area than to seize it,(79) there is perhaps an additional lesson here for the NATO alliance.(80)

It is difficult to visualize, of course, precisely what form a future war in Europe would take. According to one writer quoted by Liddell Hart, "Before a war military science seems like a real science, like astronomy, but after a war it seems more like astrology."(81) A reasonable guess, therefore, would appear to be that whatever the form, it is highly likely to be quite different from that currently planned for and commonly anticipated. If the prescient Warsaw banker, I. S. Block,(82) were around today, his demonstrated common sense and capacity for logical reasoning might well prompt him to deduce that with the first bang of a nuclear warhead, human nature being what it is, soldiers everywhere on the battlefield would probably start digging in with an unprecedented enthusiasm. The only army ever to fight knowingly under the nuclear threat, the PLA, certainly adopted this course of action in Korea. It was no accident that Chinese Communist trenches and field fortifications were much deeper and more extensive than those of the UN; the PLA seriously believed that nuclear weapons might be used against them.

Of course, a similar situation to this could conceivably develop in Europe regardless of the nuclear threat. In 1976, the German Infantry School at Hammelburg fired artillery and mortars, with the intensity prescribed by Soviet doctrine, on various field positions in which infantrymen were represented by dummies. Results of the tests showed that infantry, prone in the open, would suffer 100 percent casualties. Men in trenches without overhead protection could expect 30 percent casualties, while those dug in with overhead protection would encounter 10 percent casualties.(83) Obviously, the only effective infantry field defense in a general European war would have to be one properly dug in and camouflaged. The time required to complete such works, however, would be basically incompatible with current mobile-warfare concepts (though admittedly, it would be compatible with more positional concepts). Heavy initial casualties caused by the greatly increased effectiveness of conventional artillery and antiarmor firepower could in any case result in armies digging in as abruptly as they did in the Great War.(84)

Apart from signs that immobility rather than mobility might characterize the future European battlefield, there are concrete indications that fast-moving armored operations will be further inhibited by an electronic communications vacuum.(85) Given this possibility and Fuller's warning that "nothing is more dangerous in war than to rely upon peace training,"(86) current tactical emphasis on energy-consuming mobile armored warfare in Europe may be as misplaced as the attention focused on mobile cavalry operations prior to the Great War. At that time, even among the most pessimistic, there was a feeling that military operations of such magnitude had to end in a rapid decision one way or another. By the time the soldier began to joke that "the first seven years will be the worst,"(87) however, the short war was over, and all but the generals had lost faith in the mobile-war scenario. Today, there appears to be a striking similarity between the advent of the improved infantry antitank weapon and the cavalry-defeating machine gun. The impact of urbanization, as previously pointed out, further militates against total reliance on mobile armored forces.

On the basis of the foregoing argument, it becomes patently obvious that the foot soldier is likely to remain a force to be reckoned with in the fields, forests, and urban settlements of Europe. Reinvigorated with a power to hit harder and farther than ever before, it is unlikely that the infantryman with his traditional skills will become "merely an appendage . . . to the armoured corps,"(88) when, with less capability and battlefield credibility, he easily endured an earlier "shock of armour." While Mother Earth will doubtless continue to be the foot soldier's best friend, the combination of MICV and APC--if properly used--should greatly increase the general mobility and overall security of the infantry. While the exact use of these vehicles in the final stage of an attack is difficult to determine precisely, they most definitely limit the likelihood of infantry being killed offhandedly by shellings and air attacks. The helicopter, of course, has additionally blessed the infantry arm with an even greater mobility than the tank.(89) As well, though the soldier on foot is still probably the most vulnerable man on the open battlefield, he is at the same time the most resilient and versatile of all arms. In the setting of a European war, the NATO foot soldier would doubtless perform best in an urban or forest environment where he would have virtually unlimited scope to disorganize, disrupt, and paralyze an enemy. Training infantry for fighting a mobile battle at distance, without preparing for close-in forest and urban combat situations, might therefore be a serious error in NATO doctrine. Given a long war rather than a short one, and immobility instead of mobility, there will be a need for light infantrymen fleet of mind and foot. To arrive at such a conclusion, one has but to look beyond deterrence, whether in Sinai or in Europe.

NOTES

1. Burns, *Manpower in the Canadian Army*, p. 92. In the Normandy campaign, Canadian casualties were incurred as follows: infantry 76 percent; armored corps 7 percent; and artillery 8 percent. Stacey, *The Victory Campaign*, p. 284. According to Bidwell, small-arms fire, including bullets of rifle caliber delivered by machine guns, accounted for 30 percent of all casualties in both the Great War and World War II; artillery shells, as distinct from bombs, accounted for 60 percent in World War II. Bidwell, *Modern Warfare*, p. 55. Sidorenko, using German figures, attributes 53 percent of Great War casualties to small-arms fire; Soviet figures for World War II show small arms causing 49 percent of casualties during the first year and 39 percent during the fourth year. A. A. Sidorenko, *The Offensive (A Soviet View)* (Washington: U.S. Government Printing Office, 1970), p. 20. British figures for Northwest Europe during the period D-day (June 6, 1944)

to VE-day (May 8, 1945) show that infantry officer casualties amounted to 15.3 percent of their total taking part in operations. This compared with percentages of 8.9 for armored officers and 4.5 for artillery officers. In the case of other ranks, on the same basis, percentages were: infantry, 10.4; armored, 4.0; and artillery (including antitank), 2.2. For Italian operations, August 26, 1944 to May 2, 1945, comparative figures for officers and other ranks by corps were: infantry, 8.1 and 6.5 percent, respectively; armored, 6.2 and 2.1 percent; and artillery, 1.5 and 0.7 percent. "Battle Casualties," Canadian Army Training Memorandum, no. 56 (1945), p. 15.

2. Clausewitz, On War, pp. 358, 361, and 484.

3. Brown, Strategic Mobility, p. 199. Analysis of the Normandy campaign shows that despite an almost absolute air superiority, Allied attacks rarely succeeded unless the attacking troops had a superiority of more than five to one in fighting strength. In some cases, such as "Operation Bluecoat" (the break-out attempt by British Second Army near Caumont on July 30, 1944), attacks failed with odds of ten to one in their favor. On the Russian front, the defense often repelled attacks delivered with a superiority of seven to one or even more. Liddell Hart, Deterrent or Defence (London: Stevens, 1960), pp. 97-98, 106-9 and 179-80. It is indeed significant that fewer and fewer troops have been required over the years to defend the same amount of space. The three to one advantage currently prescribed as the force ratio required for an attack to overcome a defense is therefore questionable. Miksche, for one, relates it to numbers only, maintaining that a six to one superiority in firepower is required before the average attack will succeed. Miksche, Atomic Weapons and Armies, pp. 105-6 and 114.

4. Deitchman, Limited War, pp. 15-18.

5. Samuel Rolbant, The Israeli Soldier: Profile of an Army (Cranbury: Thomas Yoseloff, 1970), p. 108.

6. Edward Luttwak and Dan Horowitz, The Israeli Army (London: Allen Lane, 1973), pp. 62, 65-66, 68, 91, 118, 130-2 and 149; and Messenger, The Art of Blitzkrieg, pp. 223-6. Laskov was influenced by Liddell Hart's theories on the use of tanks and the "indirect approach." Yadin and Dayan, though firm believers in the latter concept, were not initially enamored of Liddell Hart's ideas on armored warfare. Bond, Liddell Hart, pp. 245-6, 249-52, 254-5, 259-66, and 269.

7. Abbreviated APC. The development of the Israeli Merkava "spaced armour" tank commenced around 1971 with Tal as its "chief architect." A heavy tank, the Merkava features armor first, mobility second, and firepower third. It can also carry six to eight infantrymen (it has a door in the rear, as the engine is in front) if 25 of the tanks's 64 rounds of ammunition are removed from the rear compartment. The Merkava can thus be used as an APC or rescue vehicle. Richard Cornblum, "Israel's New Tank: the Merkava," Canadian Defence Quarterly, 19 (1979); 34-7.

8. Luttwak, The Israeli Army, pp. 143-53, 186-9, 192 and 363-8. Of the 350 tanks of the armored corps in the Sinai Campaign, only Ben Ari's brigade had a full complement of two tank battalions. The rest were distributed to other formations by squadrons and battalions. Ibid., p. 132.

9. Luttwak, The Israeli Army, pp. 240, 246-9 and 289-96; Messenger, The Art of Blitzkrieg, pp. 232-4; and Macksey, Tank Warfare, p. 257. Gunther Rothenberg states that the "all-tank" concept never became "official doctrine" but admits that Israeli armor commanders generally neglected the infantry arm. Gunther E. Rothenberg, The Anatomy of the Israeli Army (New York: Hippocrene, 1979), pp. 159-60.

10. Luttwak, The Israeli Army, pp. 191, 215, 240-1, 258 and 295.

11. Edgar O'Ballance, The Sinai Campaign of 1956 (New York: Praeger, 1959), p. 196.

12. Rolbant, The Israeli Soldier, p. 101; and Luttwak, The Israeli Army, pp. 95-96, 100, 187, 363-5 and 370. The other functional commands are training, Gadna (pioneer, paramilitary youth), and Nahal (youth, military agricultural corps). Ibid., pp. 95-96.

13. Luttwak, The Israeli Army, pp. 95-96, 117, 108-12, 157, 177-8, 191, 215-16, 277, and 370; and Rolbant, The Israeli Soldier, p. 62.

14. Insight Team of the London Sunday Times, The Yom Kippur War (New York: Doubleday, 1974), pp. 164-5, 170-2, 191-5, and 489; Colonel J. M. E. Clarkson, "Spark at Yom Kippur: Many Surprises in an Eighteen-Day War," Canadian Defence Quarterly 3 (1974): pp. 11-13 and 21; "Lessons From the Arab/Israeli War," Report of a Seminar Held at the Royal United Services Institute for Defence Studies (London: R.U.S.I., 1974), pp. 2-3 and 5-6; and Captain L. W. Bentley and Captain D. C.

McKinnon, "The Yom Kippur War as an Example of Modern Land Battle," Canadian Defence Quarterly 4 (1974): pp. 14-18.

15. Weeks, Men Against Tanks, p. 185; and Insight Team of the London Sunday Times, The Yom Kippur War, pp. 171-2.

16. Captain (later Brigadier) Orde Wingate exerted a more direct and pervasive influence on the Israelis than Liddell Hart or any other military thinker. He emphasized personal example in leadership, practical meticulous discipline, thorough planning and preparation, concentration of force, surprise, and mobility, and ideological motivation. The Israelis called him Hayedid ("Friend"). Yigal Allon, The Making of Israel's Army (London: Valentine, Mitchell, 1970), p. 10; Bond, Liddell Hart, pp. 246-7; and Captain Luigi Rossetto, "Brigadier-General Orde Wingate and the Development of Long Range Penetration" (M.A. thesis, Royal Military College, 1978), p. 21.

17. Major General Chaim Herzog, The War of Atonement (Boston: Little, Brown and Company, 1975), p. 271.

18. Luttwak, The Israeli Army, pp. 139, 355, 365-72 and 376. In 1956, Sharon's men took three police fortresses (at Rahwa, Ghirandal, and Husan) by night. The tactics he devised for seizing a stronghold consisting of concentric lines are worth noting. Rejecting standard fire and movement action, Sharon had his men approach the trench system in darkness, walking slowly and in absolute silence until fired upon. When discovered, they ran forward as fast as possible, firing on the move, while barbed-wire fences were breached by bangalore torpedoes. On reaching the trench line, the men split into small assault groups, and without pausing to clear the fire trenches, burst into the communication trenches, running and shooting all the way to the center and out again. Sweeping one trench line after another, the assault teams kept moving until all defenders were captured or killed. The essence of Sharon's tactics was the shock effect of relentless movement and surprise to confuse the enemy and break down his resistance. The method was, of course, highly vulnerable to enemy counterattack; for this reason, Sharon prescribed isolating all strongholds prior to attack and blocking them all-round. Ibid., pp. 113-15.

18. Luttwak, The Israeli Army, p. 365, and Herzog, The War of Atonement, pp. 191 and 270.

19. For Zavah Haganah LeYisrael (Israel Defense Army) formed May 31, 1948. Allon, The Making of Israel's Army, p. 35. The official translation is "Israeli Defense Forces" as it includes air and naval forces. Luttwak, The Israeli Army, p. 37.

20. Originally established under Wingate's guidance to counter Arab terrorism by offensive measures.

21. Allon, The Making of Israel's Army, pp. 9, 11, 16, and 18-21.

22. Interestingly, in 1970, 25 percent of all Israeli career officers had university degrees or the equivalent, 42 percent had a full secondary education, and 31 percent a partial secondary education. Luttwak, The Israeli Army, p. 182; and Rothenberg, The Anatomy of the Israeli Army, p. 30.

23. Luttwak, The Israeli Army, pp. 21, 54, 73-74, and 81-86. Rabin was the first ex-Palmach Chief of Staff, appointed after Ben Gurion's resignation. Ibid., p. 177. For details of the Palmach field training program for the individual soldier and NCOs, see Allon, The Making of Israel's Army, pp. 125-30.

24. S. L. A. Marshall, Sinai Victory (New York: William Morrow, 1958), pp. 239-40; and O'Ballance, The Sinai Campaign, p. 75. Three sections plus a small headquarters with a 60-millimeter mortar and perhaps "bazooka" made up an Israeli platoon. Three platoons plus a weapons platoon and headquarters comprised a company. Four rifle companies (later three), a weapons company, and a headquarters made up a battalion of roughly 700 all ranks. An infantry brigade consisted of three infantry battalions plus a headquarters and a number of support units that varied in size and composition. An Israeli brigade is commanded by a colonel; the second in command is a lieutenant colonel, as is the commanding officer of a battalion. Rolbant, The Israeli Soldier, pp. 103 and 108; Luttwak, The Israeli Army, pp. 90 and 176; O'Ballance, The Sinai Campaign, p. 58; and Rothenberg, The Anatomy of the Israeli Army, pp. 101-2

25. Allon, The Making of Israel's Army, pp. 127 and 265. Allon was opposed to military college and other direct entry officers. He argued that any officer who served a stint as a corporal would "gain experience no military academy can ever give. . . . Whatever respect one may have for military colleges and the general and technical training they give, no military

270   A PERSPECTIVE ON INFANTRY

college graduate is fit to bear the title and responsibility of 'officer' before he has served for a period as a section-leader." Ibid., p. 266.

S. L. A. Marshall, "Why the Israeli Army Wins," Harper's Magazine, October 1958, pp. 39 and 41-43; and Marshall, Sinai Victory, pp. 21 and 233-4. An NCO candidate received a "silver rating" if he completed in less than 11 minutes an obstacle course consisting of: an initial 400-meter run (including the scaling of a 6-foot wall and the walking of a 20-foot parallel bar); a second 400-meter lap (including a 15-yard crawl under barbed wire, crossing three ditches filled with running water and walking a plank over a fourth, ten feet deep, to throw a grenade at a target); and a final 600-meter dash. Marshall, Sinai Victory, p. 235.

27. Allon, The Making of Israel's Army, pp. 91-92 and 258-9; Marshall, Sinai Victory, pp. 22-23, 90, and 226; and Rolbant, The Israeli Soldier, pp. 33, 175-6, and 331. The proportion of teeth to tail in the 1956 IDF was about 50 percent. O'Ballance, The Sinai Campaign, p. 74.

28. Allon, The Making of Israel's Army, pp. 91-92 and 258-9; Marshall, Sinai Victory, pp. 22-23, 90 and 226; and Rolbant, The Israeli Soldier, pp. 33, 175-6 and 331. Virtually every U.S. soldier who managed to survive his tour in Vietnam received a Bronze Star. This state of affairs has resulted in the very sensible recommendation that no medals be awarded for actions unconnected with combat. Gabriel and Savage, Crisis in Command, pp. 15 and 135. It is interesting to note, as well, the Israeli "principles of war": maintenance of the aim, initiative, surprise, concentration, economy of force, protection, cooperation, flexibility, and consciousness of purpose or cause. Allon, The Making of Israel's Army, p. 44.

29. Rolbant, The Israeli Soldier, pp. 109-10; and O'Ballance, The Sinai Campaign, p. 196.

30. Herzog, The War of Atonement, pp. 248-50.

31. General William E. DePuy, "Implications of the Middle East War on U.S. Army Tactics, Doctrine and Systems," mimeographed copy of a presentation by Commander, U.S. Army Training and Doctrine Command, 1976, pp. 1-2, 10-11, 13, 16 and 18.

32. In the 1973 Yom Kippur War, the Egyptians used the BMP as directed by Soviet tacticians and suffered very heavy losses. D. M. O. Miller, "The

Infantry Combat Vehicle: An Assessment," <u>Military Technology and Economics</u>, 3 (May-June 1979): 32.

33. C. N. Donnelly, "Tactical Problems Facing the Soviet Army," <u>International Defense Review</u>, 11 (December 1978):1406; and Phillip A. Karber, "The Soviet Anti-tank Debate," <u>Survival</u> 18 (May-June 1976):105-6 and 108-9. The BMP is assessed as one of the finest armored personnel carriers in the world. Amphibious and fast, it also permits troops to fire their weapons from within it while on the move. Jeffrey Record, <u>Sizing up the Soviet Army</u> (Washington: The Brookings Institution, 1975), p. 24.

34. See, for example, Joseph D. Douglass, Jr., <u>Soviet Military Strategy in Europe</u> (New York: Pergamon, 1980), pp. 76-79 and 164-5.

35. Michael McGwire, "Soviet Strategic Weapons Policy," in <u>Soviet Naval Policy: Objectives and Constraints</u>, eds. Michael McGwire, Ken Booth, and John McConnell (New York: Praeger, 1976), pp. 488-91 and 498-500; and Marshal V. D. Sokolovsky, ed., <u>Military Strategy: Soviet Doctrine and Concepts</u> (New York: Praeger, 1963), pp. 194-204 and 277-95.

36. Y. Novikov and F. Sverdlov, <u>Manoeuvre in Modern Land Warfare</u> (Moscow: Progress Publishers, 1972), p. 8. The classic model chosen was the advance of Sixth Guards Tank Army against the Japanese army in Manchuria in 1945.

37. Donnelly, "Tactical Problems Facing the Soviet Army," pp. 1405 and 1410. For design details of the BMP see Brigadier Richard E. Simpkin, <u>Mechanized Infantry</u> (Oxford: Brassey's, 1980), pp. 32-37 and 74-75.

38. This is perhaps stretching the point technically, as the French AMX-VCl, which entered service in 1956, had firing ports to enable the infantry to fight from inside. Miller, "The Infantry Comfat Vehicle," p. 32.

39. Karber, "The Soviet Anti-Tank Debate," pp. 107-8; Donnelly, "Tactical Problems Facing the Soviet Army," p. 1405; and David R. Jones, <u>Soviet Armed Forces Review Annual, Vol. 4, 1980</u> (Gulf Breeze: Academic International Press, 1980), pp. 91-92. The mechanized infantry offensive in depth had first been proposed in the 1930s by Tukhachevskii. Interestingly, and contrary to Western expectations, it took the Soviets 15 years following the introduction of the T-62 to deploy a new main battle tank (the T-72) in Central Europe. Yet over

the same period, they designed, developed, and deployed five major battlefield air defense systems, five major artillery systems, and numerous armored personnel vehicles, including the BMP. Ironically, the ATGM threat to the BMP has actually raised the status of the tank in Soviet eyes. Karber, "The Soviet Anti-Tank Debate," pp. 107-9.

40. Karber, "The Soviet Anti-Tank Debate," p. 109; Donnelly, "Tactical Problems Facing the Soviet Army," pp. 1406-7; and Sidorenko, The Offensive (A Soviet View), pp. 145-7. Recent Soviet studies have shown that only one infantryman in six can hit a moving target when firing his automatic assault rifle from inside a BMP. Donnelly, "Tactical Problems Facing the Soviet Army," p. 1410. The Soviets fully realize that night, mist, snowfall, smoke, and dust all impede the use of ATGMs, even with night vision devices. Nonetheless, they are judged "the strongest anti-tank weapon at maximum and medium ranges [more than 1,000 to 1,500 meters], in open country, under favourable conditions of observation." Major General G. Biryukov and Colonel G. Melnikov, Antitank Warfare, trans. David Myshne (Moscow: Progress, 1972), pp. 83-84.

41. Donnelly, "Tactical Problems Facing the Soviet Army," p. 1407; and Karber, "The Soviet Anti-tank Debate," pp. 107 and 110-11. Soviet commentators often make the point that NATO ground forces, particularly the West German and American armies, are not geared to a prepared defense in depth, which would make the most favorable use of antitank weapons, but remain committed to the doctrine of a mobile defense based on brigade and division-level armored counterattacks. Thus, while the West still seems to believe that the tank is the best means of antitank defense, Soviet military writers are beginning to stress the offensive use of antitank weapons. Ibid., p. 111. Interestingly, the tank only gained the reputation for being the best antitank weapon because it was, for much of World War II, the sole dependable means of directly taking on an enemy tank. In the Red Army, antitank artillery accounted for 66 percent of enemy tank losses. Using artillery in a direct mode against German tanks was common practice for the Soviets. All modern Russian artillery is recognized as having an antitank capability. Biryukov, Antitank Warfare, pp. 36, 52-55, and 73; and Sidorenko, The Offensive (A Soviet View), p. 123.

42. In Soviet usage, suppression means the infliction of 25 percent casualties on enemy personnel and equipment so that he is incapable of action during

the period of bombardment and for a short time after it has ceased. Destruction, on the other hand, means the infliction of over 60 percent casualties and the consequent inability of the enemy to recover for a long time after the attack. Donnelly, "Tactical Problems Facing the Soviet Army," p. 1410.

43. Karber, "The Soviet Anti-Tank Debate," pp. 109-10. An additional lesson of the Yom Kippur War was that unprotected batteries firing in line were extremely vulnerable. The fact that NATO counterbattery fire (computerized) could be expected within six to ten minutes of the opening of a Soviet bombardment gave impetus to the procurement of self-propelled, and protected, artillery pieces. Donnelly, "Tactical Problems Faced by the Soviet Army," pp. 1407-9. The Soviets are quite aware that parceling out artillery has certain disadvantages, among them increased difficulty in coordinating massed artillery fire and greater strain on logistics systems. As previously mentioned, infantry guns (75-millimeter and 150-millimeter howitzers attached to regiments) were peculiar to German infantry. Developed from 1927 onward to meet a requirement for rapid close-support fire, they could fire both high explosive and armor-piercing ammunition. A disadvantage of this system was that the Germans were never able to concentrate their gun fire as well as the Allies could. Davies, German Army Handbook, pp. 56-57 and 111. That the Germans generally fought better overall, however, might make this system worth examining. General Senger und Etterlin thought "on the basis of the number of batteries engaged, the German artillery was not really inferior to that of the opponent." Senger und Etterlin, Neither Fear Nor Hope, p. 235.

44. Sidorenko, The Offensive (A Soviet View), p. 137.

45. The Marder, mounting a 20-millimeter cannon, a Milan ATGM (range 2,000 meters), and two machine guns, has a range of roughly 600 kilometers (the same as the Leopard tank) and can reach speeds of up to 70 kilometers per hour. Major General Fritz Birnstiel, "German Combat Troops in Action." Infantry, 57 (November-December, 1977): p. 27. It carries six riflemen and a crew of four. Twice as heavy as the BMP, it is not amphibious. The Russians regard the Germans as "having the most experience in using tanks and, therefore, the military-strategic plans and considerations of the Bundeswehr deserve special attention." Biryukov, Antitank Warfare, pp. 15 and 29; and Miller, "The Infantry Combat Vehicle: An Assessment," p. 31.

274   A PERSPECTIVE ON INFANTRY

46. For "Tube-launched, Optical-tracked, Wire-command link" missile. It weighs 172 pounds, can be carried by four men, and has an effective range of 300 to 3,500 meters. The missile weighs 54 pounds. U.S. Army TC 7-24, <u>Antiarmor Tactics and Techniques for Mechanized Infantry</u>, pp. 2-5 and A-2.

47. The Soviets clearly recognize that "West German military theorists . . . assume that the modern defense is first the battle against the enemy tanks; therefore, it should first be organized as an antitank defense." Sidorenko, <u>The Offensive (A Soviet View)</u>, p. 53.

48. Birnstiel, "German Combat Troops in Action," pp. 26-27; Canadian Liaison Staff, Headquarters II (German) Corps, "The New Structure of the German Armed Forces," <u>Canadian Defence Quarterly</u> 4 (1975), p. 16; and Weller, <u>Weapons and Tactics</u>, pp. 148-9. The Germans have given their armored force the responsibility for the doctrine and training of <u>Panzergrenadiers</u>. Light infantry remains separate at the Infantry School at Hammelburg. General William E. DePuy, "One-up and Two-Back?," <u>Army</u> 30 (1980), p. 24.

49. Birnstiel, "German Combat Troops in Action," pp. 27-28. The Germans consider the employment of "pure" battalions (i.e., tank or infantry) to be the best method of operating. General William E. DePuy, "The U.S. Army: Are We Ready For The Future?," <u>Army</u>, 28 (September 1978), p. 28.

50. According to DePuy, this organization will also call for more officers and NCOs. Interestingly, Gabriel and Savage state that in World War II and Korea the percentage of U.S. Army officers to men was 7-9 percent, respectively; by the end of Vietnam, officers constituted roughly 15 percent of total strength. Both authors contend that there is a correlation between officer strength and cohesion, that a swelling of the officer corps corresponds with a decline in quality and cohesion. As stated in chapter three, in World War II, the German officer corps, made up of carefully selected officers virtually all with education superior to the average German, constituted only 2.96 percent of overall army strength. In the 1951 French army, officers comprised but 4.9 percent of the total. Gabriel and Savage, <u>Crisis in Command</u>, pp. 10, 31, 34-35 and 69.

51. DePuy, "The U.S. Army: Are We Ready For the Future?," pp. 22-23, 25-26, and 29. The current American infantry company tends to be larger than those

of the Germans, Israelis, and Russians and is deemed "functionally complex." The TOW ATGM was originally given to the American infantry supposedly as a "tag-along" weapon; it is frequently improperly employed because it is tied too closely to infantry companies which prefer close country and shorter ranges. Ibid., pp. 24-25.

52. Liddell Hart in 1950 recommended the "radical reorganization of the division on a five battalion basis. It would then be under the more immediate control of the divisional commander, without any interposing brigade headquarters." Liddell Hart, Defence of the West, pp. 300-2.

53. Major General Frank Kitson, "The New British Armoured Division," R.U.S.I. Journal 122 (March 1977): 17-18. In the American "pentomic" division of the 1950s, each battle group of 1,427 men consisted of five big maneuver companies commanded by captains. The pentomic reorganization, directed by General Maxwell D. Taylor, was based on the concept that tactical operations on the widely dispersed nuclear battlefield would be conducted by smaller, faster moving, hard-hitting units. DePuy, "The U.S. Army: Are We Ready for the Future?," p. 23. Liddell Hart originally suggested this pattern of organization and was pleased to see the Americans adopt it. Liddell Hart, Memoirs, vol. 1, p. 220. The new British "double brigadier" system of tactical command bears an uncanny resemblance to the two "combat commands" of the 1942 American armored division, each command "a sub-headquarters under a brigadier general, to either of which the division commander might assign such forces as he chose for specific tactical missions." Greenfield, The Organization of Ground Combat Troops, p. 323.

54. DePuy, "The U.S. Army," p. 26.

55. The Germans only dismount six riflemen per MICV; recent American opinion appears to suggest five. Lieutenant Colonel Robert G. Chaudrue, "Requiem for the Infantry," Infantry 68 (May-June 1978):30; and DePuy, "One-up and Two Back?," p. 25. Within British Army of the Rhine platoons, sections currently dismount but six men each leaving two in the APC to drive and man the gun. Major T. A. Coutts Britton, "The Assault," British Army Review, no. 61 (1979), pp. 53-56. Coutts Britton recommends that section commanders maneuver their sections through the objective in two groups rather than pairs. Ibid., p. 55.

56. Miller, "The Infantry Combat Vehicle: An Assessment," p. 32.

57. Company frontage will be 450-600 meters; platoon frontage, 150-200 meters; and section/squad frontage, 50-70 meters. Six to eight meters will be maintained between soldiers. Sidorenko, The Offensive (A Soviet View), p. 84.

58. Novikov and Sverdlov, Manoeuvre in Modern Land Warfare, pp. 52-53; and Donnelly, "Tactical Problems Facing the Soviet Army," pp. 1405-7 and 1412. It is permissible for battalions to attack in either one or two echelons or waves; tanks are always in the first echelon. The basic task of a second echelon is to "step up efforts in the direction of the main attack," shifting as necessary from one area to another. Leap-frogging a second echelon over a first in World War II usually ended in failure; far from "stepping up the pressure," the leap-frogging second-echelon troops instead stopped the advance of the first. Ibid., pp. 52-53 and 64-65. On the matter of the effectiveness of ATGM, the Soviets estimate that in World War II it took between two to three minutes, and eight to ten aimed shots on average, to destroy one attacking tank; they now estimate 30 seconds minimum, given correct conditions and, in any case, a second shot hit by ATGM. Donnelly, "Tactical Problems Facing the Soviet Army," pp. 1410-11.

59. Captain J. S. Labbé, "Future Canadian Infantry Mobility in Europe: Wheeled or Mechanized," Canadian Defence Quarterly 9 (1979):19-20; and Miller, "The Infantry Combat Vehicle: An Assessment," pp. 30, 32-36, and 38. "Commander-in-Chief's Order" in early 1978 effectively curtailed immediate introduction to service of the new American IFV, the XM 723. The General Accounting Office found the prototype slow, noisy, smoky, expensive, and too vulnerable. Captain Luigi Rossetto, "Armoured Infantry?," The Army Quarterly 109 (1979):148. The Americans are now introducing the XM 2 into service.

60. Trythall, "Boney" Fuller, pp. 93 and 168.

61. Paul Bracken, "Models of West European Urban Sprawl as an Active Defence Variable," in Military Strategy and Tactics: Computer Modeling of Land Warfare Problems, eds., Reiner K. Huber, Lynn F. Jones, and Egil Reine (New York: Plenum, 1975), pp. 219-22 and 229; Major B. M. Archibald, "Urbanization and NATO Defence," Command and Staff Review, Canadian Forces Staff College (1979), pp. 102-5; and Birnstiel, "German Combat Troops in Action," p. 26.

62. Paul Bracken, "Urban Spread and NATO Defence," Survival, 218 (1976):255-6.

63. Major A. E. Hemesley, "MOBA--Too Difficult?", R.U.S.I. Journal, 122 (1977):24. MOBA stands for "Military Operations in Built-up Areas." Another acronym is MOUT (Military, sometimes Mechanized Operations in Urban Terrain).

64. Bracken, "Urban Sprawl and NATO Defence," pp. 255 and 257; and Birnstiel, "German Combat Troops in action," p. 26. Bracken claims a typical armored brigade defensive position would contain some 85 villages; Birnstiel estimates that 30-40 built-up areas, "most of which will be small villages" (population 3,000), can be expected in a brigade defensive area. As most roads run into towns, urban density is a significant factor.

65. Trythall, "Boney" Fuller, pp. 92-93. While Simpkin argues that the introduction of compound armor has set the infantryman's antitank capability "back again to something between what it was in the Molotov cocktail and Panzerfaust eras," he does admit that this is "no more than a swing of the attack-defense pendulum." Simpkin, Mechanized Infantry, pp. 3 and 82.

66. Bracken, "Urban Sprawl and NATO Defence," pp. 254 and 257-8. According to the Bonn government's 1979 defense White Paper, West Germany's narrow width (500 kilometers or less) makes surrender of even a single inch of territory unthinkable. The heavy population density near the Warsaw Pact border further reinforces the idea that there can be no alternative to forward defense. The belt between the East-West demarcation line and the terrain 100 kilometers to the west--the zone of forward defense--encompasses 40 percent of the area, 30 percent of the population, and 25 percent of the industrial potential of the FRG. Wolfram von Raven, "The Concept of Forward Defence and the Security of Europe," Canadian Defence Quarterly 7, (1977):8.

67. Soviet strategists stress that the relatively limited depth of Western Europe makes it comparatively easy to overrun. Brown, Strategic Mobility, p. 245. Napoleon, on being shown a plan to defend France that deployed almost the whole of the Grand Army in fairly even line along the frontier, supposedly asked if it was to prevent smuggling. Adrian Hill, "Could Napoleon's Army Win Today?," R.U.S.I. Journal, 122 (1977):23.

68. Chaudrue, Requiem For the Infantry," pp. 28-29; and General Donn A. Starry, "A Tactical Evolution--FM 100-5" Military Review, 8 (1978):6-10. Based "on the history of a 1,000 tank battles," a key

feature of this doctrine is striking enemy rear areas. Breakthrough attacks are to be met through rapid concentration of units, filling denuded areas left by them with air cavalry and attack helicopter units. No large reserve or counterattack force is retained, as it is considered unaffordable. Counterattacks are to be effected in "smaller increments (the lowest level being division for formal planning) and more quickly than before . . . not cautiously, but carefully." The battle is to be fought in a series of successively deeper battle areas and positions until the attack is killed. There is to be no traditional massive counterattack to eject the enemy after this is accomplished, however. Starry, "A Tactical Evolution," pp. 6-10.

69. Hemesley, "MOBA--Too Difficult?," p. 25. The Germans do maintain some nonmechanized infantry units for employment in densely populated areas or mountainous and wooded terrain such as exists in Bavaria. However, there are but six _jaeger_ (rifle), three mountain, and three airborne brigades out of a total of 36 field brigades; the rest are armored infantry. The airborne brigades are mainly helicopter transported since the Germans place little faith in the combat jump of major units or formations. "The New Structure of the German Armed Forces," pp. 14-15; and Birnstiel, "German Combat Troops in Action," p. 29.

70. German defensive tactics in Normandy and later were a blend of static defense and dynamic defense by dispersed battle groups making sharp "finger-thrusts." They repeatedly checked Allied columns and brought them gradually to a halt. In contrast to the effect of multiple small thrusts, attempts at concentrated counterattack failed repeatedly. Liddell Hart, _Deterrent or Defence_, p. 181. The trouble with the American "active defence" is that it proposes no static or positional framework.

71. The idea is Fullerian, though it has its roots in history. Partisan operations were, of course, part of Soviet and Yugoslav efforts in World War II. The regular German army deprecated the idea of the people in arms and therefore the Werwolf threat never materialized, though it was seriously feared by the Allies. The Germans have a greater tradition for irregular warfare than is supposed; the _Freicorps_, for example, modeled on the example of the _Spanish_ guerrillas, go back to the Napoleonic Wars. Heilbrunn, _Conventional Warfare in the Nuclear Age_, pp. 56 and 107-9. Fuller considered the British formation of the Home Guard a sensible step in World War II; London, unlike Paris, appears to have been ready to turn itself

into a giant Stalingrad. The modern German Territorial Army could conceivably be used to the same end. Sir John Slessor was one of the first to suggest employing "a highly-trained German semi-static Home Guard" armed with antitank weapons. Slessor, The Great Deterrent, p. 309.

72. R. G. Smith, "The Soviet Armoured Threat and NATO Anti-Armour Capabilities," The Army Quarterly 109 (1979):158.

73. C. N. Donnelly, "Fighting in Built-up Areas: A Soviet View--Part II," R.U.S.I. Journal, 122 (1977):63-65; Hemesley, "MOBA--Too Difficult?", p. 25; and Bracken, "Urban Sprawl and NATO Defence," pp. 255 and 258-9. The Soviets have been more traditionally concerned with urban warfare in the defense; offensive urban combat doctrine generally stresses staying out of cities if at all possible. During World War II, the Russians preferred to use two other methods, often in combination, to reduce cities. One was to blow the central core of a city to smithereens by prolonged and intense bombardment; the second was to encircle the city and by cutting off its supplies compel it to capitulate. P. H. Vigor, "Fighting in Built-up Areas: A Soviet View--Part I", R.U.S.I. Journal, 122 (1977):39 and 41-46; and Bracken, "Models of West European Urban Sprawl," p. 222. Against NATO cities, the Soviets could try a third method--gas. Described by Fuller as the ideal weapon for reducing cities, it would be guaranteed to produce millions of refugees and doubtless another major problem for NATO forces. (A 1962 French estimate placed the normal expected number of refugees moving west to be at three to five million East Europeans and Germans.) Edward L. King, The Death of an Army: A Pre-Mortem (New York: Saturday Review, 1972), p. 137. The Red Army in Europe possesses a frighteningly large arsenal of chemical agents. (This could, of course, be for purposes of deterrence, as the Russians suffered terribly heavy gas casualties in the Great War.) J. S. Finan, "Soviet Interest in and Possible Tactical Use of Chemical Weapons," Canadian Defence Quarterly 4 (1974): 11-12 and 15.

74. Major A. E. Hemesley, "Soviet Military Operations in Built-up Areas," Infantry, 67 (1977):34.

75. Marshal Vasili I. Chuikov, The End of the Third Reich, trans. Ruth Kisch (London: MacGibbon and Kee, 1967), pp. 130-1, 162, and 184-5. A street would be tackled by an entire regiment, one battalion allotted to each side of the street and the third in reserve. Vigor, "Fighting in Built-up Areas: A Soviet View--Part I," pp. 42-43.

280   A PERSPECTIVE ON INFANTRY

76. "Storm groups" assaulting Berlin were composed of 20 to 25 riflemen or submachine gunners, two or three machine guns, two or three antitank rifles, one or two guns for direct fire, an engineer detachment of from three to five men, a detachment of two to four men with smoke and incendiary devices, some man-pack flamethrowers, a wireless, and, when available, one or two tanks or self-propelled guns. Other storm groups included more engineer detachments, and sometimes a platoon or company of infantry, in addition to the numbers detailed above. "Storm detachments" were larger organizations, containing up to a battalion of infantry, a company of engineers, a platoon of flamethrowers, a company each of tanks and 82-millimeter mortars, a battery each of self-propelled guns, 120-millimeter mortars, 122-millimeter howitzers, and a troop of 45- and 76-millimeter guns. Sometimes guns as large as 203 millimeters were allotted to storm detachments, the role of Soviet artillery to smash a city to pieces, house by house, in direct or indirect modes.

77. Chuikov, The End of the Third Reich, pp. 183-4.

78. Vigor, "Fighting in Built-up Areas: A Soviet View--Part I," p. 45. They learned, too, that German houses were tougher nuts to crack than their Russian or Polish equivalents. Ibid., p. 40.

79. C. N. Donnelly, "Soviet Techniques for Combat in Built-up Areas, International Defense Review 10 (April 1977):242.

80. The recent German development of the "weapon under the bed" Armbrust expendable hand-held antitank weapon is heartening in this regard. Weighing but 6.3 kilograms, Armbrust has an effective range of 200 meters or more. Its most appealing features, however, are its relative soundlessness and greatly reduced back-blast flash (0.8 meters clearance required); the latter characteristic means that it can be fired in relative safety from within enclosed rooms. Only the older British PIAT had a similar capability. Christian Eliot, "Man-Portable Anti-Tank Weapons," Military Technology and Economics, 3 (January-February 1979):22-24.

81. Liddell Hart, Thoughts on War, p. 24.

82. Writing around the turn of the century, Block foresaw clearer than most the form the Great War was eventually to take on: millions of men under arms in

entrenched positions and fortifications, collectively demonstrating the futility of mass attacks in the face of deadly modern firepower. I. S. Block, The Future of War, trans. R. C. Long (New York: Garland, 1972), pp. 36-41. According to Fuller, the "only noted soldier recorded to have troubled himself to criticize Block's views was old General Dragomirov. . . . He condemned them because they failed to prove that the bayonet was still supreme." Fuller, The Conduct of War, p. 130.

83. Chaudrue, "Requiem for the Infantry," p. 29. In World War II, the Germans did not have the proximity fuze. Patton warned that "in the next war" lots of overhead protection would be required. Patton, War As I Know It, p. 313.

84. Other scenarios have been postulated, of course. One of the most intriguing is Miksche's suggestion of a "disseminated infiltration by Russian light infantry on a large-scale moving from cover to cover" offering most unworthwhile nuclear targets. Miksche, Atomic Weapons and Armies, p. 162.

85. Lieutenant Colonel Don E. Gordon "Target: The Spoken Word," Army, vol. 29, 1979, pp. 20-22.

86. Fuller, Generalship and Its Diseases, p. 50.

87. Wolff, In Flanders Fields, p. 5.

88. Miller, "The Infantry Combat Vehicle: An Assessment," p. 34.

89. The attack helicopter, for which I see a great future in combination with other arms, I regard as essentially another arm or "weapons platform" similar to the tank. The Germans see attack helicopters and tanks working together within friendly territory. The Americans are more radical. "The Helicopter in Modern Combat," Military Technology and Economics, 3 (1979):9-10. The advantage of transport helicopters is the tremendously increased physical mobility they give to infantrymen; the weapon in this case, however, is not the helicopter but the soldier on foot. See also, Lieutenant Colonel Lynn M. Hansen, "Soviet Combat Helicopter Operations," International Defense Review 11 (1978):1244-5. The Russians, though initially skeptical of the helicopters' worth, have since built up a formidable fleet of antitank, attack, and airborne assault helicopters. Ibid., p. 1246.

9. Foul-Weather Warriors

Toward a Perfection of Infantry

The verdict of this study is that infantry has played a more significant role in twentieth-century warfare than has hitherto generally been realized and that foot soldiers will likely continue to occupy an extremely important place in any future conflict. The twin pillars of infantry strength, of course, remain technical capability and tactical prowess. As the cavalry-defeating machine gun projected the might of the "concentrated essence of infantry"(1) in the Great War, so the armor-defeating ATGM has increased the power of the foot soldier today; for the first time, we are seeing an infantry technically capable of defeating tank attacks. However, as the previous chapters have revealed, the infantry arm was primarily sustained on the field of battle through the tactical ability and applied intelligence of men. The training, motivation, and operational capacities of the soldier on foot have often decided the major issues of war. Within the infantry itself, an intellectual openness and emphasis on small groups and their leaders has normally enhanced infantry effectiveness overall. Advancing technology has increased, not lessened, the need for improved tactics and better techniques of training in these areas.

The decentralization of tactical control forced on land forces has been one of the most significant features of modern war. In the confused and often chaotic battlefield environment of today, only the smallest groups are likely to keep together, particularly during critical moments. In such circumstances, most individuals usually yield to the powers of a leader sweeping them on toward the accomplishment of a mission. That such did not occur in

France in 1940 was due to the existence of a crushing centralization that smothered all vestiges of junior leadership. Small groups that could have continued to fight instead surrendered because their communications worked against them. In view of current Soviet electronic warfare capabilities, NATO nations could conceivably experience a similar disruption in any future struggle. The more sophisticated the army, apparently the more vulnerable it is to paralysis(2) and the strains of Clausewitzian "friction," wherein things that can will go wrong. The lesson, then, is for an army not to become overly centralized. Small groups and their leaders must be capable of going it alone like so many forlorn hopes. In this regard, it is noteworthy that the Yom Kippur War taught "the incisive lesson that ground forces must be capable of dealing with all problems without depending in any way on the Air Force."(3) Judging from similar lessons in World War II in France and Russia, an infantry that has neither the means nor inclination to fight on its own is hardly worthy of the name.

A major theme of this work has been to stress the vital importance of small units and basic infantry tactics. As we have seen in these pages, battles are essentially "fought by platoons and squads"(4); in fact, superior infantry performance appears to have been related in many instances to the actions of such primary groupings. There is thus substantial reason to suspect that the section commander is a prime determinant of operational effectiveness in the infantry battalion team.(5) It follows then that far greater attention should be paid to primary-group training within units. A positive step in this direction would be to replace the outdated Jominian term "minor tactics" with a more dynamic nomenclature such as "battle patterns" and accord them a higher priority.(6) Much more time could also be devoted to making section and platoon training as interesting and challenging as rifle-company training.(7) Good sections make good platoons and companies, which, in turn, make the difference on the battlefield. Exactly how much difference is lucidly explained by a Canadian commanding officer who fought a battalion in Korea:

> The success of Kapyong was due mainly to high morale and to good company, platoon, and section commanders. . . . That is something that we should never overlook in our military training. Too much officer training is aimed at high levels of command and not enough at the company and platoon level. With a modicum of experience at the lower levels, anyone can take over at the higher. Many poor commanders have stayed in command at

brigade and divisional level in consequence of having a good staff. At the platoon or company level the poor commander is discovered the first day. So I say that a division can survive with a poor commander for a while, but I'll be damned if a platoon or company can. Therefore, play only superficially at moving divisions on . . . maps with fingers spread making right and left hooks. Concentrate on section, platoon and company tactics. Learn from experience what human beings can endure and still fight. Learn to do your jobs properly at your own levels of command. It is surprising how easy it is to command a battalion when you have had success in commanding a company.(8)

By placing stronger emphasis on section training, the general level of platoon and company effectiveness could be much improved. Treating the section commander more like the commissioned officer-commander to whom he reports would be an enlightened and practical way of contributing to this. Additional time spent on demanding and realistic section training would not only strengthen primary-group cohesion but would as well better condition individuals for the lonely reality of the battlefield. Under present training methods, the soldier grows accustomed to the presence of great numbers of men and material around him. He sees strength on parade, and he is rarely left alone in the field. The more he sees of the strength of his army, the greater grows his confidence. He is scarcely aware that it has become a factor in his morale, that is, until the day the desolate emptiness of the battlefield "chills his blood and makes the apple harden in his throat." Expecting to see action, he sees nothing.(9) He is also, more often than not, alone on the ground. Not conditioned to operating in a small group, he is not even sure if his friends are around.

According to Marshall, it is "one of the simplest truths of war that the thing which enables an infantry soldier to keep going with his weapons is the near presence or the presumed presence of a comrade."(10) It follows naturally, then, that foot soldiers should be trained to expect that immediate comrades may often be as invisible as the enemy on the field of battle. More extensive small-group training would certainly be one way of getting this message across; it would also do much to break the gregarious habit. As mentioned earlier, Allied commanders during World War II were continually plagued with the problem of overcoming the tactically undesirable herd instinct. A strong recommendation that emerged as a consequence of this concern was that "all infantry training should aim at instilling into every man more confidence in his ability to move alone at night." It was at the same time stressed, however, that such confidence could "only be gained after detailed and long instruction in small groups."(11) For armies the world over, the lesson is clear: small groups must be permitted to develop the inherent cohesion--the mutual professional interdependence--necessary to sustain themselves in war. The Germans in World War II appeared to have understood this:

> The replacement system of the the Wehrmacht operated to the same end. The entire personnel of a division would be withdrawn from the front simultaneously and refitted as a unit with replacements. Since new members were added to the division while it was out of line they were thereby given the opportunity to assimilate themselves into the group; then the group as a whole was sent forward. This system continued until close to the end of the war and helped to explain the durability of the German Army in the face of the overwhelming numerical and material superiority of the Allied forces.(12)

As far as individual training in general is concerned, the modern infantryman should spend much more time and energy practicing stalking skills and improving weapon handling. Only the very highest standard of personal camouflage should be accepted with exercises as a rule being conducted in scenarios of enemy air superiority. Although infantrymen should be capable of holding an enemy off and fighting at a distance, they must also be prepared to withstand the rigors of close-quarter combat and, moreover, prevail in such circumstances. Given the likelihood of forest and urban engagements in a general European war, it would be shortsighted not to train infantrymen to fight in these environments. In this regard, foot soldiers should be taught how to attack and destroy fortified strongpoints. Every infantryman, for example, should know how to prepare a pole charge and be generally better trained in demolitions, which have application in city fighting. He should additionally be capable of destroying tanks at close range. Though he may not often have to use this combat skill, such training would help to overcome the foot soldier's innate fear of tanks and promote an overall offensive spirit. On the Eastern Front, "tankophobia" was well recognized by the armies of both sides as a serious disease with which to be reckoned.(13)

In an associated vein, conditioning the soldier to withstand the often traumatic noise of the battlefield also appears to be a matter of major military concern. To quote one World War II infantry officer, "a soldier's nerves should be trained for battle as carefully as his brain and body. If during his training he could be introduced to the crash of bombs, the whine of bullets, the whistle and crump of shells, he would find it easier to withstand the real thing when he encountered it."(14) Had the French infantry been so conditioned in 1940, they may not have collapsed so ignominiously. To more experienced veterans, however, "battle inoculation in which you are shot at with the intention of being missed . . . [was] of little consequence." A better system was to teach the soldier how to recognize various weapons by their sounds, to tell the difference between a Bren or a Spandau, and to learn by the crack of a bullet whether it missed you by inches or yards.(15) The real aim of all such training, of course, was to make it psychologically easier for the average infantryman to return enemy fire. To be able "to plaster . . . opponents with fire," as Rommel described it, was to gain a moral victory that was half the battle.

    In Marshall's judgment, the general reluctance of troops to fire at the enemy pointed to a serious training deficiency in addressing the problem.(16) "The act of willingly firing upon the enemy," he argued, should be recognized as "an instance of high initiative on the battlefield" and not as "commanders have long considered it . . . simply a natural derivative of sound training." Stressing the doctrine of fire discipline early in training was to put the cart before the horse; far better to start with uninhibited soldiers and rein them in, "controlling" their fire instead of trying to get it started. And who, having seen the confused fire actions of harried sections on field training, could fail to agree with Marshall? Yet let sections under their commanders wander the field for a while in their little primary groups, firing their weapons as their common sense would have had them do on any hunting expedition, and even trained snipers might be reluctant to take them on. Starting from such a basis, fire and movements skills can be expected to follow naturally since the man who has the "fire habit" will always look for ground from which to give his fire more effect. The soldier who applies correct principles of fire will always move, and as with the man so with the unit.(17) German sections, moving and chattering like so many soccer teams, appear to have developed the fire habit to a high degree. From all indications, it was a mark of superior primary-group training.

    In advocating a different approach to current infantry training, Marshall reasoned that the reality of the battlefield called for "men who can think through their situation and steel themselves for action according to the situation." He was absolutely convinced, however, that "the thinking soldier--the man who is trained for self-starting--cannot be matured in a school that holds to the vestiges of the belief that automatic action is the ideal thing in the soldier."(18) Teaching a man how to think rather than what to think is a far better method of preparing him for the unexpectedness of war. A discipline of a kind that has nothing to do with common perceptions and drills is therefore required. In their exhaustive study of the American soldier in World War II, Ginzberg et al. determined that of all factors reviewed "an adequate education" emerged as the "overriding prerequisite for effective performance in military service." While an illiterate or poorly educated person could meet the demands made on him, especially if he possessed good native intelligence, was in good health, and received appropriate assignments, he was much more vulnerable than one who had had better education.(19) The most effective soldier thus appears to be one who is mentally resourceful and capable of a certain amount of inventiveness or creativity.

To be creative is not to fit into any particular pattern but to possess the necessary independence of mind and ability to go it alone. The battlefield with its "confused alarms of struggle and flight" is an ideal medium wherein maximum creativity can be practiced. Chuikov's rejection of "tactical blue prints" and formation of storm groups at Stalingrad is but one example of such practice. To quote Marshal Malinovsky, former Soviet Minister of Defense: "The activity of commanders of all ranks includes elements of creation. The making of a decision for combat, and the fulfillment of any mission presupposes a certain measure of the innovator approach to the matter, for there are absolutely no identical combat situations."(20) The German concept of Auftragstaktik was obviously aimed at encouraging innovative and creative thinking among responsible commanders who often had to act alone. They apparently recognized that the battlefield environment was conducive to creativity and the implementation of imaginative solutions. It is, of course, important to note in this regard that creative ideas spring from individuals, not groups.(21) To promote creative thinking within the infantry, then, collectivized tendencies toward "group-think" should be tempered where possible. Fraught with potential for compromise and mediocrity, group-think represents a danger to decisive action on the part of commanders:

> Not very surprisingly it has been suggested that those most susceptible to "group-think" will tend

to be people fearful of disapproval and rejection. "Such people give priority to preserving friendly relationships at the expense of achieving success in the groups work-tasks." Conversely, the sort of person who . . . makes the best military commander--the outspoken individualist--clearly cannot give of his best in the group situation. If he fails to hold his tongue, he runs the risk of being ejected by his colleagues.(22)

    In the preceding chapters, a number of unimaginative tactical approaches have been discussed. It is clear, for example, that superior firepower or "weight of metal" has not always ipso facto won the field. The idea that modern firepower precluded few tactics below battalion level was also refuted by the notable success of German and Russian infiltration operations. The fact is that fire alone cannot decide tactical issues; it requires the subordinate function of movement to enable it to arrive at the right time and place.(23) In the development of any basic tactical doctrine or framework, therefore, the function of movement must be included. While "guarding" and "hitting" may imply that movement is necessary, they do not lend it the required emphasis. To inculcate an army with the idea of maneuver, as did the Germans, calls for special efforts. To start with, the infantry section could be made into a tactical microcosm of the field army as a whole by endowing it with a capacity for maneuver. Since "only a unit that can fight in three echelons can be considered capable of independent action,"(24) the section so endowed should be organized in three elements for guarding, hitting, or moving.(25) A common German and Israeli practice was to organize their maneuver bodies into three functional components: a suppression element, an assault element, and an exploitation element.(26) As "a located section post" in World War II was often "a death trap for the men in it,"(27) a section commander would be permitted the maneuver option in defense as well.(28) Given the modern circumstances of greater battlefield dispersion and generally higher levels of education, it is doubtful whether this would be any more of a "risky business" than that faced by Ludendorff. Once the infantry section and its commander were imbued with a sense of maneuver, the entire infantry arm would, in time, become so imbued.

    With sufficient emphasis placed on maneuver, the suicidal frontal-attack mentality might just be eliminated from an army's thinking. If this study suggests anything, it is that a commander has a moral responsibility to keep his men alive and that no commander is ever justified in launching his troops to a direct attack on an enemy firmly in position. To do so under modern conditions would be to trigger, from even a dozen enemy armed with assault rifles, an intense fire of roughly 6,000 rounds a minite during the last few hundred yards.(29) Before this curtain of fire were encountered, however, attackers would doubtless have to run a gauntlet of artillery and mortar fire. Mortars were described during World War II as "the man-killers in modern combat," and it was commonly recognized that "you simply cannot advance over ground swept by mortar fire."(30) Under such conditions, foot soldiers could not assault. Their only choice would be to maneuver into a position to kill the enemy by fire or, as suggested years before by Fuller, cut him off from his line of retreat and compel him to surrender. The advantages of infiltration techniques in such circumstances are obvious. The Russian method of digging up to an enemy position, as Sherman did during the American Civil War with effect, would also appear to make good sense. It is the bullet not the trench that dampens the offensive spirit.(31)

According to Liddell Hart, the "way to success in war is strategically along the line of least expectation and tactically along the line of least resistance."(32) While this maxim applies equally to the operations of all arms--the actions of which in combination assuredly remain the key to success in modern war--it has particular implications for infantry as a specialist arm. To exploit the foot soldier's loco-mobility to maximum advantage, infantry units should not be fettered by having to adapt their formations rigidly to the movements of tanks or artillery barrages. If they are thus restricted, they are apt to lose both their initiative and special value in battle. To limit infantry employment to the holding of ground or the guarding of tanks is to fail to profit from the foot soldier's substantial offensive capability. Masters of difficult and close terrain, infantrymen are essentially foul-weather warriors; horses flounder, tanks may grind to a halt, and aircraft may fail to fly, but the foot soldier plods ever on. Obscurity, the antidote to modern weapons, is his best friend. In fog or at night in howling storm, the infantryman is monarch of the field. Using ground principally to gain security from enemy fire and to attain surprise, the primary role of infantry remains to disrupt, psychologically dislocate, and disorganize enemy resistance in preparing the way for a decision.(33) The more modern war assumes the form of a series of local actions rather than that of a "main battle," the more modern tactics should aim at "paralyzing" rather than physically "destroying" an enemy. Chuikov attempted to accomplish this at

Stalingrad by trying to make every German soldier feel that he was under the muzzle of a Russian gun. Loss of hope rather than loss of life is the factor that really decides wars, battles, and even the smallest combats.(34)

For the infantryman to be truly effective in this role, however, he will have to be as light of foot as he is quick of thought. This will apply whether he is transported to battle in an MICV or a helicopter. Mobility is needed most of all in the clash of arms. Swift and agile movement plus rapidity and intelligent tactical flexibility are its true essentials. The link between fear and fatigue having been clearly established, extreme efforts will be required to ensure that the foot soldier is not overloaded. Staff and regimental officers must be prevented from "playing it safe" and loading the soldier with everything he can possibly need for every possible emergency. A limit must surely be established; moreover, it should be enforced by a rigid system of inspection. Marshall's suggested four-fifths of optimum training load,(35) about 40 pounds, merits serious consideration. Since overloading kills men in war, the habit must be ruthlessly stamped out in peace. Infantrymen, like mules, deserve such protection.

It is highly unlikely, however, that a lean and hard-marching infantryman can spring fully armed and ready from an army that is elsewhere rolling in fat. The tendency of armies to increase their requirements--stronger armaments, more transport, and complex communications--in order to overcome an enemy, has also tended to set up internal drags on their mobility. Western armies in particular have almost become immobile through their standards of living and insistence on high technology. Then, too, the more complex the weapons system, the greater the mathematical probability of breakdown and the more vulnerable it is.(36) Sir John Slessor expressed the fear several years ago that NATO may have "been equipping themselves with weapons that they may never be able to use." Warning against the dangers of too much modernization, he urged a return to "the mobility of the boot."(37) Liddell Hart, for his part, found the "Russians getting a much higher proportion of operational divisions out of their man-power . . . giving reason to suspect that the organization of western armies is inefficient and wasteful."(38) Again, the problem is related to a reluctance to look beyond deterrence. Yet if the true object of all military training is to prepare the soldier for the next war,(39) it is in this direction that one must look.

For armies languishing in the somnolence of peace, the critical reading of the lessons of past wars is one

of the most fruitful means of seriously preparing for
future conflict. While no one war has ever exactly
resembled another, the illumination of historical study
has often provided some useful light with which to
probe the fog of distant battle. For the infantry arm
there are important signposts worth noting. From all
indications, past and present, there will be a place
for the foot soldier and his traditional skills on the
battlefield of tomorrow. To prevail over more numerous
enemies in conjunction with other arms, however, the
Western infantryman will doubtless have to rely as much
on his brain power as on his weapon power. Like Ardant
du Picq, he will have to reject as "shameful" the
"theory of strong battalions"(40) and embrace instead
the belief that superior training, endurance, and
tactical skill can compensate for quantity. The cynical
view that anyone can be made into an infantryman
because he can stop a bullet as well as the next will
likewise have to be scorned. The modern infantryman
will have to be taught to live, not die.

NOTES

   1. Wolff, In Flanders Fields, p. 7. The other
ingredients were trenches in depth and "sworls and
loops of barbed wire." Ibid.

   2. Liddell Hart, Thoughts on War, p. 54-5 and
283. Armies rigidly controlled by radio waves could
well prove as operationally ineffective as those
formerly "trussed in telephone wire." Too much control
and "fine-tuning" can be a bad thing. Many soldiers
today think NATO armies have too many radios. The
helicopter's most pernicious contribution to the
fighting in Vietnam may have been its undermining of
the influence and initiative of small-unit commanders.
Palmer, Summons of the Trumpet, p. 142.

   3. Herzog, The War of Atonement, p. 271.

   4. Patton, War As I Knew It, p. 351.

   5. By permitting all soldiers with five years
service to automatically ascend to section commander's
rank (corporal), the Canadian army actually denigrated
that position. Upgrading section commander rank to
sergeant did little to solve the problem, as,
initially, many older sergeants considered it beneath
their dignity to command sections again. The
introduction of the new rank of master corporal
(section second in command), made it possible for
certain individuals, for ten dollars extra per month,

to exercise section command in the interim period. The lasting effect of the damage done is difficult to determine.

6. Jomini divided the conduct of war into five categories: strategy, grand tactics, logistics, engineering (seige operations), and minor tactics. Howard, "Jomini and the Classical Tradition," The Theory and Practice of War, p. 15. The term "battle patterns" was taken from Canadian Army Training Memorandum, no. 51 (1945), p. 27.

7. I have in mind such things as tactical walks, sand-table exercises and discussions, and small group hunting-shooting exercises, among others.

8. Colonel J. R. Stone, "Memoir: Kapyong," CF Infantry Newsletter, 3 (1974): 11-12. The Battle of Kapyong was fought in Korea on April 24-5, 1951 by the Second Battalion, Princess Patricia's Canadian Light Infantry.

9. Marshall, Men Against Fire, pp. 44-7.

10. Ibid., p. 42.

11. "Infantry Training," Current Reports From Overseas, no. 70, 1945, p. 1.

12. Janowitz and Shils, "Cohesion and Disintegration in the Wehrmacht in World War II," p. 185. The Russians withdrew formations as well. Ely, The Red Army Today, p. 149. Replacing personnel losses by individuals rather than by units led American troops to "easily conclude" that there would be "no end to the strain" of combat" until they 'broke' or were hit." Stouffer et al., The American Soldier, p. 88.

13. Biryukov. Antitank Warfare, pp. 135-6. "Tankophobia" often led to the disastrous mistake of opening fire too soon on tanks. Ibid. To send a soldier off to fight tanks without considering "tankophobia" is not to be serious about the military profession. Soldiers could be conditioned to overcome tank fright in several ways, from having tanks drive over their individually constructed trenches, to having the soldier mount tanks on the run in mock attacks.

14. Anthony Stewart Irwin, Infantry Officer: A Personal Record (London: B. T. Batsford, 1943), p. 26; and Stouffer et al., The American Soldier, p. 229. On the dangers of using battle inoculation as a sadistic test of nerves rather than as a training expedient to

foster the growth of military judgment, see Ahrenfeldt, Psychiatry in the British Army, pp. 199-204.

15. "Battle Impressions of a Rifle Platoon Commander," Current Report From Overseas, no. 71 (1945), p. 1.

16. Some have argued that the reason troops were reluctant to fire was that there was no visible enemy to shoot at. Soldiers supposedly trained to deliver aimed fire cannot do so without targets. However, since infantry can never hope to compete in volume of fire with other arms, it makes good sense for riflemen to concentrate on accuracy of fire. After all, the only fire that counts on the battlefield is that which arrives at the right time and right place.

17. Marshall, Men Against Fire, pp. 59 and 82-3. This is not to dismiss the requirement for soldiers to hold their fire on occasion, particularly in defense.

18. Marshall, Men Against Fire, pp. 40 and 116.

19. Ginzberg, The Ineffective Soldier; Lessons for Management and the Nations: Patterns of Performance, p. 116. Farmers, surprisingly, accounted for a disproportionate number of ineffective soldiers. Men above 28 were one and a half times as likely to be ineffective as men below 22. The least likely to break were white single men below 22, with some college education, who had had a clerical job. The major determinant, however, was education. Ibid, pp. 113-16. It was not until early 1944 that the American army adopted a system of profiling men such as that which had been instituted much earlier in the Canadian and British armies. Ginzberg, The Ineffective Soldier; Lessons for Management and the Nation: Lost Divisions, p. 44. Significantly, the modern Red Army prides itself on being "the best educated" and "the best read army in the world." Herbert Goldhamer, The Soviet Soldier: Soviet Military Management at the Troop Level (New York: Crane, Russak, 1975), p. 21.

20. Colonel James Mrazek, The Art of Winning Wars (New York: Walker, 1968), pp. 32, 46 and 58.

21. Mrazek, The Art of Winning Wars, pp. 18 and 101.

22. Dixon, On the Psychology of Military Incompetence, p. 400. According to Dixon, symptoms of "group-think" include: collective attempts to rationalize away items of significance that might cause

the group to reconsider its decision; collective self-adulation and feeling of invulnerability; "mind guarding" to protect the group from within and from adverse information; and the shared illusion that silence means consent. Ibid., pp. 397-400.

23. F. O. Miksche, The Failure of Atomic Strategy (London: Faber and Faber, 1959), p. 173. To the Germans, the chief value of the tank was its mobility not its firepower. Miksche, Blitzkrieg, p. 102.

24. Miksche, Blitzkrieg, pp. 132-3.

25. It should be pointed out that in Commonwealth organization the platoon is the maneuver element, not the section. Under this system, however, only company and platoon commanders are expected to maneuver, not their NCO section commanders. As the last occupy the sole established "command" position for NCOs, I can only see their number being inculcated with a sense of maneuver through accident; yet they often end up commanding for real in war. Interestingly, Liddell Hart advocated the small section of four or five men "to operate a light machine gun in action." He claimed that experience showed that larger groups tended to lose several men quickly, but then, "when thus reduced in bulk," they were able to "make a continued advance under fire with little or no further loss." In 1938, he suggested a platoon of six sections of four men each, or three double-sections, which would "enable a tactical relay system." Later, he toyed with the idea of organizing by fives at platoon level, but fives are more complicated than threes for inexperienced commanders to handle. Liddell Hart, Defence of the West, pp. 304-6; and Thoughts on War, p. 257. Organization by threes thus appears to be optimum at lower tactical levels, preferable at least to fours, which would likely invariably result in unimaginative "two by two" posturing. The question is, Where should maneuver capability begin? I personally believe there is a good case for starting at section level. The group of three or four around a light machine gun should be the basic element. Miksche considers the eight-man section too small and recommends tripling its strength under an "alteration of relative tactical value of units" to three squads of seven men each, plus one officer and three extra men, for a total of 25. Miksche, Atomic Weapons and Armies, pp. 175-6. In 1940, he proposed a 16-man section, as it could sustain casualties and left-out-of-battle shortages better than an eight-man section, which he expected to fall below section strength in short order. Miksche, Blitzkrieg, pp. 132-3. British section strength in World War II was

eventually increased to 11 all ranks. For a defense of organization by fours see Gavin, Airborne Warfare, pp. 163-5.

26. DePuy, "One-up and Two-back?," pp. 22-3. This article identifies the advantages of attacking with two fire elements in direct fire support and one element assaulting. Whereas two elements attacking and one providing fire support only pierced a test defense 25 percent of the time, the two shooting and one moving combination penetrated 87 percent of the time. The test also demonstrated the superiority of the "parapet" foxhole, from which entrenchment the soldier did not fire to the front but rather covered to his left and right protecting the foxhole of his comrades. This was the system used by the Japanese in World War II and explained in chapter seven.

27. "What an Infantry Subaltern Really Is--Part I," Canadian Army Training Memorandum, no. 37 (1944), p. 42; and Senger und Etterlin, Neither Fear Nor Hope, p. 184.

28. This was precisely what Liddell Hart advocated in 1921.

29. Lieutenant Colonel A. J. Jeapes, "Letter to the Editor," British Army Review, no. 68 (1979), p. 73. One of Patton's tactical counsels was "not to try a sneak frontal attack at night or in daytime against a dug-in enemy who has been facing you for some time." Patton, War As I Knew It, p. 299.

30. Canadian Army Training Memorandum, no. 48 (1945), p. 20.

31. B. H. Liddell Hart, Sherman: Genius of the Civil War (Ernest Benn, 1930), p. 136. At Dien Bien Phu, General Giap pushed his own trenches closer and closer to overcome the dug-in French. Jac Weller, Fire and Movement (New York: Thomas Y. Crowell, 1967), p. 43.

32. Liddell Hart, Thoughts on War, p. 242.

33. Liddell Hart, Thoughts on War, pp. 55, 201, 261, 285-6 and 302; and Miksche, Blitzkrieg, p. 61. John Weeks lists the "tasks" of infantry as: first, to hold ground against enemy armor and infantry attacks and provide a firm pivot for counterattacks or other maneuvers; second to dominate and control the close country; third, to close with the enemy and clear his defensive positions; and finally, to provide observation, reconnaissance and early warning. John

Weeks, "The Modern Infantryman," *Military Technology and Economics* 3 (May-June 1979): 23-4.

34. Liddell Hart, *Deterrent or Defence*, p. 183; and *Defence of the West*, p. 263.

35. German and British studies showed that optimum marching load was roughly one-third of a man's weight (about 48 pounds). Lothian, *The Load Carried by the Soldier*, pp. 55-7; and Marshall, *The Soldier's Load*, pp. 31-2, 53, 60, 65, 71 and 74. Like Marshall, I am convinced that the soldier must not be loaded for tomorrow and that the army logistic system should adapt to this.

36. Marshall, *The Soldier's Load*, p. 83; Liddell Hart, *Thoughts on War*, pp. 193-5; and Bidwell, *Modern Warfare*, pp. 151-2.

37. Slessor, *The Great Deterrent*, pp. 282 and 309. He was specifically concerned about the "enormous quantities of motor transport (and fuel to keep it moving) which in fact is such a drain on the mobility of modern armies." Ibid., p. 282.

38. Liddell Hart, *Deterrent or Defence*, p. 142.

39. Fuller, *Memoirs*, p. 462.

40. Ardant du Picq, *Battle Studies*, p. 131.

Bibliography

PRIMARY SOURCES

Official Histories

Appleman, Roy E. <u>United States Army in the Korean War: South to the Naktong, North to the Yalu.</u> Washington: Office of the Chief of Military History, 1961.

Barclay, Brigadier C. N. <u>The First Commonwealth Division: The Story of British Commonwealth Land Forces in Korea, 1950-1953.</u> Aldershot: Gale and Polden, 1954.

Edmonds, Brigadier General Sir James E., and Becke, Major A. F. <u>British Official History of the Great War: Military Operations, France and Belgium, 1918.</u> London: Macmillan, 1937.

Ellis, Major L. F. <u>The War in France and Flanders, 1939-1940.</u> HM Stationery Office, 1953.

_____. <u>Victory in the West.</u> 2 vols. London: HM Stationery Office, 1960.

Erskine, David. <u>The Scots Guards 1919-1955.</u> London: William Clowes, 1956.

Frank, Benis M., and Shaw, Jr., Henry I. <u>Victory and Occupation: History of U.S. Marine Corps Operations in World War II.</u> Washington: Historical Branch, G-3 Division, Headquarters U.S. Marine Corps, 1968.

Greenfield, Kent Roberts; Palmer, Robert F.; and Wiley, Dell I. <u>The Organization of Ground Combat Troops: The United States Army in World War II; The Army Ground Forces.</u> Washington: Department of the Army, 1947.

Hermes, Walter G. <u>United States Army in the Korean War:</u>

Truce Tent and Fighting Front. Washington: Office of the Chief of Military History, 1966.

Milner, Samuel. *United States Army in World War II: Victory in Papua.* Washington: Office of the Chief of Military History, 1957.

Montross, Lynn, and Canzona, Captain Nicholas A. *U.S. Marine Operations in Korea, 1950-1953. The Chosin Reservoir Campaign.* Washington: Historical Branch, G-3 Division, Headquarters U.S. Marine Corps, 1957.

Montross, Lynn; Kuokka, Major Hubard D.; and Hicks, Major Norman W. *U.S. Marine Operations in Korea, 1950-1953: The East-Central Front.* Washington: Historical Branch, G-3 Division, Headquarters U.S. Marine Corps, 1962.

Nicholson, Lieutenant Colonel G. W. L. *The Canadians in Italy, 1943-1945.* Ottawa: Queen's Printer, 1956.

Stacey, Colonel C. P. *The Victory Campaign: The Operations in North-West Europe, 1944-1945.* Ottawa: Queen's Printer, 1960.

Stanley Clarke, Major E. B., and Tillott, Major A. T. *From Kent to Kohima: The History of the 4th Battalion The Queen's Own Royal West Kent Regiment (T.A.), 1939-1947.* Aldershot: Gale and Polden, 1951.

Stevens, G. R. *Princess Patricia's Canadian Light Infantry, 1919-1957.* Montreal: Southam, 1958.

Wood, Herbert Fairlie. *Strange Battleground: The Operations in Korea and their Effects on the Defence Policy of Canada.* Ottawa: Queen's Printer, 1966.

Manuals and Pamphlets

Airborne Operations: A German Appraisal. Washington: Department of the Army, Pamphlet No. 20-232, 1951.

The Army Lineage Book, Volume II: Infantry. Washington: Office of Military History, Department of the Army, 1953.

Beyer, Major James C, ed. *Wound Ballistics.* Washington: Office of the Surgeon General, Department of the Army, 1962.

Blau, George E. *The German Campaign in Russia: Planning and Operations (1940-1942).* Washington: Department of the Army, Pamphlet No. 20-261a, 1955.

BIBLIOGRAPHY

Canadian Army Training Memoranda. Ottawa: King's Printer, 1939-1945.

Combat in Russian Forests and Swamps. Washington: Department of the Army, Pamphlet No. 20-231, 1951.

Effects of Climate on Combat in European Russia. Washington: Department of the Army, Pamphlet No. 20-291, 1952.

German Army Handbook, April 1918. London: Arms and Armour Press, 1977.

The German Campaign in Russia, Planning and Operations (1940-1942). Washington: Department of the Army, Pamphlet No. 20-261a, 1955.

German Defence Tactics Against Russian Break-throughs. Washington: Department of the Army, Pamphlet No. 20-233, 1951.

Handbook on the Soviet Army. Washington: Department of the Army, Pamphlet No. 30-501, 1958.

Infantry, Airborne Infantry and Mechanized Infantry Rifle Platoons and Squads. Washington: Department of the Army, Field Manual FM 7-15, 1962.

Kennedy, Major Robert M. *The German Campaign in Poland (1939).* Washington: The Department of the Army, Pamphlet No. 20-255, 1956.

Military Improvisations During the Russian Campaign. Washington: Department of the Army, Pamphlet No. 20-201, 1951.

Operations of Encircled Forces: German Experiences in Russia. Washington: Department of the Army, Pamphlet No. 20-235, 1952.

Operations in Sicily and Italy. West Point: U.S.M.A. Department of Military Art and Engineering, 1950.

Rear Area Security in Russia: The Soviet Second Front Behind the German Lines. Washington: Department of the Army, Pamphlet No. 20-240, 1951.

Small Unit Actions. Washington: War Department Historical Division, 1946.

Small Unit Actions During the German Campaign in Russia. Washington: Department of the Army, Pamphlet No. 20-269, 1953.

300   BIBLIOGRAPHY

Small Unit Tactics Infantry. Harrisburg: The Military Service Publishing Company, 1948.

Tactics and Technique of Infantry, 2 vols. Harrisburg: The Military Service Publishing Company, 1949.

Terrain Factors in the Russian Campaign. Washington: Department of the Army, Pamphlet No. 20-290, 1951.

Tindall, Major Richard G. et al. Infantry in Battle. Washington: The Infantry Journal, 1934.

U.S. Army TC7-24, Antiarmor Tactics and Techniques for Mechanized Infantry.

The War in Eastern Europe. West Point: U.S.M.A. Department of Military Art and Engineering, 1949.

War Office. Current Reports From Overseas.

War Office. WO1852. Field Service Pocket Book, 1914. London: HM Stationery Office, 1914.

War Office. WO2232. Field Service Regulations, Part 1: Operations. London: HM Stationery Office, 1909. With Amendments to 1914.

War Office. WO 8847. Field Service Regulations, Volume II: Operations. London: HM Stationery Office, 1924.

War Office. WO2052. Infantry Training: Company Organization. London: HM Stationery Office, 1914.

War Office. Manual 1447. Infantry Training: Training and War (1937). London: HM Stationery Office, 1937.

War Office. WO2227. Musketry Regulations, Part 1. London: HM Stationery Office, 1909. With Amendments to 1914.

Semi-Official Manuals and Associated Documents

Bond, Colonel P. S., ed. Military Science and Tactics: Infantry Advanced Course; A Text and Reference of Advanced Infantry Training. Washington: Bond, 1944.

Canadian Land Forces Command and Staff College. 1978 Course Package. Restricted.

The Canadian Infantry 1986-1995. Land Forces Combat

Development Study 77-8-1. Committee Draft, 1 March, 1978. NATO Secret.

_____. Land Forces Combat Development Study Final Report, 22 June, 1978. NATO Secret.

Fry, Major General James C. Assault Battle Drill. Harrisburg: The Military Service Publishing Company, 1955.

Harper, Lieutenant General Sir G. M. Notes on Infantry Tactics and Training. London: Sifton Praed, 1921.

Kearsey, Lieutenant Colonel A. Simple Tactics, Aldershot: Gale and Polden, 1951.

Kinsman, Lieutenant Colonel J. H., Tactical Notes. Dublin: E. Ponsonby, 1914.

Langford, Lieutenant Colonel R. J. S. Corporal to Field Officer: A Ready Reference for all Ranks in Peace and War. Toronto: Copp Clark, 1941.

Lothian, Major N. V. The Load Carried by the Soldier. London: John Bale, Sons and Danielsson, circa 1920.

Ney, Virgil. Organization and Equipment of the Infantry Rifle Squad from Valley Forge to ROAD. Fort Belvoir: U.S. Army Combat Operations Research Group Memorandum 194, January, 1965.

_____. The Evolution of the Armored Infantry Rifle Squad. Fort Belvoir: U.S. Army Combat Operations Research Group Memorandum 198, March 19, 1965.

Thompson, Lieutenant Colonel Paul W. et al. How the Jap Army Fights. New York: Penguin, 1943.

Turner, Brigadier A. J. D. Valentine's Sand Table Exercises. Aldershot: Gale and Polden, 1955.

Valentine, Captain A. W. Sand Table Exercises. Aldershot: Gale and Polden, undated.

Memoirs, Journals, and Accounts

Allon, Yigal. The Making of Israel's Army. London: Valentine, Mitchell, 1970.

Ben-Porat, Yeshayahu, and Carmel, Hezi, et al. Kippur. Translated by Louis Williams. Tel Aviv: Special Edition, 1973.

Caputo, Philip. *A Rumor of War.* New York: Holt, Rinehart and Winston, 1977.

Carrington, Charles Edmund [Charles Edmonds]. *A Subaltern's War.* London: Peter Davies, 1929.

Chapman, Guy. *A Passionate Prodigality.* New York: Holt, Rinehart and Winston, 1966.

Chuikov, Marshal Vasili I. *The Beginning of the Road.* Translated by Harold Silver. London: MacGibbon and Kee, 1963.

———. *The End of the Third Reich.* Translated by Ruth Kisch. London: MacGibbon and Kee, 1967.

Dunkelman, Ben. *Dual Allegiance.* Toronto: Macmillan, 1976.

Emmrich, Kurt [Peter Bamm]. *The Invisible Flag.* Translated by Frank Herrman. London: Faber and Faber, 1957.

Fuller, J. F. C. *Memoirs of an Unconventional Soldier.* London: Ivor Nicholson and Watson, 1936.

Graves, Robert. *Good-bye to All That.* London: Jonathan Cape, 1929.

Groom, W. H. A. *Poor Blood Infantry.* London: William Kimber, 1976.

Guderian, General Heinz. *Panzer Leader.* Translated by Constantine Fitzgibbon. London: Michael Joseph, 1952.

Herbert, Anthony B. *Soldier.* New York: Dell, 1973.

Irwin, Anthony Stewart. *Infantry Officer: A Personal Record.* London: B. T. Batsford, 1943.

Jack, Ben J. L. *General Jack's Diary.* Edited by John Terraine. London: Eyre and Spottiswoode, 1964.

Junger, Ernst. *The Storm of Steel: From the Diary of a German Storm-Troop Officer on the Western Front.* London: Chatto & Windus, 1929.

Kennedy, Major General Sir John. *The Business of War.* London: Hutchinson, 1957.

Kesselring, Field Marshal Albert, *Memoirs.* London: William Kimber, 1953.

Kippenberger, Major General Sir Howard. *Infantry Brigadier.* Oxford: University Press, 1949.

Liddell Hart, B. H. Memoirs. 2 vols. London: Cassell, 1965.

Lowry, Major M. A. An Infantry Company in Arakan and Kohima. Aldershot: Gale and Polden, 1950.

Ludendorff, General Erich von. My War Memories. 2 vols. London: Hutchinson, 1919.

MacDonald, Charles B. Company Commander. New York: Ballantyne, 1947.

Majdalany, F. The Monastery. London: John Lane, 1945.

Manstein, Field Marshal Erich von. Lost Victories. Translated by Anthony G. Powell. London: Methuen, 1958.

Martel, Lieutenant General Sir Giffard le Q. An Outspoken Soldier. London: Sifton Praed, 1949.

Masters, John. The Road Past Mandalay. New York: Bantam, 1979.

Mellenthin, Major General F. W. von. Panzer Battles, 1939-1945. Translated by Betzler. London: Cassell, 1955.

Mowat, Farley. The Regiment. Toronto: McClelland and Stewart, 1977.

Moynihan, Michael, ed. A Place Called Armageddon: Letters From the Great War. London: David and Charles, 1975.

Patton, General George S. War As I Knew It. New York: Pyramid, 1970.

Pope, Lieutenant General Maurice A. Soldiers and Politicians. Toronto: University Press, 1962.

Richards, Frank. Old Soldiers Never Die. London: Faber and Faber, 1966.

Ridgway, General Matthew B. The Korean War. New York: Doubleday, 1967.

Reitz, Deneys. Commando: A Boer Journal of the Boer War. London: Faber and Faber, 1929.

Sajer, Guy. The Forgotten Soldier. Translated by Lily Emmet. London: Weidenfeld and Nicolson, 1967.

Samwell, Major H. P. An Infantry Officer With the Eighth Army. London: William Blackwood and Sons, 1945.

Sasson, Siegfried. Memoirs of An Infantry Officer. London: Faber and Faber, 1930.

Senger und Etterlin, General Frido von. Neither Fear nor Hope. Translated by George Malcolm. London: Macdonald, 1960.

Simson, Ivan. Singapore: Too Little, Too Late. London: Leo Cooper, 1970.

Slim, Field Marshal Sir William. Defeat into Victory. London: Cassell, 1956.

Speidel, General Dr. Hans. Invasion 1944: Rommel and The Normandy Campaign. Chicago: Henry Regnery, 1950.

Strategy and Tactics of the Soviet-German War by Officers of the Red Army and Soviet War Correspondents. London: Hutchinson, circa, 1942.

Swinton, Major General Sir Ernest D. Eyewitness. London: Hodder and Stoughton, 1932.

Thomason, Captain John W., Jr. Fix Bayonets. New York: Charles Scribner, 1926.

Upton, Major General Emory. The Armies of Asia and Europe. New York: D. Appleton, 1878.

Zhukov, Marshal Georgy Konstantinovich. Memoirs. London: Jonathan Cape, 1971.

SECONDARY SOURCES

Books

Addington, Larry H. The Blitzkrieg Era and The German General Staff, 1865-1941. New Brunswick: Rutgers University Press, 1971.

Ahrenfeldt, Robert H. Psychiatry in the British Army in the Second World War. London: Routledge and Kegan Paul, 1958.

Allen, Ralph. Ordeal by Fire: Canada, 1910-1945. Toronto, Popular Library, 1961.

Anders. General Wladyslaw. Hitler's Defeat in Russia. Chicago: Henry Regnery, 1953.

Ardant du Picq, Colonel Charles J. J. J. Battle Studies. Translated by Colonel John N. Greely and Major

Robert C. Cotton. Harrisburg: The Military Service Publishing Company, 1947.

Balck, Colonel William. Tactics. 2 vols. Translated by Walter Krueger. Fort Leavenworth: U.S. Cavalry Association, 1911.

_____. Development of Tactics--World War. Fort Leavenworth: The General Service Schools Press, 1922.

Barclay, Brigadier C. N. The New Warfare. London: William Clowes, 1953.

Barker, A. J. British and American Weapons of World War II. London: Arms and Armour Press, 1969.

_____. German Infantry Weapons of World War II. London: Arms and Armour Press, 1969.

_____, and Walter, John. Russian Infantry Weapons of World War II. New York: Arco, 1971.

Barnett, Correlli. The Swordbearers. London: Eyre and Spottiswoode, 1963.

Baynes, John. Morale: A Study of Men and Courage. London: Cassell, 1967.

Beaumont, Roger A. Military Elites. Indianapolis: Bobbs-Merrill, 1974.

Becke, Captain A. F. An Introduction to the History of Tactics 1740-1905. London: Hugh Rees, 1909.

Bek, Alexander. Volokolamsk Highway. Moscow: Foreign Languages Publishing House, 1944.

Bernhardi, General Friedrich von. The War of the Future in the Light of the Lessons of the World War. Translated by F. A. Holt. London: Hutchinson, 1920.

Bidwell, Shelford. Modern Warfare: A Study of Men, Weapons and Theories. London: Allen Lane, 1973.

Biryukov, Major General G. and Melnikov, Colonel G. Antitank Warfare. Translated by David Myshne. Moscow: Progress, 1972.

Blaxland, Gregory. Destination Dunkirk: The Story of Gort's Army. London: William Kimber, 1973.

Bloch, I. S. The Future of War. Translated by R. C. Long. New York: Garland, 1972.

Blumenson, Martin, ed. The Patton Papers. 2 vols. Boston: Houghton Mufflin, 1974.

Bond, Brian. Liddell Hart: A Study of His Military Thought. London: Cassell, 1977.

Brodie, Bernard. War and Politics. New York: Macmillan, 1973.

_____, and Brodie, Fawn. From Crossbow to H-Bomb. New York: Dell, 1962.

Brown, Neville. Strategic Mobility. London: Chatto and Windus, 1963.

Buchan, John. The Courts of the Morning. London: Hodder and Stoughton, 1929.

Burns, Major General E. L. M. Manpower in the Canadian Army 1939-1945. Toronto: Clarke, Irwin, 1956.

Caiden, Martin. The Tigers are Burning. New York: Hawthorn, 1974.

Cameron, Colonel J. M. The Anatomy of Military Merit. Philadelphia: Dorrance, 1960.

Campbell, Arthur. The Seige: A Story from Kohima. London: George Allen and Unwin, 1956.

Carell, Paul. Hitler Moves East, 1941-1943. Translated by Ewald Osers. Boston: Little, Brown and Company, 1964.

Carrington, Charles Edmund. Soldier From the Wars Returning. London: Hutchinson, 1965.

Carver, Field Marshal Sir Michael, ed. The War Lords. Boston. Little, Brown and Company, 1976.

Chapman, Guy, ed. Vain Glory. London: Cassell, 1937.

Clark, Alan. Barbarossa: The Russian-German Conflict, 1941-1945. London: Hutchinson, 1965.

Clausewitz, Carl von. On War. Edited and translated by Michael Howard and Peter Paret. Princeton: University Press, 1976.

Coox, Alvin D. The Anatomy of a Small War: The Soviet-Japanese Struggle for Changkuteng/Khasan, 1938. London: Greenwood, 1977.

Creveld, Martin van. Supplying War: Logistics From

Wallenstein to Patton. Cambridge: University Press, 1977.

Davies, W. J. K. German Army Handbook, 1939-1945. New York: Arco, 1974.

Davis, Burke. Marine! The Life of Lt. Gen. Lewis B. (Chesty) Puller. Toronto: Bantam, 1964.

De Gaulle, General Charles. The Army of the Future. Philadelphia: Lippincott, 1941.

_____. The Edge of the Sword. Translated by Gerard Hopkins. New York: Criterion, 1966.

Deighton, Len. Blitzkrieg. London: Jonathan Cape, 1979.

Deitchman, Seymour. Limited War and American Defense Policy. Cambridge: The M.I.T. Press, 1964.

Dixon, Norman. On the Psychology of Military Incompetence. London: Jonathan Cape, 1976.

Doby, John T.; Boskoff, Alvin; and Pendleton, William W. Sociology: The Study of Man in Adaptation. Lexington: D.C. Heath, 1973.

Douglass, Joseph D. Jr. Soviet Military Strategy in Europe. New York: Pergamon, 1980.

Dupuy, R. Ernest, and Dupuy, Trevor N. Military Heritage of America. New York: McGraw-Hill, 1956.

Dupuy, Colonel T. N. A Genius for War: The German Army and General Staff, 1807-1945. London: Macdonald and Jane's, 1977.

Earle, Edward Mead, ed. Makers of Modern Strategy. Princeton: University Press, 1944.

Ellis, Chris, and Chamberlain, Peter. Handbook of the British Army 1943. London: Arms and Armour Press, 1975.

Ely, Colonel Louis B. The Red Army Today. Harrisburg: The Military Service Publishing Company, 1949.

Erfurth, General Waldemar. Surprise. Translated by Dr. Stefan T. Possony and Daniel Vilfroy. Harrisburg: The Military Service Publishing Company, 1943.

Erickson, John. The Soviet High Command. London: Macmillan, 1962.

_____. *The Road to Stalingrad.* London: Weidenfeld and Nicolson, 1975.

Falls, Cyril. *Ordeal by Battle.* London: Methuen, 1943.

_____. *Caporetto 1917.* London: Weidenfeld and Nicolson, 1966.

Farago, Ladislaw. *Patton: Ordeal and Triumph.* New York: Dell, 1975.

Farrar-Hockley, Anthony. *Infantry Tactics.* London: Almark, 1976.

Foertsch, Colonel Hermann. *The Art of Modern War.* Translated by Theodore W. Knauth. New York: Oskar Piest, 1940.

Freidin, Seymour and William Richardson, eds. *The Fatal Decisions.* Translated by Constantine Fitzgibbon. London: Michael Joseph, 1956.

Fuller, J. F. C. *The Reformation of War.* London: Hutchinson, 1923.

_____. *Sir John Moore's System of Training.* London: Hutchinson, 1924.

_____. *On Future Warfare.* London: Sifton Praed, 1928.

_____. *Lectures on F.S.R. II.* London: Sifton Praed, 1931.

_____. *Generalship; its Diseases and their Cure: A Study of the Personal Factor in Command.* London: Faber and Faber, 1933.

_____. *Machine Warfare.* London: Hutchinson, 1941.

_____. *Armored Warfare: An Annotated Edition of Lectures on F.S.R. III (Operations Between Mechanized Forces).* Harrisburg: The Military Service Publishing Company, 1943.
_____. *Thunderbolts.* London: Skeffington and Son, 1946.

_____. *The Conduct of War.* London: Eyre and Spottiswoode, 1961.

Furse, Colonel George Armand. *The Art of Marching.* London: William Clowes, 1901.

Gabriel, Richard A., and Savage, Paul L. *Crisis in Command.* New York: Hill and Wang, 1978.

Garder, Michel. A History of the Soviet Army. London: Pall Mall, 1966.

Garthoff, Raymond L. Soviet Military Doctrine. Glencoe: The Free Press, 1953.

Gavin, Major General James M. Airborne Warfare. Washington: Infantry Journal Press, 1947.

Gawne, Captain, trans. A Summer Night's Dream (Colonel Meckel) and The Defence of Duffer's Drift (E. D. Swinton). Kansas: Hudson, circa 1909.

George, Alexander L. The Chinese Communist Army in Action: The Korean War and its Aftermath. New York: Columbia University Press, 1967.

Ginzberg, Eli et al. The Ineffective Soldier: Lessons for Management and the Nation. 3 vols. New York: Columbia University Press, 1959.

Goerlitz, Walter. History of the German General Staff, 1657-1945. Translated by Brian Battershaw. New York: Praeger, 1953.

Goldhamer, Herbert. The Soviet Soldier: Soviet Military Management at the Troop Level. New York: Crane, Russak, 1975.

Goutard, Colonel A. The Battle of France, 1940. Translated by Captain A. R. P. Burgess. London: Frederick Muller, 1958.

Griffith, Samuel B., II. The Chinese People's Liberation Army. New York: McGraw-Hill, 1967.

Hackett, Sir John W. The Profession of Arms. London: Times Publishing, 1963.

Halperin, Morton H. Limited War. Cambridge: Harvard University Center for International Affairs, 1962.

Hayashi, Saburo, and Coox, Alvin D. Kogun: The Japanese Army in the Pacific War. Quantico: The Marine Corps Association, 1959.

Heilbrunn, Otto. Conventional Warfare in the Nuclear Age. New York: Praeger, 1965.

Herlin, Hans. Udet: A Man's Life. Translated by Mervyn Savill. London: Macdonald, 1960.

Herzog, Major General Chaim. The War of Atonement. Boston: Little, Brown and Company, 1975.

BIBLIOGRAPHY

Hittle, Lieutenant Colonel J. D. *The Military Staff: Its History and Development.* Harrisburg: The Military Service Publishing Company, 1949.

Horne, Alastair. *The Price of Glory: Verdun 1916.* London: Macmillan, 1962.

_____. *To Lose a Battle.* London: Penguin, 1979.

Howard, Michael, ed. *The Theory and Practice of War.* London: Cassell, 1965.

Huber, Reiner K.; Jones, Lynn F.; Reine, Egil. *Military Strategy and Tactics: Computer Modeling of Land War Problems.* New York: Plenum, 1975.

Insight Team of the London *Sunday Times*, *The Yom Kippur War.* New York: Doubleday, 1974.

Irving, David, *The Trail of the Fox.* New York: Avon, 1977.

Jackson, W. G. F. *The Battle for Italy.* London. B. T. Batsford, 1967.

Janowitz, Morris, ed. *The New Military.* New York: Russell Sage, 1964.

_____. *Military Conflict: Essays in the Institutional Analysis of War and Peace.* Beverly Hills: Sage, 1975.

Jessel, Major R. G. *G, A and Q.* Aldershot: Gale and Polden, 1947.

Jones, David R. *Soviet Armed Forces Review Annual, Vol 4, 1980.* Gulf Breeze: Academic International Press, 1980.

Keegan, John. *The Face of Battle.* New York: Viking, 1976.

Kerr, Walter. *The Russian Army: Its Men, Its Leaders, and Its Battles.* London: Victor Gollancz, 1944.

King, Edward L. *The Death of the Army: A Pre-Mortem.* New York: Saturday Review, 1972.

Knorr, Klaus. *On the Uses of Military Power in the Nuclear Age.* Princeton: University Press, 1976.

Laffin, John. *Jackboot.* London: Cassell, 1965.

Leckie, Robert. *Conflict: The History of the Korean War, 1950-1953.* New York: G. P. Putnam's Sons, 1962.

Leeb, Field Marshal General Ritter von. Defense. Translated by Dr. Stefan T. Possony and Daniel Vilfroy. Harrisburg: The Military Service Publishing Company, 1943.

Leed, Eric T. No Man's Land; Combat and Identity in World War I. Cambridge: University Press, 1979.

Lester, J. R. Tank Warfare. London: George Allen and Unwin, 1943.

Lindsay, Major General G. M. The War on Civil and Military Fronts. Cambridge: University Press, 1942.

Liddell Hart, B. H. A Science of Infantry Tactics Simplified. London: William Clowes, 1923.

_____. Paris, or the Future of War. London: Kegan Paul, Trench, Trubner, 1925.

_____. The Remaking of Modern Armies. London: John Murray, 1927.

_____. Sherman: The Genius of the Civil War. London: Ernest Benn, 1930.

_____. The British Way in Warfare. London: Faber and Faber, 1932.

_____. The Future of Infantry. London: Faber and Faber, 1933.

_____. Europe in Arms. London: Faber and Faber, 1937.

_____. Thoughts on War. London: Faber and Faber, 1944.

_____. The Revolution in Warfare. New Haven: Yale University Press, 1947.

_____. Defence of the West. London, Cassell, 1950.

_____. Deterrent of Defence. London: Stevens, 1960.

_____. History of the Second World War. London: Pan, 1978.

_____. The Other Side of the Hill. London: Pan, 1978.

Liddell Hart, B. H., ed. The Rommel Papers. Translated by Paul Findlay. London: Collins, 1953.

_____. The Soviet Army. London: Weidenfeld and Nicolson, 1956.

Light, Donald Jr., and Keller, Suzanne. *Sociology.* New York: Alfred A. Knopf, 1979.

Lindsay, Major General G. M. *The War on the Civil and Military Fronts.* Cambridge: University Press, 1942.

Lloyd, Alan. *The War in the Trenches.* New York: David McKay, 1976.

Lloyd, Colonel E. M. *A Review of the History of Infantry.* London: Longmans, Green, 1908.

Lucas, James. *War on the Eastern Front: The German Soldier in Russia.* London: Jane's, 1979.

Lucas, Lieutenant Colonel. *The Evolution of Tactical Ideas in France and Germany During the War of 1914-1918.* Translated by Major P. V. Kieffer. Paris: Berger-Leorault, 1925.

Luttwak, Edward, and Horowitz, Dan. *The Israeli Army.* London: Allen Lane, 1973.

Luvaas, Jay. *The Education of an Army: British Military Thought, 1815-1940.* Chicago: University Press, 1964.

MacDonald, Charles B. *The Mighty Endeavor: American Armed Forces in the European Theatre in World War II.* New York: Oxford University Press, 1969.

McGwire, Michael; Booth, Ken; and McConnell, John, eds. *Soviet Naval Policy: Objectives and Constraints:* New York: Praeger, 1976.

McKee, Alexander. *Caen: Anvil of Victory.* London: Souvenir Press, 1964.

_____. *The Race for the Rhine Bridges.* New York: Stein and Day, 1971.

Mackintosh, Malcolm. *Juggernaut: A History of the Soviet Armed Forces.* New York: Macmillan, 1967.

Macksey, Kenneth. *Crucible of Power: The Fight for Tunisia 1942-1943.* London: Hutchinson, 1969.

_____. *Tank Warfare: A History of Tanks in Battle.* London: Rupert Hart-Davis, 1971.

_____. *The Guinness History of Land Warfare.* Enfield: Guinness Superlatives, 1973.

_____. *Guderian: Panzer General.* London: MacDonald and Janes, 1975.

McLean, Donald B., ed. *German Infantry Weapons.* Forest Grove: Normount Armamant Company, 1968.

Maguire, T. Miller. *Notes on the Austro-Prussian War of 1866.* London: Hugh Rees, 1904.

Majdalany, F. *The Battle of Cassino.* Boston: Houghton Mifflin, 1957.

Manchester, William. *American Caesar: Douglas MacArthur 1880-1964.* Boston: Little, Brown and Company, 1978.

Marshall, S. L. A. *Blitzkrieg.* New York: William Morrow, 1940.

——. *Armies on Wheels.* New York: William Morrow, 1941.

——. *Men Against Fire.* New York: William Morrow, 1947.

——. *The Soldier's Load and the Mobility of a Nation.* Washington: The Combat Forces Press, 1950.

——. *The River and the Gauntlet.* New York: William Morrow, 1953.

——. *Pork Chop Hill.* New York: William Morrow, 1956.

——. *Sinai Victory.* New York: William Morrow, 1958.

——. *Battle at Best.* New York: William Morrow, 1963.

——. *The Officer as a Leader.* Harrisburg: Stackpole, 1966.

Maude, Colonel F. N. *Notes on the Evolution of Infantry Tactics.* London: William Clowes, 1905.

Messenger, Major Charles. *The Art of Blitzkrieg.* London: Ian Allan, 1976.

Miksche, F. O. *Blitzkrieg.* London: Faber and Faber, 1941.

——. *Atomic Weapons and Armies.* London: Faber and Faber, 1955.

——. *The Failure of Atomic Strategy.* London: Faber and Faber, 1959.

Moran, Lord. *The Anatomy of Courage.* London: Constable, 1967.

Morgan, J. H. Assize of Arms. New York: Oxford University Press, 1946.

Mowat, Farley. The Regiment. Toronto: McClelland and Stewart, 1977.

Mrazek, Colonel James. The Art of Winning Wars. New York: Walker, 1968.

Necker, Wilhelm. Hitler's War Machine and The Invasion of Britain. Translated by H. Leigh Farnell. London: Lindsay Drummond, circa 1940.

Nelson, Harvey W. The Chinese Military System: An Organization Study of the Chinese People's Liberation Army. Boulder: Westview, 1977.

Newman, Bernard. The Cavalry Went Through. Other details unspecified.

Nickerson, Hoffman. The Armed Horde, 1793-1939: A Study of the Rise, Survival and Decline of the Mass Army. New York: G. P. Putman's Sons, 1940.

_____. Arms and Policy, 1939-1944. New York: G. P. Putman's Sons, 1945.

Novikov, Y., and Sverdlov, F. Manoeuvre in Modern Land Warfare. Moscow: Progress, 1972.

O'Ballance, Edgar. The Sinai Campaign of 1956. New York: Praeger, 1959.

_____. The Red Army. London: Faber and Faber, 1964.

_____. Korea: 1950-1953. London: Faber and Faber, 1969.

Ogorkiewicz, R. M. Armoured Forces. London: Arms and Armour Press, 1970.

Osgood, Robert Endicott. Limited War: The Challenge to American Strategy. Chicago: University Press, 1957.

Palit, Major General D. K. War in the Deterrent Age. London: MacDonald, 1966.

Palmer, Dave Richard. Summons of the Trumpet. San Rafael: Presido, 1978.

Palmer, Michael. Warfare. London: B. T. Batsford, 1972.

Pardieu, Major M. F. de. A Critical Study of German Tactics and the New German Regulations. Translated by

Captain Charles F. Martin. Fort Leavenworth: U.S. Cavalry Association, 1912.

Pitt, Barrie. 1918: The Last Act. London: Cassell, 1962.

Possony, Stefan T., and Pournelle, J. E., The Strategy of Technology: Winning the Decisive War. Cambridge: Dunellen, 1970.

Preston, Richard A.; Wise, Sydney F.; and Werner, Herman O. Men in Arms. New York: Praeger, 1962.

Record, Jeffrey. Sizing Up the Soviet Army. Washington: The Brooking's Institution, 1975.

Reinhardt, Colonel G. C., and Kintner, Lieutenant Colonel W. R. Atomic Weapons in Land Combat. Harrisburg: The Military Service Publishing Company, 1953.

Richardson, Major General F. M. Fighting Spirit: Psychological Factors in War. London: Leo Cooper, 1978.

Richardson, William and Freidin, Seymour, eds. The Fatal Decisions. Translated by Constantine Fitzgibbon. London: Michael Joseph, 1956.

Rolbant, Samuel. The Israeli Soldier: Profile of an Army. Cranbury: Thomas Yoseloff, 1970.

Rommel, General Field Marshal Erwin. Infantry Attacks. Translated by Lieutenant Colonel G. E. Kidde. Washington: The Infantry Journal, 1944.

Ropp, Theodore. War in the Modern World. New York: Collier, 1962.

Rosinski, Herbert. The German Army. Washington: The Infantry Journal, 1944.

Rothenberg, Gunther E. The Anatomy of the Israeli Army. New York: Hippocrene, 1979.

Rowan-Robinson, Major General H. The Infantry Experiment. London: William Clowes, 1934.

Ryabov, V. The Soviet Armed Forces Yesterday and Today. Moscow: Progress, 1976.

Ryan, Cornelius. A Bridge Too Far. London: Book Club Associates, 1975.

Saxe, Marshal Maurice de. Reveries on the Art of War.

Edited and translated by Brigadier General Thomas R. Phillipp. Harrisburg: The Military Service Publishing Company, 1944.

Schlieffen, General F. M. Count Alfred von. Cannae. Fort Leavenworth: The Command and General Staff School Press, 1936.

Schram, Stuart. Mao Tse-tung. New York: Simon and Schuster, 1966.

Seaton, Albert. The Russo-German War 1941-45. New York: Praeger, 1970.

_____. The Battle for Moscow. London: Rupert Hart-Davis, 1971.

_____. The Soviet Army. Reading: Osprey, 1972.

Seeckt, General Hans von. Thoughts of a Soldier. Translated by Gilbert Waterhouse. London: Ernest Benn, 1930.

Shirer, William L. The Rise and Fall of the Third Reich. New York: Simon and Schuster, 1960.

Sidorenko, A. A. The Offensive (A Soviet View). Washington: U.S. Government Printing Office, 1970.

Simpkin, Brigadier Richard E. Mechanized Infantry. Oxford: Brassey's, 1980.

Slessor, Sir John. The Great Deterrent. London: Cassell, 1957.

Sokolovsky, Marshal V. D., ed. Military Strategy: Soviet Doctrine and Concepts. New York: Praeger, 1963.

The Soviet Army. Translated by Vladimir Talmy. Moscow: Progress Publishers, 1971.

Stouffer, Samuel A. et al. The American Soldier: Combat and Its Aftermath. Princeton: University Press, 1949.

Swettenham, John. To Seize the Victory: The Canadian Corps in World War I. Toronto: Ryerson, 1965.

Swinson, Arthur. Kohima. London: Cassell, 1966.

Swinton, Sir Ernest D. [Ole Luk-oie]. The Green Curve and Other Stories. London: William Blackwood, 1915.

Thompson, Lieutenant Colonel Paul W.; Doud, Lieutenant

Colonel Harold; Scofield, Lieutenant John; and Hill, Colonel Milton A. How the Jap Army Fights. New York: Penguin, 1943.

Thompson, W. Scott, and Frizzell, Donaldson D., eds. The Lessons of Vietnam. New York: Crane, Russak, 1977.

Trythall, Anthony John. "Boney" Fuller: The Intellectual General, 1878-1966. London: Cassell, 1977.

Tuchman, Barbara W. The Guns of August. New York: Macmillan, 1962.

———. Stilwell and the American Experience in China, 1911-45. New York: Bantam, 1979.

Turney, Alfred W. Disaster at Moscow: Von Bock's Campaigns, 1940-1942. Albuquerque: University of New Mexico Press, 1970.

Warner, Philip, and Youens, Michael. Japanese Army of World War II. Reading: Osprey, 1973.

Weeks, John. Men Against Tanks. New York: Mason Charter, 1975.

Weller, Jac. Weapons and Tactics. London: Nicholas Vane, 1966.

———. Fire and Movement. New York: Thomas Y. Crowell, 1967.

West, Captain Francis J. Small Unit Action in Vietnam, Summer 1966. Washington: Historical Branch, G-3 Division, Headquarters U.S. Marine Corps, 1967.

White, D. Fedotoff. The Growth of the Red Army. Princeton: University Press, 1944.

Wilmot, Chester. The Struggle for Europe. London: Collins, 1974.

Wintringham, Tom. Deadlock War. London: Faber and Faber, 1940.

———. Weapons and Tactics. London: Faber and Faber, 1943.

Wolff, Leon. In Flanders Fields. New York: Viking, 1958.

Wood, Herbert Fairlie. The Private War of Jacket Coates. Toronto: Longmans, 1966.

_____. Vimy! London: Corgi, 1972.

Wynne, Captain G. C. If Germany Attacks: The Battle in Depth in the West. London: Faber and Faber, 1939.

Young, Desmond. Rommel. London: Collins, 1950.

Young, Brigadier Peter. World War 1939-45. London: Pan, 1966.

_____, ed. Decisive Battles of the Second World War: An Anthology. London: Arthur Barker, 1967.

Articles, Periodicals, and Unpublished Papers

Alfoldi, Dr. Lazlo M. "The Hutier Legend." Parameters 5 (1976): 69-74.

Altrichter, Colonel "Problems of Infantry Attack Tactics." Infantry Journal 47 (1940); 223-38.

Archibald, Major B. M. "Urbanization and NATO Defence." Command and Staff Review, Canadian Forces Staff College (1979), pp. 99-115.

Bentley, Captain L. W. and McKinnon, Captain D. C. "The Yom Kippur War as an Example of Modern Land Battle." Canadian Defence Quarterly 4 (1974): 13-19.

Bidwell, Shelford. "The Pleasures of Defeat." British Army Review, no. 61 (1979). pp. 5-7.

Birnstiel, Major General Fritz. "German Combat Troops in Action." Infantry 67 (1979): 24-29.

Bracken, Paul. "Urban Sprawl and NATO Defence." Survival 18 (1976): 254-65.

Canadian Liaison Staff, Headquarters II (German) Corps, "The New Structure of the German Armed Forces." Canadian Defence Quarterly 4 (1975): 13-17.

Chadrue, Lieutenant Colonel Robert G. "Requiem for the Infantry." Infantry 68 (1978): 28-31.

Clarkson, Colonel J. M. E. "Spark at Yom Kippur: Many Surprises in an Eighteen-Day War." Canadian Defence Quarterly 3 (1974): 9-22.

Cornblum, Richard. "Israel's New Tank: The Merkava." Canadian Defence Quarterly 19 (1979): 34-38.

Coutts, Britton Major T. A. "The Assault." British Army Review, no. 61 (1979), pp. 53-58.

Creveld, Martin van. "Supplying an Army: An Historical View." Journal of the Royal United Service Institution, 123 (1978): 56-63.

Croft, Lieutenant Colonel W. D. "Second Military Prize Essay for 1919." Journal of the Royal United Service Institution 65 (1920), 443-76.

Depuy, General William E. "Implications of the Middle East War on U.S. Army Tactics, Doctrine and Systems." Mimeographed. Presentation by Commander, U.S. Army Training and Doctrine Command, 1976.

──────. "11 Men, 1 Mind." CF Combat Arms School Infantry Journal 6 (1976-1977): 2-9.

──────. "The U.S. Army: Are We Ready for the Future?" Army 28 (1978): 22-29.

──────. "One-Up and Two-back?" Army 30 (1980): 20-25.

Donnelly, C. N. "Soviet Techniques for Combat in Built-up Areas." International Defense Review 10 (1977): 238-42.

──────. "Fighting in Built Up Areas: A Soviet View--Part II." Journal of the Royal United Service Institution 122 (1977): 63-67.

──────. "Tactical Problems of Facing the Soviet Army." International Defense Review 11 (1978): 1405-12.

Eliot, Christian. "Man-Portable Anti-Tank Weapons." Military Technology and Economics, 3 (1979): 18-26.

English, J. A. "Confederate Field Communications." MA thesis, Duke University, 1964.

"An Era of Empire Ends at Singapore." Life, February 23, 1942, pp. 17-19.

Finan, J. S. "Soviet Interest in and Possible Tactical Use of Chemical Weapons." Canadian Defence Quarterly 4 (1974): 11-15.

Fuller, J. F. C. "The Application of Recent Developments in Mechanics and Other Scientific Knowledge to Preparation and Training for Future War On Land--Gold Medal (Military Prize Essay for 1919)." The Journal of the Royal United Service Institution 65 (1920): 239-74.

_____. "The Introduction of Mechanical Warfare on Land and Its Possibilities in the Near Future." The Royal Engineers Journal 33 (1921): 1-13.

"The Fuller-Liddell Hart Lecture." Dialogue held at RUSI, London, October 1978, between Brigadier A. M. Trythall and Mr. Brian Bond.

Gabriel, Major Richard A. "Some Implications for Combat Effectiveness." Military Review 58 (1978): 27-30.

Gordon, Lieutenant Colonel Don E. "Target: The Spoken Word." Army 29 (1979): 20-24.

Hansen, Lieutenant Colonel Lynn M. "Soviet Combat Helicopter Operations." International Defense Review 11 (October 1978): 1242-6.

Hart, Major General T. S. "Determination in Battle." Paper presented to Canadian Land Forces Command and Staff College, Kingston, 1979.

Hartness, Captain Harlan N. "Germany's Tactical Doctrine." Infantry Journal 66 (1939): 249-50.

"The Helicopter in Modern Combat." Military Technology and Economics. 3 (1979): 9-14.

Hemesley, Major A. E. "MOBA--Too Difficult?" Journal of the Royal United Service Institution 122 (1977): 24-26.

_____. "Soviet Military Operations in Built-up Areas." Infantry 67 (1977): 30-34.

Hill, Adrian. "Could Napoleon's Army Win Today?" Journal of the Royal United Service Institution 122 (1977): 20-23.

Hopkins, Dr. N. J. "A Look Ahead to the Land Warfare Weaponry of the Mid-Eighties, Part 1: Infantry and Armoured Vehicle Weapons." Canadian Defence Quarterly 6 (Winter 1977): 19-24.

"Infantry in Battle." Canadian Defence Quarterly 12 (1934): 68-75.

"Infantry Tactics, 1914-1918." The Journal of the Royal United Service Institution 64 (1919): 460-9.

Janowitz, Morris, and Moskos, Charles B. "Five Years of All-Volunteer Force: 1973-1978." Armed Forces and Society 5 (1979): 16-33.

Jeapes, Lieutenant Colonel A. J. "Letter to the Editor." British Army Review, no. 68 (1979), p. 73.

Karber, Phillip A., "The Soviet Anti-Tank Debate." Survival, 18 (1976): 105-11.

Kitson, Major General Frank. "The New British Armoured Division," Journal of the Royal United Service Institution 122 (1977): 17-19.

"Lessons from the Arab/Israeli War." Report of a Seminar Held at the Royal United Services Institute for Defence Studies on Wednesday, 30 Jan. 1974. London: RUSI, 1974.

Liddell Hart, B. H. "The 'Ten Commandments' of the Combat Unit--Suggestions on Its Theory and Training." The Journal of the Royal United Service Institution 64 (1919): 288-93.

_____. "Suggestions on the Future Development of the Combat Unit--the Tank as a Weapon of Infantry." The Journal of the Royal United Service Institution 64 (1919): 660-6.

_____. "The 'Man-in-the-Dark' Theory of Infantry Tactics and the 'Expanding Torrent' System of Attack." The Journal of the Royal United Service Institution 66 (1921): 1-22.

_____. "A Science of Infantry Tactics," The Royal Engineers Journal 33 (1921): 169-82 and 215-23.

_____. "The Soldier's Pillar of Fire by Night: The Need for a Framework of Tactics." The Journal of the Royal United Service Institution 66 (1921): 618-26.

_____. "The Development of the 'New Model' Army," The Army Quarterly 9 (1924): 37-50.

Loomis, Lieutenant Colonel D. G. "On Conflict." M.A. thesis, Royal Military College of Canada, 1969.

_____. "The Regimental System." CF Mobile Command Letter, Special Supplement, 1975.

Marshall, S. L. A. "Why the Israeli Army Wins." Harpers' Magazine, Vol. 227, No. 1301 (1958), 38-45.

Miller, D. M. O. "The Infantry Combat Vehicle: An Assessment." Military Technology and Economics 3 (1979): 35-38.

O'Neill, R. P. H. "The Case for a Single Combat Arm." *Journal of the Royal United Service Institution* 122 (June 1977): 35-38.

Pogue, Dr. Forrest C. "Liddell Hart's Last Testament." *Air University Review* 23 (1972): 73-76.

Raven, Wolfram von, "The Concept of Forward Defence and the Security of Europe." *Canadian Defence Quarterly* 7 (1977): 6-10.

Rossetto, Captain Luigi. "Armoured Infantry." *The Army Quarterly* 109 (1977): 147-52.

_____. "Brigadier-General Orde Charles Wingate and the Development of Long Range Penetration." M.A. thesis, Royal Military College of Canada, 1978.

Rowan-Robinson, Major General H. "The Danger of Catch-Words and Phrases," *The Journal of the Royal United Service Institution.* 66 (1921): 430-9.

_____. "Lessons of a Blitzkreig." *Infantry Journal* 67 (1940): 210-22.

Simmonds, Captain G. G. "The Attack." *Canadian Defence Quarterly* 16 (1939): 379-90.

Smith, Field Officer R. G. "The Soviet Armoured Threat and NATO Anti-Armour Capabilities." *The Army Quarterly* 109 (1979), 153-61.

Starry, General Donn A. "A Tactical Evolution--FM 100-5." *Military Review* 58 (1978): 2-11.

Stolfi, R. H. S. "Equipment for Victory in France." *History* 55 (1970): 1-20.

_____. (ed.), Major L. O. Ratley, and O'Neill, Major J. F. *German Disruption of Soviet Command, Control, and Communications in Barbarossa, 1941.* Prepared for Director, Net Assessment, Office of the U.S. Secretary of Defense (1980).

Stone, Colonel J. R. "Memoirs: Kapyong." *CF Combat Arms School Infantry Newsletter* 3 (1974): 3-12.

Vigor, P. H. "Fighting in Built-up Areas: A Soviet View--Part I." *Journal of the Royal United Service Institution* 122 (1977): 39-47.

Weeks, John. "The Modern Infantryman." *Military Technology and Economics* 3 (1979): 23-26.

Weigley, Russell F. "To the Crossing of the Rhine: American Strategic Thought to World War II." Armed Forces and Society 5 (1979) 301-20.

Wendt, Lieutenant Colonel Robert L. "Army Training Development--A Quiet Revolution." Military Review 58 (1978): 74-83.

Yelshin, Lieutenant Colonel N. "Soviet Small Arms." Soviet Military Review, no. 2 (1977), pp. 15-17.

Index

Aachen, 262
Abu Agheila, 242
Abyssinia, 159
Achtung! Panzer!, 52
active defense, 132-136, 261
  Afrika Korps, 160, 161
  Ainse River, 99
  Alam Halfa, 160
  Albert Canal, 98
  Alexander, Gen. Sir Harold, 156, 175
Allied armies, 14-15, 86, 173-186 and passim; [in Great War] tactics, 15; [in WW II] air supremacy of, 173, 181; reluctance to camouflage, 181-182; assessment of German tactics and performance, 183; German assessment of Allied infantry performance, 185; Allied infantry compared with German infantry, 181-182, 182-183; infantry modifies tactics in bocage, 179 logistics and divisional "slices," 176-177; outfought by Germans, 101, 102-103, 173, 181-183; shortage of infantry, 175, 186; Sicily landing, 173; actual strength in Battle of France, 86; tendency toward technological solutions, 155, 186; long vehicle columns in Italy, 181; weaknesses of infantry, 183-184, 186; "weight of metal" tactical bias, 175
Allon, Gen. Yigal, 247-248, 249
"all-tank" idea, 40, 51-52, 73, 91, 240, 242, 244-245
American army, 163-173 and passim; [pre-WW II] French army model, 164; in Great War, 164-165; [in WW II] armored division reorganization, 180-181; divisional "slice," 176; rapid expansion of, 168; lack of junior leaders, 169; manuals on minor tactics, 169; replacement problem, 175-176; tactical methods, 169-172; teething problems in Tunisia, 172;

325

triangular organization 165; [in Korea] longest retreat in American military history, 223; myth of "push-button" war, 226; replacement system, 226; weakness of combat arms' fighting spirit, 219, 223; [in NATO] active defense concept, 261; Division Restructuring Study, 254, 255-256; [units and formations] Third Army, 4, 176, 180; Eighth Army, 218, 223, 227; Second Armored Division, 180; Third Armored Division, 180; Seventh Armored Division, 181; Seventy-seventh Division, 170; 101st Airborne Division, 181, 186; Sixteenth Infantry, 178; 350th Infantry, 170
American infantry, 165-168 and passim; [in WW II] basic organization, weapons, and equipment, 165; motorized infantry, 168; overloading of individual soldier, 178-179; Patton's opinion of, 181; reluctance to fire, 183-184; tactical methods, 169-170, 171-172; shortage of infantry, 176; [in Korea] Chinese assessment of, 219; weapons and tactical organization, 224-226
American Civil War, 2, 7, 164, 289
Amiens, 102
Anglo-Americans: military approach, 155; tendency toward oversupply, 176, 177; learn lesson of relevance of infantry, 175
Angriffesgruppe, 95
antitank defense, 65, 67, 103, 140, 161-162, 163, 172, 179, 206, 219, 244; best antitank defense, 70, 262; effectiveness of antitank guns in World War II, 160, 180; first effective hand-held antitank weapon, 122; impact of the antitank guided missile (ATGM), 245, 250, 252-253, 259, 264; Russian methods, 137; "sword and shield" tactics, 39, 161
Antonov, Col. G. I., 115
Anzio, 173, 174
Arab-Israeli wars, 240-250
Arakan, 202
Ardant du Picq, Col. Charles J. J. J., 221, 291
Ardennes, 181, 182, 183, 185
Arnhem, 186
Arras, 103, 156
artillery, 7, 11, 15, 17, 219, 227 and passim; contributes to paralysis of French army in 1940, 101; dominance of in Great War, 14, 17, 21; double-edged nature of, 17, 175; effect of modern artillery, 264; Feuerwalze, 23; French faith in artillery-infantry array, 67; "creeping" barrage, 15, 21; Soviet decentralization of, 253; "standing" barrage, 15
"Atlantic," Operation, 180
"Atlantic Wall," 179

"Attack in Trench
  Warfare, The," 23
Attu, 211
Aufrollen, 93, 95, 96
Auftragstaktik, 87,
  95-96, 287
Australian Corps, 78
Australians, 160, 185, 201
Austro-Prussian War, 2

Balck, Col. William, 6
Balkans, 130
Barbarossa, Operation,
  111, 120
barrage (see "creeping"
  barrage)
Bastogne, 181, 186
"battle drill," 41, 45,
  157, 158, 158, 170,
  179, 247; reducing
  pillboxes, 179;
  soldier's pillar of
  fire by night, 42
"battle principles," 42
"battle procedure," 157
"battle schools," 156, 174
battlefield, face of, 14,
  284; invisibility of
  enemy, 6
battlefield mobility, 21,
  26, 124, 203, 207-208,
  209-210, 222, 241,
  265, 288, 290 and
  passim
battalion establishments,
  4-5, 9-10, 67, 71, 78,
  89, 91, 126, 167-168,
  206, 215, 220-221, 224,
  254-257
"bazooka" antitank
  weapon, 122, 167, 179,
  215
Beck, Gen. Ludwig, 52,
  65, 169; urges gradual
  expansion of
  Reichsheer, 63; view
  of infantry as mass of
  decision lends balance
  to panzer organiza-
  tion, 52
"beehive" charge, 174, 179
Belgium, 98

Ben Ari, Lt. Col. Uri, 241
Ben Gurion, Prime
  Minister David, 247
Berlin, storm of, 262, 263
Bialystok-Minsk pocket,
  111
Biddulphsberg, 6
Bidwell, Brig.
  Shelford, 88
blitzkrieg, 37-38, 52,
  62, 86, 113, 114, 158,
  242 and passim; based
  on infiltration, 96;
  1940 conquest of
  France, 98-103;
  founders in Russia,
  117-118, 118-121, 129,
  130, 135; vital role
  of infantry in, 86,
  96, 99; methodology
  and ingredients of,
  37, 93-96, 100-101;
  weaknesses of, 102, 132
"blob" defense, 21 (see
  also "elastic" defense)
Bloch, I. S., 263
BMP problem, 250-251,
  251-253
bocage, 161, 179
Bock, Field Marshal Fedor
  von, 96, 111
Boer War, 6, 10;
  British reluctance to
  dig, 6; Boer tactics
  and firepower, 6, 76
Bond, Brian, 52
"Bore War," 96
Borisov, 120
Bougainville, 214
Bren gun, 76, 77-78, 175,
  285
British army, 4, 9-10,
  11, 17, 26, 49, 52,
  73-78, 155-164,
  173-202, 203, 206-208,
  210, 256-257 and
  passim; [in Great War]
  introduces "creeping"
  barrage, 15, 21;
  failure to comprehend
  concept of defense in
  depth, 21, 76;
  improved marksmanship

of, 13; superior methods of Australians and Canadians, 78; "leap-frogging" tactics of, 21; wave tactics of, 15; [between wars] "all-tank" school, 40-41, 51, 74; infantry organization, 74-75; ignores "soft spot" infiltration tactics, 76; continues to place emphasis on linear frontal methods of attack and defense, 78; basic organizational problem related to maneuver and the role of the LMG, 74-75; attempt at rejuvenation, 76-78, 156; shortcomings of training and maneuvers, 75-76; attitude toward tank, 73; [in WW II] armored division reorganization, 53, 161; in Burma theater, 206-211; victory over Italians, 159; judged too "rigidly methodical" by the Germans, 158; performance in Malaya, 201, 203; post-Dunkirk new tactical ideas, 51, 156-157; narrow regimental loyalty of British soldier, 89; problem of tank-infantry cooperation, 158; tank-to-tank battles erroneously considered key to modern land operations, 158; "teeth to tail" ratio, 176-177; [in NATO] British Army of the Rhine reorganization, 185-186; [units and formations] Fifth Army, 26; Eighth Army, 158, 250; Fourteenth Army, 211, VIII Corps, 180; 50th (Northumbrian) Division, 179; First Canadian Division, 174, 175; Fifth Canadian Armored Division, 175; Second Canadian Infantry Brigade, 174; Second Battalion, First Parachute Brigade, 186; Regiment de la Chaudière, 182

British Expeditionary Force, 155

British infantry: company organization in World War II, 162; cooperation with tanks, 158; Dunkirk performance of, 155; in Operation "Goodwood," 180; German assessment of, 185; at "Leichenfeld von Loos," 15; period between wars, 74-76; post-Dunkirk reforms, 155-157; 1936 reorganization of, 76-78, 155; shortage of British and Canadian infantry, 175; tactical revitalization in Burma, 209-211; tactical inadequacies in Burma, 206-208; tactical methods in Tunisia, 162-163

Brooke, Gen. Sir Alan, 155

Browning Automatic Rifle (BAR), 165, 170 and passim

Bruchmuller, Col. Georg, 23

Bryansk, 118

Burma, 201-202, 206-208, 209, 211 and passim

Caen, 180
Cambrai, 26, 48
Canadian Army: WW II divisional "slice,"

176-177; at Ortona, 174, 186; shortage of infantry in World War II, 175; First Canadian Division, 174, 175; Fifth Canadian Armored Division, 175; Second Canadian Infantry Brigade, 174; Regiment de la Chaudière, 182
Canadian Corps, 78
Cannae, 62, 96, 111 (see also Kesselschlacht)
Caporetto, 26, 71
Cardin-Loyd tankettes, 50
Carlson, Major Evans F., 212-214
Case Yellow, 96-98
Cassino, 174-175 and passim; compared to Passchendaele, 175
Cauldron, the, 160
cavalry, 7, 14, 40, 50, 62, 264
centers of resistance, 21, 132, 135, 263 and passim
Central Pacific Theater, 211, 212, 215 and passim; American tactical methods in, 211
Chauvineau, Gen. Narcisse, 67
Cherbourg, 179
China, 202, 203, 204, 212-213 and passim
"China problem," 203
Chinese army (see People's Liberation Army)
Chuikov, Gen. Vasili I., 132-133, 135, 136, 250, 262, 289 and passim
Chunkufeng, 202
Citadel, Operation, 136
city fighting, lessons of, 132-136, 263
city "hugging" tactics, 262
Clausewitz, Gen. Carl von, 93, 155, 240 and passim; on the "friction" of war, 282-283
combat team, limitations of, 257
combat unit, 41, 43
Commonwealth forces, British, 158, 186, 226 (see also British Army)
communications disruption in future European war, 254
Communist Chinese army, 212 (see also People's Liberation Army)
company column, 2, 4, 9
company establishments, 6, 9, 65, 67, 71-73, 74, 76-77, 89-90, 126, 162, 165-168, 184, 204-206, 215, 220-221, 224, 254-255
"contracting funnel" concept of defense, 45-47
cooperation between arms, 50, 53, 62, 65, 86, 89, 91-92, 112, 124, 158-159, 161
"corner posts," 43, 49
Corps Expéditionnaire Français, 186
counterattack as central to elastic defense, 18-20, 21, 76
"crabs," 175
"creeping" barrage, 15, 21
Crimean War, 1, 2
Currie, Gen. Sir Arthur, 24
Cyrenaica, 160

Danish War, 2
Dayan, Gen. Moshe, 241
D-day, 178 (see also Normandy)
decentralization of tactical control, 5, 18, 45-46, 93, 95, 282; "a risky business," 19; need for in city fighting,

263; modern Soviet trend toward 253-254
"decisive" arm, 48
"decisive" attack, the, 11; difference between German and French approach to, 11-12
defense, superiority of the, 14-15, 46, 70, 121, 129, 132, 136-137, 173, 180, 186, 210, 211, 227, 240, 250-251, 257-258, 260, 261-262, 282
"Defensive Battle, The," 18
de Gaulle, Gen. Charles, 52, 112; suggests armée de métier, 67
density of force (to space), 4, 9-10, 10-11, 13, 101, 136-137, 180, 253
De Puy, Gen. William E., 250, 257
Der Kampfwagenkrieg, 52
desert warfare, relevance of lessons learned, 161-162, 175, 179, 250
deterrence, conceptual flaw in Western concept of, 251, 265
"developing" attack, the, 11
digging (see entrenchment)
Dinant, 98
dive bomber (see Sturzkampfflugzeuge)
divisional "slices," 176-177
Doolittle Report, 226
Douhet, Gen. Giulio, 112
Dragomirov, Gen. M. I., 8, 120
Dunkirk, 51, 155
Dupuy, Col. T. N., 182
Dyle River, 155

Eastern Front, 23, 110-144, 181 and passim (see also Ostfront and Soviet Union)

Eben Emael, 98
effort principal, 93
Egyptian forces, 242-243, 244-245
Eimmansberger, Gen. Ludwig R. von, 52
Eisenhower, Gen. Dwight D., 176, 259
El Alamein, 160
"elastic" defense in depth, 17-21, 70, 261-262 and passim
Elbe river, 177
entrenchment, 6, 7, 10, 11, 21, 78, 162; British reluctance to dig in Boer War, 6; entrenching tool uselessness, 14; German decision to dig in after the Marne, 14; Japanese tactical methods in World War II, 204, 209; need for digging in a future war in Europe, 204; PLA experience in Korea, 227, 263-264; the Russian as "champion digger," 121-122, 127-128; superiority of the well dug in defense, 186; trenches as death traps, 21, 47, 288
Erfurth, Gen. Waldemar; 91, 111
Eritrea, 159
Etude sur l'attaque, 22
"expanding torrent" concept of attack, 70-72, 47, 49 and passim
extended line or open order tactics, 1, 4, 7; inflicts fewer casualties, 1; risks loss of control, 4; versus close order, 1

Falkenhayn, Gen. Erich von, 61

Fall Gelb (see Case Yellow)
Faustpatrone, 174, 263
Feuerwalze, 23
fighting unit, 6, 19
Finnish (Winter) War, 114-121, 124
fire team, 214-215
fire unit, 4, 9, 43, 68
firing-line tactics, 4, 6-7, 8, 9, 10-11, 42-43
Flachen und Lukentaktik, 95
Flanders, Battle of France and, 155
Foch, Marshal Ferdinand, 11; argument for offensive, 11-12
form of future war in Europe, 263-264
Formosa, 202
Fortress Europe, 179
forward defense, NATO strategy of, 260-261
France, 4 and passim; Allied invasion of, 177-181; Corps Expeditionnaire Français, 186; 1940 German conquest of, 86-103
Franco-Prussian War, 2
Frederick Charles, Prince, 96
French army, 4, 7, 8, 9, 14, 23, 65-71, 96-98, 101-102 and passim; difference between French frontal and German flanking doctrines before Great War, 11; French pre-Great War emphasis on mass and the offensive à outrance, 13; [in Great War] commitment to the offense, 14; method of defending in depth, 21; infantry relegated to "mopping up," 14; on infiltration, 76; small unit tactics, 15; [between wars]
General Kennedy's assessment of, 66; belief in the supreme value of the defensive, 67, 70; emphasis on centralized control of operations, 70, 95; failure to fortify villages and cities, 70, 129, 261-262; field army organization, 67; French-German doctrinal differences, 70, 93, 95; groupe de combat basic tactical organization, 68-70; rated best army in Europe, 65, 261; role of tanks, 40-41, 70-71; worship of firepower over mobility, 41, 67, 70; [in WW II] could have checked Germans, 101-102, 129, 286; dispositions of Second and Ninth Armies, 101; artillery communications contribute to paralysis, 101-102; infantry allowed to stagnate, 101
"friction" of war, 282-283
Fritsch, Gen. Werner von, 52
frontages, 2, 9-10, 11, 12, 13-14, 15, 23, 24, 70, 73, 100, 113, 126, 169, 172, 174, 214, 258
frontal attacks and attitudes toward, 7, 11-12, 39, 43-45, 62, 70, 75, 78, 114, 127-128, 162-163, 169-170, 183, 208, 211, 215-217, 221-222, 245, 249, 288-289; Zhukov orders halt to, 130
Frost, Lt. Col. John, 186
Frunze, M. V., 112
Fry, Col. J. C., 170

332 INDEX

Fuller, Maj. Gen. J. F. C., 38-40, 42, 50, 51, 62 and passim; on Cassino battles, 174; "Father of Blitzkrieg," 38; grasp of infantry tactics, 38; on greatest tank obstacle, 70, 260; influence on Germans, 52; influence on Russians, 112-113; Lectures on F.S.R. II, 112; On Future Warfare, 112; paralysis theory, 38, 62; "Plan 1919," 38; on minor tactics of the Red Army, 127; on motorized guerrilla warfare, 39; tactical model of "sword and shield," 38-39; thinks in terms of tactical elements rather than arms, 38; on the "true attack," 38

Future of Infantry, The, 47-50

future of the foot soldier, 264-265

Gavin, Gen. James M., 186
Gazala Line, 160
German army, 8, 9, 11, 14, 17-21, 21-26, 86-111, 117-118, 120-121, 122-123, 129-131, 137-144, 181-185, 254-255, 261 and passim; [in Great War] abundance of NCOs, 19; backbone of elastic defense, 70; decentralization of tactical control, 22, 23; decision to dig in after Battle of the Marne, 14; elastic defense in depth, 18-21; encouragement of intellectualism and debate, 18-19, 21; gruppe as tactical battle unit, 19, 24; high standard demanded of junior leadership, 24; role of infantry expanded, 21-22; emphasis on principle of surprise, 23; tactics more sparing of own soldiers' lives, 15; tactical solution to impasse of trenches found in "soft spot" infiltration tactics, 21-26; theory of the "unlimited" objective, 24; "triangular" structure adopted, 17; [between wars] development of "all arms" doctrine, 51-52, 53, 65, 89; balance in infantry-tank organization, 91-92; blitzkrieg development, 52, 62, 96; elevation of status of section commander, 88; advocacy of flanking movements, 11, 63-64, 93, 95; last 300 yards, 63-64; military excellence of, 61, 63; noncommissioned officers, 87; officer education, 61; officer-man relationships, 87; panzer organization, 91, 161; on the role of the tank, 62, 63, 65; emphasis on surprise, maneuver, and disruptive infiltration tactics, 62, 93; standard tactical field organization, 65, 89-91; view of battle as series of local actions, 93-95; [in World War II] assessment of Red Army, 110; assault on France, 86,

INDEX 333

98; attacks Soviet Union, 110-112, 117; breakdown of blitzkrieg in Russia, 117-118, 118-121, 129, 130, 135-136; blitzkrieg technique, 98-103; casualties on Eastern Front, 118, 131; divisional "slice," 176-177; faith in small-unit actions and counterattack, 183, 184-185; fighting skill, 138-144, 182-184; immobilization in Russian rasputitsa, 120, 129; "institutionalization" of military excellence, 182; Kameradschaft, 87; reaction to Red Army infantry assaults, 184-185; use of reverse slopes, 140-141; rear-echelon troops in combat, 141-142; superior training of junior commanders, 142-143; primary group integrity, 143; [in NATO] new army structure, 254-255; military strategy and tactics, 260-261; [units and formations] Army Group Center, 111, 129-130, 136; Army Group North, 111; Army Group South, 111, 136; Army Group "A," 97, 98; Army Group "B," 96-97, 98; Army Group "C," 96; Fourth Army, 98; Eighth Army, 25, Twelfth Army, 98; Sixteenth Army, 98; Afrika Korps, 160, 161; IV Corps, 155; Forty-eighth Panzer Corps, 128; Fifty-seventh Panzer Corps, 118; Panzergruppe Kleist, 98, 102; "Group Menny," 160; Ninth SS Panzer Division, 181; Tenth SS Panzer Division, 183; First Panzer Division, 98, 99-100; Second Panzer Division, 98; Fifth Panzer Division, 98; Sixth Panzer Division, 98; Seventh Panzer Division, 98; Eighth Panzer Division, 98; Tenth Panzer Division, 98, 185; Grossdeutschland Division, 140; SS Totenkopf Division, 103; 98th Division, 118; 112th Division, 30; First Parachute Division, 174; 21 SS Panzergrenadier Regiment, 185; First Rifle Regiment, 99-100

German infantry: antitank training for Russia, 140-141; breaks in Russia, 130; company strengths reduced by combat, 129, 131, 184; compared with Russian infantry, 139-140; comprises 70 percent of Russian field force, 143-144; German traditional predilection for infantry arm, 50, 91; marching ability, 88, 99, 139; emphasis on marksmanship, 64; minefield breaching methods in Russia, 141; motorized division organization, 91; standard organization, weapons, and equipment, 89-90, 95, 184; organized for offense, 86, 101, 156; only arm capable of advancing in Russia, 129; patrolling methods, 163; WW II performance in France,

99, 102-103, 156; pre-World War II organization, 65; radical reformulation of tactics and reorganization in Great War, 22-26; role in blitzkrieg, 86, 96, 99-101; weaknesses of, 120, 122-123, 132
Germany, Federal Republic of (FRG), 259 and passim
Geyer, Captain, 23, 24
Gilbert Islands, 211
Ginsberg, Eli, et al., 287
Golan Heights, 244
Goltz, Gen. Colmar von der, 4
Gomel, 118
"Goodwood," Operation, 180, 186
Goumiers, 186
Grandmaison, Col. Louis de, 13
Great War, 1, 2, 4, 7, 9, 13-17, 37 and passim; wave tactics in, 15, 24
Great Patriotic War, 127, 261 (see also Eastern Front)
"Group Menny," 160
"group think," dangers of, 287-288
groupe de combat, 21, 68-70
gruppe organization, 19, 24, 70
Guadalajara, Battle of, 65
Guadalcanal, 202, 217
Guam, 211
Guderian, Gen. Heinz, 47, 98, 99 and passim; infantry breaks in Russia, 130; development of tactical thought and "all arms" theory, 52-53

Haganah, 246, 248
Haig, Gen. Sir Douglas, 7
Hamilton, Gen. Sir Ian, 7

Hammelberg, 264
"hedgehog" defense, 136, 144, 261
"hedgerows, battle of the," 179
helicopter, 265 and passim
Herzog, Gen. Chaim, 246
Hindenburg, Field Marshal Paul von, 94
Hindenburg Line, 48
Holland, 98
Hoth, Gen. Hermann, 98
Hungarian army, 136
Hutier, Gen. Oskar von, 26

Ignatieff, Col. Nicholas, 177
Il Duce 71, 159
Imphal-Kohima (see Kohima)
Indo-China, 202, 240
Indonesia, 202
Indian Army, 201, 209, 217 and passim; daily divisional tonnage in WW II, 210
indirect approach, 249
Infantry Greift an, 47
infantry: ability to fight alone, 117-118, 127, 181-182, 283; adaptability, 259, 265; ascendancy over cavalry, 2; casualties incurred compared with other arms, 175-176, 240; combat vehicle debate, 245; dominant role of, 240, 282; effectiveness in Yom Kippur War, 244; enhanced capability of modern infantry, 250, 260, 264-265; true function of, 48, 186, 290; junior commanders and primary group importance, 19, 24, 26, 88, 142-143, 156-157, 183-184, 221, 247-249, 283-285; loco-mobility of, 51, 127-129, 222, 259,

INDEX   335

289-290; modern tactical dilemma, 257-258; as the "tactical" arm, 49; as "queen of fortresses," 39; as "queen of battle," 14, 39, 40, 50, 124; regains movement in Great War, 21; twin pillars of infantry strength, 282; tactics in Tunisia, 161-162
"infantry tank" idea, 40
Infantry Training, British manual, 41, 49, 75
infiltration "soft spot" tactics, 22-36, 43, 44, 47, 49, 62, 88, 96, 99-101, 156, 174-175, 182-183, 206-209, 220, 221-223; ignored by the French and British, 23, 70, 75-76; influence of Captain Laffargue, 22; Russian infantry infiltration tactics, 127-128;
Israeli Defense Forces (IDF), 240-250 and passim; "all tank" trend, 240-245; "functional" commands, 243, "indirect approach" of, 249; "mobile infantry school," 241; motivation and Kameradschaft, 249; organization of field forces, 241, 242, 247; Palmach influence, 246-248; emphasis on small unit training and tactics, 247-249; Seventh Armored Brigade, 241, 245; 190th Armored Brigade, 244; Golani Brigade, 244
Israeli infantry: mechanized, 241-244; parachute, 244; reluctance to dismount, from vehicles, 244, 250;
status of section commander, 247-248; weaknesses of, 242-244, 245
Italian army, 71-73, 159; in Battle of Guadalajara, 65; fitness of soldiers not enough, 71, 159; German assessment of, 65, 159-160; infantry, 71-73, 159; neglects military essence, 159; in Russia, 136; tactics and organization pre-WW II, 73
Italian campaign, 173-175; battles compared to Great War, 174; mobility of Juin, 173; great relevance of infantry, 175; shortage of infantry, 175
Italy, 71, 161-162, 173-174 and passim
Iwo Jima: casualties compared with Passchendaele, 202

Japan, 202 and passim
Japanese army, 6-7, 9, 201-217 and passim; "banzai" attack, 211; disposition of forces in World War II, 202; not jungle orientated, 202, 203; Mongolian-Manchurian clashes with Red Army, 114-115; refusal of soldiers to surrender, 202; in Russo-Japanese War, 6-7, 208; regarded as second-rate by British, 201, protracted struggle with China, 202; tactics in World War II, 206-209
Japanese infantry, 203-208; camouflage and digging, 204; poor marksmanship, 204; mobility and spartanism

of, 203-204; weapons, equipment, and tactical organization in World War II, 204-206
Java, 208
Jerusalem, 242, 250
Jewish Brigade, 246
Jomini, Antoine Henri Baron de, 2, 164, 283
Juin, Gen. Alphonse P., 173, 186
Junger, Ernst, 21

Kameradschaft, 87, 143, 249
Karelia, 114
Katyusha rockets, 138
Keegan, John, 136
Keil und Kessel, 111
Kennedy, Gen. Sir John, 66
Kesselring, Field Marshal Albert, 159
Kesselschlacht, 62, 63, 94
Khalkhin Gol, 114, 203 (see also Nomonhan)
Kharkov, 131
Khasan, Lake, 114
Kiev pocket, 111
Kitchener divisions, 15
Kleist, Gen. Ewald von, 98, 99, 102
Kohima-Imphal, 202, 209, 210
Korea, 202, 217, 218, and passim
Korean War, 217-227
Kuriles, 202
Kursk, Battle of, 136-137, 142

Laffargue, Capt. André, 22, 37
"landship concept," 40
Laskov, Gen. Haim, 241
"last 300 yards," problem of the, 63-64, 253
Lebanon, 246
Lectures on F.S.R. II, 112
Leeb, Field Marshal Wilhelm Ritter von, 96, 111
"Leichenfeld von Loos," 15

Leningrad, 111 and passim
Lewis gun, 15, 74-75, 76
Libyan campaign, 161
Liddell Hart, B. H., 6, 26, 41-51, 52-53, 73, 74, 75, 76, 127, 156, 163, 169, 249 and passim; argument for accurate as opposed to volume fire, 47, 50-51; "contracting funnel" concept of defense, 45-47; conversion to mechanized warfare school, 50, 51; "expanding torrent" concept of attack as tactical solution to riddle of the trenches, 41, 43-45, 47-49; on forms of attack, 49; Future of Infantry, The, 47-50; on the infantry section, 157; influence on the Red Army, 112-113; influence on the German, British, and American armies, 47, 49-50, 52; influenced by German storm tactics, 47; on "land marines," 50; on light infantry skills, 48, 50; "Man in the Dark" framework for tactics, 41-43,; Memoirs, 47; "New Theory of Infantry Tactics," 41; on night attacks and attacks in fog, 48; Paris, or the Future of War, 50; reformulation of infantry tactics, 41; Remaking of Modern Armies, The, 50; "'Ten Commandments' of the Combat Unit," 41; training and equipment of infantry, 48-49, 50-51; limitations of

armor in city fighting, 263
Lin Piao, 223
"Little Stalingrad," 174 (see also Ortona)
long war scenario, 260, 264
Lossberg, Col. Fritz von, 18-19, 22, 261 and passim; conversion to tactical theories of his juniors, 18; gives his name to elastic defense in depth, 18
Ludendorff, Gen. Erich, 18, 19, 21, 24, 87, 288 and passim; on "flabby" infantry unable to shoot straight and "fight from a distance," 22; interest in basics of infantry, 18; on development of the "group," 19; emphasis on shooting skill, 64; lack of strategic grasp, 26; promotes tactical controversy, 18-19
Luftwaffe, 131, 132, 182
Luxembourg, 98
Lvov, 118

MacArthur, Gen. Douglas, 211, 215, 217 and passim; critical of Central Pacific Theater methods, 211; on "island-hopping," 215
Macksey, Kenneth, 63, 121
McKee, Alexander, 181
McNair, Lt. Gen. Lesley, 165, 172
machine guns, introduction and impact of, 7, 10, 13-14 and passim; differentiation between weapon of stability and weapon of mobility, 75; "Leichenfeld von Loos," 15; as "queens of battle," 14

"machine gun destroyer," 37, 40
Maginot Line, 67, 96, 101
Malaya, 201, 202, 203, 204, 208 and passim
Malinovsky, Marshal R. 174, 287
Manchuria, 202, 203, 217 and passim
maneuver (movement), 17, 21, 23, 26, 39, 41, 49, 52, 62, 67, 74-76, 91-93, 100, 112, 116-117, 126, 130-131, 133, 156, 162-163, 170, 183, 206-208, 212, 213, 215, 217, 221-223, 249, 251-252, 253, 254, 261, 288-289, 290; in defense, 18-22, 45-47; on fire preferred to movement, 67, 70, 224, 288; Liddell Hart's concept of fixing and maneuvering, 42-45; at section level, 45, 47, 157; true attack based on, 39
Mannerheim Line, 114, 115, 116, 142
Manstein (Gen. Erich von) Plan, 96
"Manual of Infantry Training for War," 18
Mao Tse-tung, 219, 220
marching performance, 71, 78, 101, 209-210, 248-249, 290; Chinese, 218-219; German, 88, 99, 139; Japanese, 204; Puller's Marines, 224; Russian, 123; under fire, 11-12
Marco Polo Bridge, 202
Marcks (Gen. Erich) Plan, 110
marksmanship, 2, 6, 10, 13, 63-64, 70, 71, 115, 116, 121,

135-136, 186, 204, 248-249, 253, 254-255; argument for accurate small-arms fire, 48, 50-51, 260; Boer shooting ability, 6; in British army, 9, 13; casualties inflicted by small arms, 240; failure of troops to fire, 183-184; effect of "misses" of the beaten zone, 4; Ludendorff on, 21, 64

Marne, 14

Marshall, Brig. Gen. S. L. A., 178, 182, 183-184, 217, 227, 248, 284, 286, 290 and passim; on overloading the soldier, 177-179, 217, 290; on soldiers' reluctance to fire, 183-184, 286; on "thinking" soldier, 182, 286-287

Marshall, Gen. George C., 169

Martel, Gen. Gifford le Q., 40, 52; impressions of Red Army, 124

Martel-Morris tankettes, 50

Maxse, Gen. Sir Ivor, 41, 74

mechanization and motorization, 38-39, 49-53, 62, 65, 91, 99-100, 112-113, 129, 203, 245-246, 252, 254-257, 263, 289-290; "army of chauffeurs," 174; light infantry in mechanized army, 48; on mechanized-motorized troops, 39-40; motorized infiltration, 96, 98-99; no panacea for improved mobility or military performance, 5, 37, 158-159, 173, 209-210, 218, 220

Mellenthin, Gen. F. W., 53, 127 and passim; comments on British rigid method of waging war, 161

Meuse River, 98, 99, 101, 102

Miksche, F. O., 93, 96 and passim; advocates "web"-style defense, 129

military operations in built-up areas (MOBA), 260-264

minor tactics, 26 and passim; elimination of term suggested, 283; Fuller on Soviet, 127

Minsk, 118

"mission tactics," 95 (see also Auftragstaktik)

mobility, 13, 48, 51, 75, 86-87, 91-93, 112, 124, 129, 140, 143-144, 158-159, 161, 203, 210, 218-219, 222, 226, 241, 264-265, 282-283, 288 and passim; "mobility of the boot," 290

modern armies, 218, 223-224, 226, 254, 264; road-bound immobility of, 173, 206, 210, 220, 290

Modder River, 6

Moltke, Count Helmuth von (the Elder), 2, 52, 95

Mons, Battle of, 13

Monte Cassino, 174

Montgomery, Gen. Bernard Law, 156, 160, 186

Monthermé, 98, 102

Moran, Lord, 143

Mort Homme, 66

Moscow, Battle of, 111, 118, 129-130, 136-137, 142 and passim

motti fighting, 115

"mouse-holing," 174

Mueller, Col. Wolfgang, 138

Mukden, 7
Mussolini, Benito, 71

Namur, 97
New Britain, 214
New Guinea, 214
New Zealand forces, 186
Nicaragua, 212
Nicholson's Nek, 6
Nomonhan, 202, 203 (see also Khalkhin Gol)
noncommissioned officer (NCO), 19, 24, 26, 42, 45, 74, 77, 87-88, 135, 143, 156-157, 164, 165-166, 168, 170, 182, 185, 212-215, 221, 224-226, 248, 262-263, 283-284
Normandy, 177, 178, 179, 181 and passim
North Africa, 159-164, 165, 172, 180, 250
North Atlantic Treaty organization (NATO), 251-252, 253, 254, 260 and passim
North Italian War of 1859, 2

O'Connor, Gen. Richard N., 159, 180
offensive à outrance, 13
offensive power of a division, 184-185
Okinawa, 211
Omaha beach, 178
On Future Warfare, 112
Ortona, 174, 186, 250
Ostfront, 118, 127; contrasted with Western Front, 138 (see also Eastern Front)

Pacific, 201-202, 211-212, 215-216, 217, 223 and passim
Paget, Gen. Sir Bernard, 156
pakfront, 137, 241-242
Palmach, 246-247, 248

Panfilov, Gen. I. V., 130
panje wagon, 129
Panzer Leader, 52
panzer organization, 91, 161
Panzerfaust, 89, 141, 174, 179, 244
Panzerkeil, 137
parachute forces, 113, 174, 175, 186, 242, 244, 245
paralysis, vulnerability of an army to, 38, 101-102, 254, 283
Pardieu, Maj. M. F. de, 7
Paris, 70
Paris, or the Future of War, 50
Passchendaele, 15, 173, 175, 202 and passim; Cassino compared with, 174-175
Patton, Gen. George S., Jr., 121, 181 and passim; on poor company officers, 169; compares German and American infantry, 181; advocates marching fire, 170-171; replacement requirements, 176
Pavlov, Gen. D. G., 113
Pavlov, Sgt. Jacob, 135
Pearl Harbor, 117, 203
Peiper, Col. Joachim, 121
Peleui, 217 and passim
Pendleton, Camp, 214
People's Liberation Army (PLA), 218-223, 226-227; appreciation of American infantry weakness in Korea, 219-220; superior defenses in Korea, 227; marching ability, 218; first army to fight knowingly under nuclear threat, 227, 263-264; rations, 219; tactics, 218, 220-221, 222-223;

tactical organization, 220-221; weapons and equipment, 218-219; Eighth Route Army, 218
Percival, Lt. Gen. A. E., 201
Peronne, 102
Petain, Marshal Henri Philippe, 14
Philippines, 201, 202, 203 and passim
Philippine Sea, Battle of the, 202
"phoney war," 101
"Plan 1919," 63
plane defense, 21
platoon organization, 2, 9, 24, 41, 42-47, 65, 67-70; American army, 164-168, 170, 171-172, 226; British, 74-75, 76-78, 155-158, 162, 210; in defense, 162-163; Chinese, 220-221, 221-222; German, 89-91, 128, 183; Japanese, 204-205, 206-208; Israeli, 247; Italian, 71-73; Soviet, 124-126; U.S. Marines, 212, 215-217
Polish campaign, 101, 114, 129 and passim
Porsche Tiger tank defect, 137
primary groups, 143, 283 and passim
psychological dislocation, 48, 62, 99, 186, 265, 290
Puller, Lt. Gen. Lewis B., 217, 223 and passim; "man over weapons" philosophy, 223-224

"Queen of the Desert," 160

Rabin, Yitzhak, 244
"rates of wastage," 175-176
rasputitsa, 120, 129
recommendations on infantry training: conduct all field maneuvers in scenarios of enemy air superiority, 285; introduce better battle inoculation training, 286; increase demolition training for infantrymen, 285; promote "fire habit" through less restrictive shooting control, 286; discourage frontal attack mentality, 288-289; select substitute term for "minor tactics," 283; emphasize maneuver at lowest tactical levels, 288-289; keep sections together for longer periods, 285; enhance tactical status and training of the section commander, 282-284, 288-289; devote more time to dynamic small group field training, 283-285; encourage soldiers to think and be creative, 286-287; place greater stress on stalking skills, camouflage, and weapon handling, 285; condition infantry to engage tanks at pointblank range, 285; rigidly limit weight of the soldier's load, 290
Red Army, 110-144, 250-254, 258, 262-263 and passim; ATGM threat to BMP, 250-251, 252-253; attitude toward blitzkrieg, 113, 114; "bridgehead" tactics

in Great Patriotic War, 127; current tactical theory and debate, 251-254, 258; defeats blitzkrieg, 117-118, 121, 129-130, 132-137, 262; divisional "slice" in World War II, 176-177; Finnish War and reforms, 114-117; Mongolian-Manchurian clashes with Japanese, 114-115; tactical recklessness of, 130; refusal to abandon ground, 121-122; effect of Stalinist purges, 112, 113; strength on Nazi invasion, 110-111; tactical theory and doctrinal development before Finnish War, 112-114; formulates "web"-style defense in Great Patriotic War, 121, 129, 136, 261; experience in urban warfare, 131-136, 262-263

Red Army formations: Seventh Army Group, 114; Forty-third Army, 118; Sixty-second Army, 132; Second Guards Tank Army, 263; V Airborne Corps, 118; 316th (later Eighth Guards) Division, 130

Red Army infantry, 117, 120-128, 132, 133, 137-139, 144 and passim; in defense and forest tactics of, 120-123; small-unit tactics of, 124-128; spartanism of, 123-124; basic tactical organization in World War II, 124-126; weapons and equipment, 123-124

Red Chinese (see People's Liberation Army)
refugees, 99
Reichswehr, 67, 169
Reichscheer, 63 (see also German army)
Reinhardt, Gen. Georg-Hans, 98
Remaking of Modern Armies, The, 50
reserves, proper use of, 26
replacement systems: American, 176, 226; German, 143-144, 285
Rhine Army, 256-257
Ridgway, Gen. Matthew B., 223, 226
rifle, superiority of, 1-4, 10 and passim; argument for accuracy of fire, 48, 50-51, 260; chassepôt, 4; effect of defensive fire, 3; indirect fire of, 10; musketry training of French and German armies before and during Great War examined, 63-64; "needle gun," 2, 4; reluctance to fire, 183-184, 286
Riga, 26
Rommel, Gen. Erwin, 47, 160-161, 168, 180, 181, 186, 249 and passim; compares American and British tactical handling of forces, 180-181; use of antitank weapons, 161; emphasis on basic tactics, 160; on British and Anzac infantry, 160, 185-186; on the encounter battle, 100; on the primary role of infantry, 160
Rostov, 130

Rowan-Robinson, Maj. Gen. H., 40
Royal Air Force (RAF), 182
Royal Engineers, 163
RPG-7, 244
Rumania, 61
Rumanian army, 136
Runstedt, Field Marshal Gerd von, 96, 111
Russia, 111-112, 118-120 and passim (see also Soviet Union)
Russian army, 9 (see also Red Army)
Russo-Japanese War, 6-7, 208
rushing tactics, 2, 4, 7, 9, 11, 15, 75-76, 162, 170 and passim; Patton opposed to, 170

Saar offensive, 101
Saipan, 211
Sakhalin, 202
Saxe, Marshal Maurice de, 140, 163
Schlieffen, Count Alfred von, 52
Schlieffen Plan, 62, 96
Schwerpunkt, 93-95, 96, 97 and passim
section, 2, 9, 24, 43, 45, 252, 254-255, 258, 288; American army, 164-168, 169-170, 224-225; British, 74-75, 76-77, 156-157; Chinese, 221, 224; French, 21, 67-68; German, 19, 24, 70, 88-89, 89-90, 143, 182-184; groupes de tirailleurs, 22; Israeli, 247-248; Italian, 71-72; Japanese, 204-206, 206-208; not maneuver unit in British army, 45, 157; Soviet, 124-126, 126-127, 133, 252, 262-263; U.S. Marines, 212-215, 224

section commander, as prime determinant of operational effectiveness, 283-284
Sedan, 26, 97, 98, 99, 101 and passim;
Seeckt, Gen. Hans von, 62, 86, 87
Senger und Etterlin, Gen. Frido von, 182, 184, 186
Shanghai, 212
Sharon, Gen. Ariel, 242, 245
Sherman, Gen. William Tecumseh, 289
shortage of infantry, 90, 141, 155, 175-177, 180, 184, 186-187
Sicily, invasion of, 173
Sidorenko, Col. A. A., 254
Sinai, 241, 265
Sinai Campaign of 1956, 241-242, 249
Singapore, 201, 203 and passim
"Sitzkrieg," 96
Six Day War (June 1967), 242, 246, 249; Egyptians compared to Russians at Kursk, 242
skirmishing, 2-3, 4, 169-170 and passim;
Slessor, Sir John, 290
Slim, General W. J., 206, 208 and passim; opposition to special forces, 210-211; on prowess of Japanese, 209; enlightened training methods and tactics, 209-210
small group combat, 19, 21, 42, 62, 67-68, 88, 116, 143, 182-184, 221; defensive techniques of, 91-92, 162-163, 255; dominance of in city fighting, 133-136, 262-263; influence of in modern war, 156-157, 248, 282-283,

284; impact on large-unit effectiveness, 184-185; battle as series of local actions, 93, 95, 122, 124, 183; neglected area deserving of more study, 283-285; replaces long lines, 9; stormtroop tactics, 24-26; Soviet small-unit infiltration, 127-128
Smolensk, 113, 120
Smith, Gen. Holland M., 215
"snail" offensive, 142, 210
"soft spot" tactics (see infiltration tactics)
soldier's load, 14, 48, 123-124, 173, 177-179, 290
Somme, Battle of the, 15, 17, 173-174 and passim
South African War, 9 (see also Boer War)
Southwest Pacific Theater, 211-212
Soviet Union: invasion of, 110-112, 117; geography not enough to save nation in Great War, 118; modern strategic theory of, 251
Spanish Civil War, 71, 76; confirms French view that tanks should only be used in infantry support role, 65; Russian experience, 113-114, 117, 129
Special Night Squad, 246
Speidel, Dr. Gen. Hans, 182
St. Vith, 181
Stalin, 112, 113, 114 and passim
Stalingrad, 131-136, 137, 174, 182, 220, 221 and passim; Great War storm tactics applied in battle for, 136; tanks unsuited for city fighting, 131
Stilwell, Gen. "Vinegar Joe," 217
storm detachments, 263
storm group tactics and organization, 133-135, 263
stormtroops, 23-24 and passim (see also "elastic" defense and infiltration)
Stosstruppen, 23, 89
strongpoint defense, 132, 135 (see also "web" defense)
strongpoints, 21, 132, 135, 263 and passim
Sturzkampfflugzeuge (Stukas), 98, 99, 100, 101 and passim
Sturmtruppen, 20, 23-24, 135
Suez Canal, 241, 244; 245
Suez City, 250
Sung Shin-lun, Gen. 224
Suomussalmi, 115
Swinton, Maj. Gen. Sir Ernest, 6, 15
Syria, 245, 246

tactical arthritis, 6
tactical methods and theory: blitzkrieg, 99-100; blitzkrieg theory, 93-96; British in defense in World War II, 155-156, 157, 160, 162-163, 185; city fighting, 131-136, 173-174, 262-263; "contracting funnel" defense, 45-47; defense in depth, 17-21, 76, 136, 180, 226-227, 261-262; "expanding torrent" attack, 41-45; German defensive tactics in World War II, 183, 184-185; "hedgehog

defense," 140, 144, 261; "hugging" tactics, 132, 209, 220, 262; Japanese defensive tactics, 209; MacArthur's technique, 215-216; maneuver problem, 74; U.S. Marine methods, 215-217; "mouse-holing," 174; attacking pillboxes, 179-180, 215; PLA tactics, 221-223; "roadblock" tactics, 208-209, 222; resistance-riposte idea, 19-21, 45-47, 49; Rommel's techniques, 100, 161; Slim's methods, 209-210; "sword and shield" theory, 39, 129, 161; Soviet assault tactics, 126-129; stormtroop tactics, 22-26; tank-infantry cooperation, 158

tactical paralysis, 23

tactical unit, 4, 5, 9, 19, 24, 45

"tail to teeth" ratios, 155, 176-177, 186, 210, 219, 290

Tal, Gen. Israel, 241-244, 245

tank: "landship concept," 40; as machine gun destroyer, 37, 40; as "solution to trenches" missed by Germans, 62; "The Tank Army," 40; tank-infantry ratios, 53, 91, 161, 180-181, 252; most formidable antitank obstacle, 70, 260; limitations in street fighting, 263; "Porsche" Tiger, 137; T-34 tank, 116, 117 and passim

"tank infantry," 40

tank terror or "tankophobia," 140, 285-286

tankettes, 50

Tarawa, 215

Tel Shams, 245

Thomas, Gen. Gerald C., 214

Timoshenko, Marshal S. K., 115, 116, 120

Tobruk, 160,

Torgau, 177

trench warfare, 14-17 and passim; infantry take to shell holes and regain freedom of movement, 21; two possible solutions to "riddle of trenches," 26; trenches as death traps, 21; "troglodyte" warfare, 22;

"triangular" tactical organization, 9, 17, 165-166, 166-167, 206, 220-221; advantages of organizing by "threes," 65; Chinese "three by three" structure, 221

Trotsky, Leon, 112

Trythall, Brig. A. J., 260

Tukhachevskii, Marshal Mikail, 112-114, 117, 124 and passim

Tunisia, 161-162 and passim

Ukraine, 111, 118

Umurbrogal Ridge, 217

United Nations, forces in Korea, 218, 219, 220, 223, 227 and passim

U.S. Army (see American army)

United States Army Air Forces (USAAF), 182

United States Marine Corps, 212-217 and passim; amphibious operations and tactics in World War II, 215-217; highest casualties ever, 217;

imitate Communist Chinese basic fire group, 213-214; develop "fire team," 214-215; weapons, equipment and tactical organization in World War II, 212-215; weapons, equipment and tactical organization in Korean War, 224; Fleet Marine Force Pacific, 215; First Marine Division, 224; First Marines, 217; First Raider Battalion, 214; Second Raider Battalion, 212
"unlimited objective," theory of, 24, 93
Upton, General Emory, 5, 164
urban warfare, 131-136, 173-174, 250, 261-263; advantage of urban "hedgehogs," 261
urbanization of Europe, 259-260
USSR (see Soviet Union)

Van Fleet, Gen. James, 221
Verdun, 24, 48, 66, 71, 173 and passim; Germans' first use of infiltration techniques, 24; visit by General Currie, 24
"Verdun on the Volga," 131 (see also Stalingrad)
Vers l'armée de Métier, 52
Versailles, Treaty of, 61
Vietnam, 240
Volga, 111, 131, 136 and passim
Volksgrenadier divisions, 91

Volkov, 123
Volokolamsk, 130
volley fire, 4, 8, 63, 121
Volkes, Maj. Gen. Chris, 185
Vyazma-Bryansk pocket, 111

Waffen S.S., 50
Warsaw, stand against panzer attack, 129 and passim
"web" defense, 117-118, 129, 136-137, 261-262;
"weight of metal" doctrine, 67, 175, 288
Wehrmacht, 120, 122, 129, 130, 132, 139, 141, 142, 142 and passim
Wehrsport, 88
Werwolf guerrilla war, 262
Western Desert, 158-159
Western Desert Force, 159
Western Front, 13, 14, 17 and passim
Weygand Line, 102
Wietersheim, Gen. Gustav von, 98
Wingate, Brig. Gen. Orde, 245, 246
Wintringham, Tom, 76

Yadin, Yigael, 241, 249
Yom Kippur War, 244-246; technological implications and major lessons of, 250; Israeli neglect of night fighting, 245

Zahal, 246 (see also Israeli Defense Force)
Zaitsev, Vasili, 135
Zelva, 127
Zhukov, Marshal Georgy K., 110, 114, 129, 130 and passim
Zossen, 135

About the Author

Major John A. English holds the appointment of Staff Officer to the Deputy Chief of the Defence Staff, National Defence Headquarters, Ottawa, Canada. An infantry officer in the Princess Patricia's Canadian Light Infantry, he has seen service with the British and Canadian armies in England, Germany, Denmark, Cyprus, Canada, and Alaska.

Major English has published a number of articles on military subjects, including "The Trafalgar Syndrome," which appeared in the Naval War College Review.

The author holds a B.A. in history from the Royal Military College of Canada, an M.A. from Duke University, where he studied under Dr. Theodore Ropp, and an M.A. in war studies from the Royal Military College of Canada.

LIBR